Physiology and Biochemistry of Seeds

in Relation to Germination

In Two Volumes

1 J. D. Bewley · M. Black
Development, Germination, and Growth

Corrected Printing of the First Edition

Springer-Verlag
Berlin Heidelberg New York 1983

Dr. J. Derek Bewley, Department of Biology, University of Calgary, Calgary, Alberta, Canada T2N 1N4

Dr. Michael Black, Department of Biology, Queen Elizabeth College, University of London, Campden Hill Road, London W8 7AH, Great Britain

With 122 Figures

ISBN 3-540-08274-3 Springer-Verlag Berlin Heidelberg New York
ISBN 0-387-08274-3 Springer-Verlag New York Heidelberg Berlin

Library of Congress Cataloging in Publication Data. Bewley, J. Derek, 1943 —. Physiology and biochemistry of seeds in relation to germination. Includes bibliographical references and index. Contents: v.l. Development, germination, and growth. 1. Seeds. 2. Germination. I. Black, Michael, joint author, II. Title. DNLM: 1. Seeds — Physiology. 2. Seeds — Growth and development. Qk740 B572p QK661.B48. 582′.01′662. 77–7953.
Typesetting, printing, and bookbinding: Universitätsdruckerei H. Stürtz AG, Würzburg
2131/3130–543210

"The true writer will let his wife starve, his children go barefoot, his mother drudge for his living at seventy sooner than work at anything but his book"

Slightly amended, from G.B. Shaw's
"Man and Superman" (with apologies)

Dedicated to Christine Bewley and Marianne Black and their respective children who, despite their initial shock on discovering that they had aspiring writers as husbands and fathers, proved to be pillars of strength and encouragement.

Acknowledgements

First and foremost we owe a debt to the thousands of scientists who have studied and written about various aspects of seed biology. Many of their contributions have been drawn upon to sustain our efforts but inevitably much work receives no reference, if only because of limitations of space and time. Especial thanks are due to Drs. D.L. Laidman, M. Parker, P.I. Payne, P. Slack and D.L. Smith who allowed us to use their unpublished work and to many others for the use of published photographs and graphs: these sources are indicated with the appropriate figures. We thank Dr. P. Halmer for valuable discussions and critical advice and J. Pacey for photographic contributions. We also express our appreciation to our respective research students and other colleagues who soldiered on at their laboratory benches while we were immured in the libraries. Queen Elizabeth College, University of London, kindly gave facilities to J.D. Bewley during his sabbatical leave which was generously permitted by the University of Calgary and supported by the British Council, the Department of External Affairs, Government of Canada and the National Research Council of Canada.

Thanks to the stamina, conscientiousness and excellent typing of Catherine Wardale our manuscript was transformed into a presentable condition. We sincerely thank her and Gwen Turner, also in London, who helped out when the pressure became too high. The final typescript was produced by Laurie Dowson and Linda Dunne of Calgary for whose efforts we are most grateful.

Contents

Chapter 4. Imbibition, Germination, and Growth

Chapter 5. Biochemistry of Germination and Growth

Chapter 6. Mobilization of Reserves

**Chapter 7. Control Processes in the Mobilization
of Stored Reserves**

Contents of Volume 2

Chapter 1. Introduction

1.1. The Subject Matter of This Book

In this book we will consider the biochemical and physiological phenomena that occur in a germinating seed and the activities that are uniquely related to germination such as food mobilization and early growth of the seedling. It seems clear, furthermore, that in order to appreciate these parts of our subject we need to understand certain aspects of seed development — since what occurs here lays down much of the pattern for later events. Volume 1 of this book is restricted to the above topics. Volume 2 deals with the control of germination by internal and environmental factors, with dormancy, viability and the ecology of seeds and germination.

1.2. The Seed

With the seed the independence of the next generation of plant begins. The seed, containing the new plant in miniature, is equipped with structural and physiological devices to fit it for its role as a dispersal unit and is well-provided with food reserves which sustain the young plant until a self-sufficient, autotrophic organism can be established. The embryonic plant is protected within its coverings, its metabolic activities at an extremely low ebb often not to be re-awakened until some considerable time has passed or a particular environmental stimulus experienced; the new individuals may be dispersed in time as well as in space.

To fulfill its unique role in the plant's life history the seed possesses some special physiological and biochemical properties. Perhaps the most remarkable, and the most obvious, is that most species of seed can remain alive though dehydrated. The water content of the seed may drop to about 10% by weight, many of its cellular organelles becoming disorganized and inactive. In this state of quiescence the seed resists the vicissitudes of the environment but can resume full metabolic activity, growth and development when conditions so permit. An important, fundamental question may, therefore, be asked about the seed: how can the embryo and some of its associated structures, unlike almost all other parts of the plant, withstand desiccation and avoid death? Regrettably, the answer is still not available.

The seed is dispersed from the mother plant endowed with a store of food reserves of protein, carbohydrate and fat in a more concentrated package than occurs anywhere else on the plant. Animals exploit this property when using seeds as an extremely important part of their diet. It is also debatable that civilization began its development when man started to cultivate plants for the food that their "seeds" provided, especially the cereals—wheat and barley in the Near East and Europe, rice in Asia and maize in the Americas. It need hardly be necessary to remind the reader, moreover, that virtually all of man's exploitation of plants in agriculture depends upon seeds—that they can be stored, transported, multiplied and, most important of all, germinated!

Mention of the cereals creates a suitable moment at which to note that many dispersal units which are commonly referred to as seeds are not true seeds at all but single-seeded fruits. In these cases the pericarp remains thin and dry and may even become fused to the underlying testa as in the cereal grains. In other species, e.g. lettuce and sunflower, such fusion does not occur. Physiologically and biochemically these "dispersal units" should be considered as seeds. In this book we have commonly used the term seed, except in the case of cereals whose dispersal units we refer to as grains or kernels.

It is not surprising, therefore, that for the above reasons—biological and agricultural—the physiology and biochemistry of seeds have been intensively studied. This book is an attempt to discuss modern findings on this subject.

1.3. What is Germination?

When a viable (i.e. living) seed is wetted, water is taken up, respiration, protein synthesis and other metabolic activities begin and after a certain period of time the embryo emerges from the seed, generally radicle first: the seed has germinated. Various requirements must obviously be satisfied before these events can occur; in most cases there must be sufficient oxygen to allow some aerobic respiration, and a suitable temperature to permit the various processes to proceed at an overall adequate rate. Many species of seed nevertheless fail to germinate even when these requirements are satisfied. This is because there exists within the seed a block (or blocks) somewhere along the sequence of changes which normally would culminate in protrusion of the radicle through the surrounding structures. These impedances are overcome either by the provision before visible germination of some environmental stimulus such as light or low temperature, or by subtle changes which slowly occur in the seed with the passage of time. This condition—the failure to germinate even under apparently favourable conditions—is called dormancy. But once germination has occurred, growth of the young seedling continues, supported by the mobilization of the food reserves; eventually the plumule is carried upwards and in nature is raised out of the ground into the light where its autotrophic life can begin.

Implicit in the foregoing discussion is the definition of germination which we will use in this book. Germination consists of those processes which begin with water uptake and which successfully terminate with the emergence of the radicle or hypocotyl through the seed coverings. We have taken all events subsequent to this to be part of or associated with seedling growth. Thus, mobilization of food reserves according to our definition is not strictly a component of germination. But clearly, since it is uniquely associated with the germinated seed it is nevertheless best considered in this context.

1.4. How is Germination Measured?

We have seen above that germination can be recognized by emergence from the seed. As far as each individual seed is concerned it is therefore an all-or-nothing event — the seed has, or has not, the expressed ability to germinate. For seed populations though, it becomes possible to mark grades of germination ability (germinability) or capacity, which is simply the maximum percentage of seeds which germinate under favourable conditions. This is not necessarily the same as the germination rate, which is the germination percentage obtained after a certain time under certain stipulated conditions which may or may not be optimal, e.g. in the presence of certain chemicals or at a certain temperature, etc. This is a commonly used measure of germination; reference to the literature will reveal many instances where researchers have expressed their results as the final germination percentage after time, t. The germination rate is, of course, the reciprocal of the "time to germination", another index in common use. This can be expressed for a single seed, for a population, or for a certain fraction of the population, e.g. 50%. Now we have already seen that germination as defined above is really divisible into two parts, *viz* the biochemical preparative processes and emergence itself. Expression of the final germination percentage is informative only of the proportion of seeds reaching the stage of emergence but it reveals nothing about the time taken to reach this stage.

   An expression for the mean germination rate of a population was devised by Kotowski [11]. This expresses the so-called coefficient of velocity (C_v) (though not strictly a measure of velocity or speed but of rate) thus:

$$C_v = \frac{\Sigma G_n}{\Sigma (G_n \cdot D_n)} \cdot 100$$

where G_n = number of seeds germinated on day n; D_n = days from initial sowing. Several criticisms have been made of this approach. Heydecker [9] for example, has pointed out that the expression provides no information about the "distribution" of germination, i.e. a certain average rate results when all seeds germinate at the same time or when some germinate early and others very late. Kotowski's coefficient can, however, be transformed into a measure of the distribution of germination ("uniformity of germination") (see [3]). Another device for reducing the contribution of the few, late germinators to the mean rate uses probability graph paper, and plots the cumulative number of germinated seeds against time (see, e.g. [13]).

Realizing the desirability of expressing both the rate of germination *and* the final amount several workers have attempted to derive a single value to combine these two parameters. The "germination value," C, of Czabator [7] can be calculated from $C=pmt^{-1}$, where p is the "peak" germination percent, i.e. the point of inflexion of the curve of germination against time, $m=$ the final germination percent and $t=$ the time for the test. The major deficiency of this method, emphasized by Goodchild and Walker [8] is that C is a value only of the average rate of germination and that identical values can result from several different curves; for example, the time to reach p can be varied without altering mt^{-1}, p or C. An attempt has also been made by Timson [14] to combine rate and final level of germination from $\Sigma\, g_i\,(t-j)$ where $g_i=$ germination in time interval i (i varies from 0 to t), $t=$ total number of time intervals and $j=i-1$ [8]. Here, too, the result is not completely satisfactory since the same value may be given, for example, by two seed populations one which germinates 90% by the first day and the remaining 10% over the next 9 days, and the other which shows no germination up to the 9th day but on the 10th day 100% germination.

A number of researchers have attempted to overcome these difficulties and include the three factors — total germination, mean rate of germination and variation in the rate — into one description. Goodchild and Walker [8] found it adequate to use polygonial regression methods for curve fitting; whereas Janssen [10] describes a method using the average germination time, the standard deviation and the total accumulated sum of the normal curve. Interested readers should also refer to other papers which discuss the recording of germination data [12].

We must note that measurement of numbers of germinated seeds does not always convey the required information. In certain species radicle growth may commence before this organ visibly bursts through the testa; and in some experiments it may be necessary to pinpoint, to within just a few hours, the time when the radicles first begin to grow. In such cases, the course of germination can be followed by recording changes in fresh weight.

Inspection of the literature will reveal a certain imprecision in expressing the time factor in "germination." Generally, time is measured from first exposing the seed or seed part to water. This is sometimes referred to as the "time of imbibition," a description which is clearly undesirable if we accept imbibition as being only the initial stage of water uptake (Chap. 4). In some cases, "time of germination" is also unsatisfactory; this expression has been used, for example, for isolated cotyledons or even isolated endosperms where obviously no germination in its proper sense can occur. Moreover, according to the usage of "germination" as defined above, only the hours or days up to radicle or hypocotyl emergence from the seed should strictly be classed as germination time — all times after this cover seedling growth. These small difficulties can easily be avoided by reference to, say, the time after sowing or planting, or after the start of imbibition.

1.5. Some Comments on Our Sources

To obtain the material for this book we have consulted a great volume of published literature — research papers, reviews and general works. Each Chapter is extensively referenced, though the reference list generally represents only a fraction of the total source.

In attempting to set down the events of germination in some generally applicable pattern we have encountered certain difficulties. We mention them now not so much as an excuse for the deficiencies of this book but in the hope that researchers and other students in this field might gain from our experience. When trying to construct an overview it is necessary to compare work from different laboratories. The problems we have met in making these comparisons include the following: (1) Various species or different cultivars of the same species are used by different, or sometimes the same workers. (2) Seeds of the same species or cultivar but of different provenances or harvests show variations in behavior. (3) Frequently, the information for a particular species or cultivar is incomplete. For example, much may be known about food mobilization in a species but little about deposition of food reserves; protein breakdown may have been followed in one cultivar, starch breakdown in another and respiration in a third. An overall picture of the sequence of reserve deposition and mobilization therefore has to be constructed from several isolated pieces of information obtained from different species or cultivars. (4) Dissimilar experimental conditions are used by workers investigating the same phenomenon. For example, in studying the breakdown of reserves in isolated cotyledons or endosperm, different researchers have incubated the tissues in various amounts of water on paper, sand, vermiculite or even almost completely submerged. With such a variety of experimental conditions meaningful physiological and biochemical comparisons are rendered difficult. (5) It is often hard to relate the physiological time courses of various sets of experiments to each other; this is because some kind of "marker" has been omitted. If an "event marker" such as time of radicle emergence, increase in fresh weight, appearance of a certain enzyme, start of mobilization of a particular reserve were always included, comparisons among published results could more easily be made.

1.6. Plant Names

Throughout this book we use English and botanical names. Frequently, but not always, both names are employed together. We have assumed that the English names for "popular" plant species are well known even to those readers whose native language is not English and in these cases we often use only the English name; otherwise only the botanical name appears. We hope that the irritation caused by any inconsistencies can be relieved by reference to the glossary of plant names.

Some Articles of General Interest

1. Brown, R.: Germination. In: Plant Physiology. Steward, F.C. (ed.). New York: Academic Press, 1972, Vol. VIC, pp. 3–48
2. Crocker, W., Barton, L.V.: Physiology of Seeds. Waltham, Mass.: Chronica Botanica, 1957
3. Heydecker, W. (ed.): Seed Ecology. Proc. 19th Easter School in Agr. Sci., Univ. Nottingham. London: Butterworths, 1973
4. Kozlowski, T.T. (ed.). Seed Biology. New York: Academic Press, 1972, Vol. I, II, III
5. Mayer, A., Poljakoff-Mayber, A.: The Germination of Seeds, 2nd ed. Oxford: Pergamon Press, 1975
6. Ruhland, W. (ed.): Encyclopedia of Plant Physiology. Berlin: Springer, 1960, Vol. XV/2

References

7. Czabator, F.J.: Forest Sci. **8**, 386–396 (1962)
8. Goodchild, N.A., Walker, M.G.: Ann. Botany (London) **35**, 615–621 (1971)
9. Heydecker, W.: Nature (London) **210**, 753–754 (1966)
10. Janssen, J.G.M.: Ann. Botany (London) **37**, 705–708 (1973)
11. Kotowski, F.: Proc. Am. Soc. Hort. Sci. **23**, 176–184 (1926)
12. Nicholls, M.A., Heydecker, W.: Proc. Intern. Seed Test. Ass. **33**, 531–540 (1968)
13. Roberts, E.H.: In: Viability of Seeds. Roberts, E.H. (ed.). London: Chapman and Hall, 1972, pp. 14–18
14. Timson, J.: Nature (London) **207**, 216–217 (1965)

Chapter 2. The Structure of Seeds and Their Food Reserves

2.1. Seed Structure

In this chapter we will survey the major features of seed structure which should be understood in order to appreciate points raised in subsequent chapters. Detailed accounts of seed structure can be found in works on plant anatomy and morphology as well as in recently published studies [1, 7]. We will, however, give special, detailed attention to the food reserves of seeds, the site of accumulation of which is obviously closely associated with structure.

The seed is derived from the fertilized ovule. In almost all cases the following can be recognized as the fertilized ovule develops: (1) the testa — the product of one or both integuments of the ovule; (2) the perisperm — derived from the nucellus; (3) the endosperm — produced as a result of fusion between one male generative nucleus and the two polar nuclei to form the triploid endosperm nucleus [1]; (4) the embryo — the result of fertilization of the oosphere (ovum) by a male nucleus. The degree to which these various components continue their development or even whether or not they are all retained, leads to some of the fundamental structural differences among various types of seed (Fig. 2.1). In addition, in many species extra-ovular tissue, especially the ovary wall (pericarp), becomes closely associated with the seed during its formation. We should also note the variability in structure even in seeds produced by one plant, i.e. seed polymorphism. Variations in size, presence or absence of endosperm, colour of testa, and amounts of chlorophyll can be found in several species. The factors responsible for producing these differences are incompletely understood.

It is necessary only to remind the reader of the great range in size and shape of seeds, from the spore-like seeds of the Orchids, through the familiar seeds of crop plants (peas, beans, cereals, lettuce, tomato) to the large coconut and huge *Lodoicea* which may weigh up to 15 kg!

2.1.1. The Testa

The testa is generally a hard coat; in some cases a thinner inner testa is present formed from the inner integument. A great deal of attention has been given to the anatomy of the testa and the differences between genera and species are often exploited for taxonomic purposes. Its physiological importance arises from the presence of an outer and inner cuticle, often fatty or waxy, and one or more layers of thickened, protective cells (Fig. 2.2). These features confer upon the testa some degree of impermeability to water and/or gases, including

---

[1]   The "endosperm" of Gymnosperm seeds is haploid megagametophytic tissue.

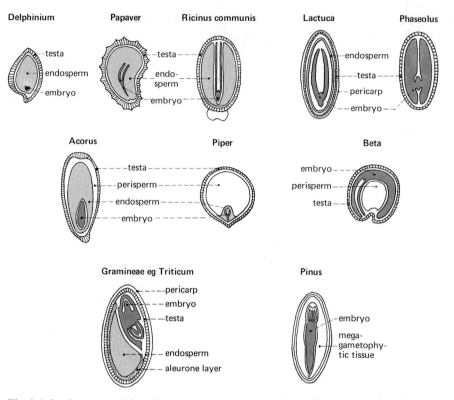

Fig. 2.1. Seed structure. Note the relative proportions of embryo, endosperm and perisperm

oxygen, so that it can consequently exert a regulatory influence over the metabolism and growth of the inner tissues and organs of the seed. In some cases, the testa may be mucilaginous and thereby play an important role in water retention and seed dispersal.

Besides its colouring (sometimes there is a mottled pattern) and texture, an obvious feature of the testa is the hilum. This is the scar, generally of different colour from the rest of the testa and of variable shape and size according to species, marking the point of attachment of the seed to the funiculus. In many seeds a small hole, the micropyle, is at one end of the hilum. The testa of some species, but not many, may have hairs or wings which aid in seed dispersal (e.g. *Epilobium, Salix, Lilium* spp.). Also situated on the testa of many species of seed are outgrowths, such as the warty growth on the hilum named the strophiole; this may be important in controlling movement of water into and out of the seed. Other outgrowths are termed arils. The aril associated with the micropyle, such as the one found in *Ricinus communis,* is termed a caruncle. Arils can take other forms—knobs, bands, ridges or cupules—and are frequently brightly coloured. An aril with which readers may be familiar is that of the nutmeg, *Myristica fragrans,* which yields the spice, mace. Arils in fact frequently contain unusual chemical compounds not found elsewhere

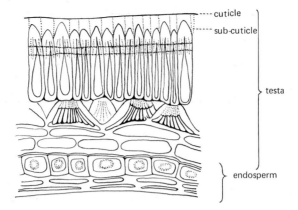

A. Melilotus alba (Leguminosae)

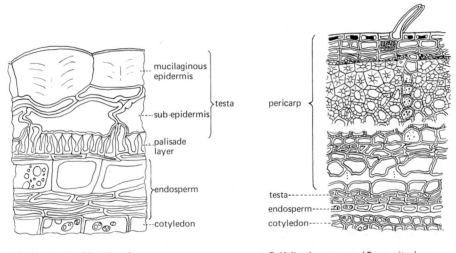

B. Sinapis alba (Cruciferae)

C. Helianthus annuus (Compositae)

Fig. 2.2 A–C. Seed coats. Note the thick testa of the legume (A), the mucilaginous Crucifer (B) and the pericarp of *Helianthus* (C). (A) After Hamly, 1932 [32]; (B) and (C) After Vaughan, 1970 [7]

in the plant. One interesting example is that of *Thaumatococcus* which has an extremely sweet-tasting protein. Arillar contents may be important in attracting animals which aid seed dispersal.

In very many species of "seed" the above features of the "testa" are apparently lacking; this is because the outer coat is not, in fact, the testa but the pericarp (Fig. 2.2 C). Many "seeds" whose biochemistry and physiology have been intensively studied are of this type and are therefore truly fruits. The sunflower (*Helianthus annuus*) and lettuce (*Lactuca sativa*) are both cypselas (a type of achene), the cereal grains are caryopses (achenes in which the pericarp and rudimentary testa are fused) and *Fraxinus* (ash) and *Ulmus* (elm) are winged achenes called samaras. In a small number of species the outer "testa" of the seed is derived from neither ovular integuments nor pericarp but from the outer layers of the endosperm (e.g. in *Crinum*).

2.1.2. Perisperm and Endosperm

The perisperm in the majority of seeds fails to pass beyond an incipient stage of development but in a few cases (e.g. *Yucca, Coffea*) this tissue becomes the major store for the seed's food reserves. In these seeds the endosperm is virtually absent, but in other species appreciable amounts of endosperm are also present (Fig. 2.1).

Seeds are described as endospermic or non-endospermic depending on the presence or absence of a well-formed endosperm (the synonyms albuminous and exalbuminous are archaic and obviously are best not applied to starchy and predominantly oily endosperms). We should note, however, that many species which cannot be classified as endospermic nevertheless do possess an endosperm, but this may be only one or few cell layers thick (e.g. *L. sativa*, Fig. 2.3B). The endosperm, when relatively large, stores the food reserves; well-known examples include members of the Gramineae, *Trigonella foenum-graecum, R. communis* (Fig. 2.4C) and *Phoenix dactylifera*. In these cases the cotyledons are relatively small in mass or else are haustorial organs. During seed development the endosperm surrounds the embryo (except in Orchidaceae where it is absent) and may persist as a relatively large tissue until the seed is fairly well advanced (e.g. *Pisum*). But when embryo growth accelerates, the endosperm may be absorbed or become depleted to a thin tissue and the embryo comes to occupy virtually the whole seed (see *P. coccineus*, Fig. 2.4B). Where the endosperm does not suffer this fate (i.e. in the endospermic seeds) its development may nevertheless be uneven. In the cereals, for example, the endosperm develops unilaterally so that in the mature "seed" (i.e. caryopsis) it resides completely on one side of the embryo (Fig. 2.5). Some seeds possess a ruminate endosperm into which there are ingrowths from the outer coverings (e.g. *Hedera helix*); the furrow in cereal caryopses may be seen as a simple form of this.

The endosperm of members of the Gramineae and certain other species (e.g. *Trigonella, Fagopyrum*) is characterized by the possession of an outer aleurone layer. As the "seed" begins to mature, the peripheral cells of the endosperm which cut off cells internally begin to divide anticlinically to form small cells of rectangular appearance in section. One or more layers of such cells may be produced; they become fairly thick-walled, develop protein bodies called aleurone grains and, most important, remain alive, unlike for example the other endosperm cells of cereals which are dead cells packed with starch and to a lesser extent protein.

As mentioned above, the true endosperm is derived from two polar nuclei of the ovule and one male nucleus. It is therefore at least a triploid tissue,

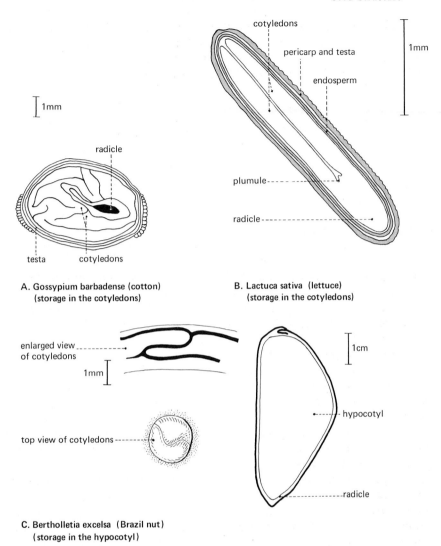

Fig. 2.3A–C. Structure of cotton seed (A), lettuce achene (B) and Brazil nut seed (C)

but in many species varying degrees of ploidy are encountered, often in different cells. This results from nuclear fusion which occurs during endosperm development. We might note finally, in this context, that the so-called endosperm of Conifers (e.g. *Pinus,* Fig. 2.4 D) is haploid, developing from megagametophytic tissue. Although it serves the same function as true endosperm, being the repository of stored reserves, we should be aware of its fundamentally different developmental origin.

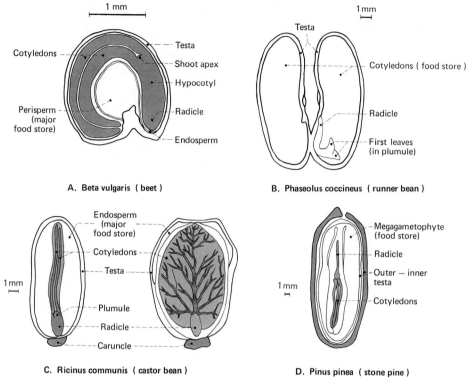

Fig. 2.4A–D. Structure of beet (A), runner bean (B), castor bean (C) and pine (D) seeds

2.1.3. The Embryo

This consists of the embryonic axis bearing one or more cotyledons. The embryonic axis is composed of the hypocotyl to which the cotyledons are attached, the radicle (which is usually difficult to delineate from the hypocotyl) and the plumule (the shoot apex with the first true leaf or leaves). Rarely, a mesocotyl (an internode between the cotyledons) is present. These parts are generally easy to distinguish in a dicot embryo but harder to identify in many monocot species, especially in the Gramineae. The single cotyledon of these embryos has largely become haustorial, i.e. the scutellum. The basal sheath of the cotyledon has elongated into the coleoptile and the hypocotyl has become modified, in some species, partly into the mesocotyl. The coleorhiza can be interpreted as the base of the hypocotyl sheathing the endogenous radicle. The epiblast has been suggested to be a cotyledonary ligule (Fig. 2.5).

The shapes of embryos and their position within the seed varies greatly among species. These features have been intensively investigated and a detailed classification has been drawn up [48]. Some of the types are illustrated in Figure 2.1.

In those dicot species which have a substantial endosperm the embryo, of course, occupies proportionately less of the seed than when the endosperm is lacking. Moreover, the cotyledons of these endospermic seeds, since they

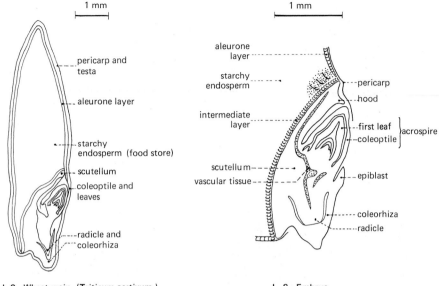

L.S. Wheat grain (Triticum aestivum)   L. S. Embryo

Fig. 2.5. Structure of the wheat caryopsis and embryo

do not store reserves, are often thin, delicate and leaf-like (Fig. 2.4C). In contrast, the cotyledons of non-endospermic seeds are very much bulkier and in many cases such as *Pisum, Vicia* and *Phaseolus* account for over 90% of the mass of the seed (Fig. 2.4B). Cotyledons of non-endospermic, epigeal (Sect. 4.4) species (e.g. *Lactuca, Cucumis*) which become leaf-like after germination are relatively not as massive as in the hypogeal types. In some species the cotyledons exhibit rather complex shapes; they are, for example, deeply-divided in *Tilia,* cupular in *Myzodendron,* and with many convolutions in *Juglans* (walnut). Rarely, the cotyledons are greatly reduced as in *Bertholletia excelsa* (Brazil nut) where they are tiny organs at the summit of a massive, swollen hypocotyl (Fig. 2.3C). The cotyledons of seeds of many parasitic species are absent. Chloroplasts are present in cotyledons of many species such as *Acer* spp, *Capsella bursa pastoris* and *Salsola* spp.

Seeds of some species such as *Linum usitatissimum, Citrus* spp, *Empetrum nigrum, Poa alpina* and *Opuntia* show polyembryony. In different species this can result from one of several causes, including cleavage of the zygote, development of one or more synergidae, existence of more than one embryo sac per nucellus and various forms of apogamy and adventitious embryony. Clearly, in some types of polyembryony haploid embryos are formed, such as in *Linum.*

Finally, we should note that seeds of some species possess greatly reduced or immature embryos. Seeds of orchids have minute, poorly-differentiated embryos and, as mentioned above, no endosperm. Other species produce seeds with embryos which are much more developed than those of the orchids but nevertheless are still immature. This happens, for example, in *Fraxinus* spp, and *Ranunculus ficaria* where development of the embryo continues after the seed (actually a samara in *Fraxinus*) has been liberated from the mother plant.

2.2. Food Reserves

Seeds characteristically contain relatively large amounts of food reserves which support growth and development of the seedling until it can establish itself as a photosynthesizing, autotrophic plant. These reserves are for the most part, but not exclusively, laid down as discrete, intracellular bodies and include lipid, protein, carbohydrate, organic phosphate and various inorganic compounds. It is, of course, these storage materials that render seeds an extremely valuable part of the animal and human diet. In just a few species though (for example in the Orchidaceae) there are almost no food reserves. Here, germination and continued growth depend upon an external supply of organic substances such as carbohydrate; this is often provided, it is thought, by relationships with saprophytic fungi.

There are great differences among seeds in their content of food reserves. Carbohydrates, mainly starch, predominate in the cereals and other Grami-

Table 2.1. Food reserve composition of various seeds

Species	Average percent composition (dry wt)			Major storage organ
	Protein	Fat	Nitrogen-free extract[a] (major component)	
Corn (*Zea mays*)	11	5	75 (starch)	Endosperm
Sweet corn (*Zea mays*)	12	9	70 (starch)	Endosperm
Oats (*Avena sativa*)	13	8	66 (starch)	Endosperm
Wheat (*Triticum aestivum*)	12	2	75 (starch)	Endosperm
Rye (*Secale cereale*)	12	2	76 (starch)	Endosperm
Barley (*Hordeum vulgare*)	12	3	76 (starch)	Endosperm
Broad bean (*Vicia faba*)	23	1	56 (starch)	Cotyledon
Flax (*Linum usitatissimum*)	24	36	24 (starch)	Cotyledon
Field Pea (*Pisum arvense*)	24	6	56 (starch)	Cotyledon
Garden Pea (*Pisum sativum*)	25	6	52 (starch)	Cotyledon
Peanut (*Arachis hypogaea*)	31	48	12 (starch)	Cotyledon
Soybean (*Glycine max*)	37	17	26 (starch)	Cotyledon
Cotton (*Gossypium* spp)	39	33	15	Cotyledon
Rape (*Brassica napus*)	21	48	19 (starch)	Cotyledon
Watermelon (*Citrullis vulgaris*)	38	48	5	Cotyledon
Brazil nut (*Bertholletia excelsa*)	18	68	6	Radicle/Hypocotyl
Oil palm (*Elaeis guineensis*)	9	49	28	Endosperm
Ivory nut (*Phytelephas macrocarpa*)	5	1	79 (galactomannan)	Endosperm
Date (*Phoenix dactylifera*)	6	9	58 (galactomannan)	Endosperm
Castor bean (*Ricinus communis*)	18	64	trace	Endosperm
Pine (*Pinus pinea*)	35	48	6	Megagametophyte

[a] Nitrogen-free extract consists of material which is not protein, fat, fiber (including cellulose) or ash (mineral nutrients). Thus starch, free sugars and dextrins are the usual components. Based on Crocker and Barton, 1957 [2] and Winton and Winton, 1932 [74]

naceous plants, though protein and lipids are also present. A large proportion of seeds, many of which are agriculturally important, store lipids as their main reserve as triacylglycerols (triglycerides); but again, substantial amounts of other reserves, especially protein, are present too. High levels of protein, together with even higher amounts of starch but little lipid, are found in a third group of seeds which includes many legumes such as peas and beans. Some species store carbohydrates as mannans (e.g. fenugreek, *T. foenum-graecum*), or nitrogen as free amino acids (e.g. *Griffonia*). In addition to the major organic reserves, phytin (salts of phytic acid) and carbohydrates such as stachyose and raffinose are frequently found.

The reserve composition of some seeds are shown in Table 2.1; more details may be found in Crocker and Barton [2] and the references therein. It should be realized, however, that no definite values for any particular species can be given since considerable variation may result from environmental conditions experienced by the mother plant during seed development or from genetic factors. The composition of seeds can of course be modified by breeding; much effort has been and is still expended by plant breeders to produce seeds with more desirable contents of carbohydrate, lipids or protein to suit human and animal dietary requirements.

2.2.1. Location of Reserves

Reserve materials may be stored in the embryo or in extraembryonic tissues (endosperm or more rarely perisperm) or both. In many species, for example some members of the Leguminosae, protein and carbohydrate are located in the bulky cotyledons which either remain underground during germination and seedling growth, e.g. pea (*P. sativum*) broad bean (*V. faba*), or are carried upwards as the hypocotyl grows, e.g. dwarf bean (*P. vulgaris*). Cotyledons are also the storage organs for lipids and proteins in seeds of many species such as *C. sativus* (cucumber), *Cucurbita pepo* (squash), *Lactuca* (lettuce), *Sinapis alba* (mustard) and *Brassica* spp. In these cases, the cotyledons are somewhat swollen but they become leafy, photosynthetic organs as seedling growth progresses. Unusually, the embryonic axis itself is the major storage region as in the Brazil nut (*B. excelsa*). Here, the cotyledons are minute and the radicle/hypocotyl is greatly swollen, containing much oil (Fig. 2.3 C).

Most of the stored carbohydrate (starch) and protein of cereals and other grasses is located in the endosperm. In some oil-storing seeds, such as the castor bean (*R. communis*) and tung (*Aleurites* spp), the endosperm is the storage tissue. Endosperm stores mannans in fenugreek (*T. foenum-graecum*) and date (*P. dactylifera*). In some cases, the perisperm (arising from the nucellus) accumulates reserves as in *Yucca* where fat and protein are stored and coffee (*C. arabica*) where mannans are laid down.

It must be realized, however, that reserves are frequently present in some measure throughout the whole seed. In lettuce, for example, though most of the protein and oil stores are located in the cotyledons, protein and oil bodies nevertheless do occur in the radicle/hypocotyl region. Additionally the endo-

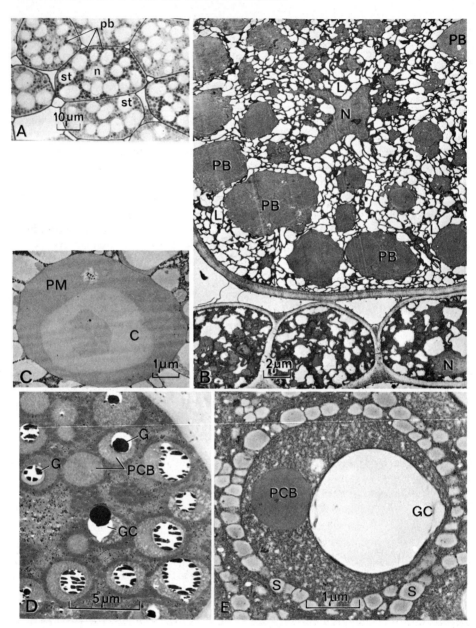

Fig. 2.6A–E. Histology of food reserves 1. (A) Cotyledon cells of mung bean (*Phaseolus aureus*). From Harris and Chrispeels, 1975 [33]. (B) and (C) Electron micrographs of mesophyll cell and epidermal cells (B) and protein body (C) of squash (*Cucurbita maxima*) cotyledons. From Lott et al., 1971 [46]. (D) and (E) Electron micrographs of a barley (*Hordeum vulgare*) aleurone cell with aleurone grains (D); and aleurone grain (E). From Jacobsen et al., 1971 [40]. *C*: crystalloid, *G*: globoid, *GC*: globoid cavity, *PCB*: protein carbohydrate body, *PB, pb*: protein body, *PM*: protein matrix, *L* and *S*: oil bodies, *N* and *n*: nucleus, *st*: starch grains

sperm, while not the major storage tissue, is rich in mannans and also contains protein bodies. Similarly in cereals we can see a distribution of protein, oil and carbohydrate throughout the kernel. The endosperm, of course, contains the bulk of the reserves; almost all the starch and protein is located here, but carbohydrates (mainly di- and trisaccharides) occur in the embryonic axis. The larger proportion of the grain's oil is in the neighbouring scutellum but small quantities occur in the endosperm, especially in the aleurone layer. The greater proportion of starch and protein is in the bulky cotyledons of leguminous seeds such as peas and beans; protein and carbohydrate (sugars) are, however, also present in the embryonic axis. And although, as stated above, by far the larger proportion of oil and protein in the castor bean is in the endosperm, the embryo also possesses some of these reserves.

These few examples will, it is hoped, illustrate three points: (1) that some reserves may be found in different seed tissues; (2) that one tissue may contain all or most of a seed's particular reserve; (3) that different reserves co-exist in the same tissue (Figs. 2.6, 2.7, 2.8).

Discontinuities in distribution of reserves can occur in the storage tissues. The central endosperm of many cereals, for example, possesses tightly-packed starch grains and relatively few protein bodies, whereas the outer cells (i.e. the subaleurone cells) have smaller starch grains and more abundant protein bodies [28]. In the cotyledons of certain legumes, starch grains are interspersed with protein bodies and/or oil bodies but again some regions may have more or less of the polysaccharide and protein.

In the seeds which accumulate oil in the endosperm or embryo high levels of protein are usually also present. Chemical differences can exist in any one species between the triglycerides of various tissues; in rapeseed (*B. napus* cv Regina) for example, the proportions of palmitic and erucic acids in the triglycerides show differences between the cotyledons and hypocotyl [9].

We will now consider the chemistry and biology of the reserves in more detail.

2.3. Protein

Osborne's classification of seed proteins [57] was based on their solubility and this property is still used today in methods for their separation. Four groups of proteins (which are now known to be heterogeneous and to vary among species) are defined: (1) Albumins are soluble in water at neutral or slightly acid pHs and are heat coagulatable. These are mainly enzymic proteins. (2) Globulins are insoluble in water but dissolve in salt solutions (e.g. 0.4 M NaCl) and are not as readily coagulatable upon heating as are the albumins. (3) Glutelins are insoluble in water, salt solutions and ethanol but can be extracted with fairly strong acidic or alkaline solutions. (4) Prolamins dissolve in ethanol (90%) but not in water.

Not all these groups of protein are necessarily found in any one species of seed (Table 2.2). The Gramineae contain prolamins (e.g. gliadin, zein and

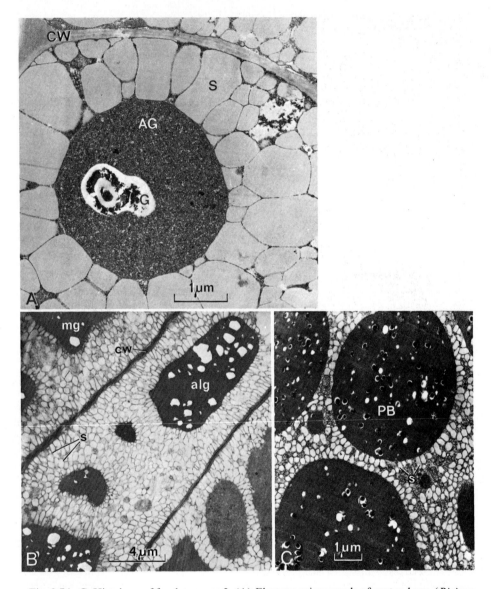

Fig. 2.7A–C. Histology of food reserves 2. (A) Electron micrograph of castor bean (*Ricinus communis*) endosperm cell. From Vigil, 1970 [72]. (B) Electron micrograph of cotyledon cell of *Sinapis alba*. From Rest and Vaughan, 1972 [62]. (C) Electron micrograph of cotyledon cell of *Sinapis alba*. From Werker and Vaughan, 1974 [73]. *AG, alg*: aleurone grains (protein bodies), *G*: globoid, *mg*: myrosin grains (presumed to contain the enzyme myrosinase), *PB*: protein body, *S, s*: oil bodies, *CW*: cell wall

hordein of *Triticum, Zea* and *Hordeum* respectively), but this group of proteins is uncommon in other seeds. Cereal grains are also rich in glutelins; the glutenin of *Triticum,* which is important in giving the structure to bread, and oryzenin

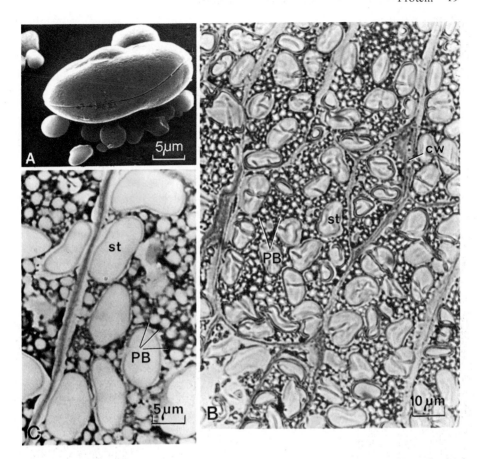

Fig. 2.8A–C. Histology of food reserves 3. (A) Scanning electron micrograph of large lenticular and small spherical wheat (*Triticum aestivum*) starch grains. From Evers, 1971 [29] (B) Transverse section of barley (*Hordeum distichon*) endosperm cells. (C) As (B) but at higher magnification. Note large and small starch grains and darkly-staining protein bodies amongst them. *cw*: cell wall, *PB*: protein bodies, *st*: starch grain. By J. Pacey

Table 2.2. Protein composition of certain seeds

Species and strain	Percent of total protein			
	Albumin	Globulin	Glutelin	Prolamin
Avena sativa	Trace	80	5	15
Zea mays (IND 260)	14	0	31	48
Zea mays (opaque-2)	25	0	39	24
Oryza sativa	5	10	80	5
Pisum sativum	40	60	0	0
Cucurbita pepo	Trace	92	Small amount	Trace

After Beevers, 1976 [15]

of *Oryza* are examples. Globulins, on the other hand, predominate in dicotyledonous seeds such as the legumes. Examples of such globulins described in the early classification of Osborne are arachin, legumin, vignin, glycinin and vicilin.

2.3.1. Storage Proteins of Legumes

Because leguminous seeds are of very considerable agricultural importance they have understandably received a great deal of attention. Together with the cereals they provide most of our biochemical and physiological knowledge of seed storage proteins.

Osborne divided the globulin, firstly of *P. sativum* then of *P. vulgaris* and *Glycine max,* into two fractions, *vicilin* and *legumin,* which precipitated differentially when saline extracts were diluted and heat treated. These two protein fractions have since been shown to make up the globulins of a wide range of leguminous seeds; legumin-type proteins are also found in certain non-leguminous species such as *H. annuus, Prunus* spp and *Brassica* spp (see [24] for a detailed, up-to-date review on these proteins).

The original separation methods have now been superseded by the modern ones of electrophoresis, ultracentrifugation, gel filtration and other techniques. These newer methods are superior since they exploit the fundamental differences in charge and molecular size of proteins. The relationships among the globulins of different species are also examined using serological methods. Application of these and other analytical techniques demonstrate that legumin and vicilin differ from species to species, though different legumins have much in common.

Separation of the leguminous globulins by ultracentrifugation generally reveals two bands with sedimentation coefficients of approximately 11–12 S (often termed 11 S) and 7–8 S (often termed 7 S) held to be legumin and vicilin respectively (Table 2.3). In some species (e.g. *Glycine*) additional proteins (2.2 S) are present,

Table 2.3. Legumin-like and vicilin-like proteins in a variety of seeds

Species	"Legumin"		"Vicilin"	
	Sedimentation coefficient(S)	MW (daltons)	Sedimentation coefficient(S)	MW (daltons)
Arachis hypogaea	12.0	350,000	8.7	190,000
Glycine max	12.2	345,000–363,000	8.0	330,000
Lupinus angustifolius	11.6	336,000	7.8	181,000
Phaseolus vulgaris	11.6	340,000	6.8	151,000
Pisum sativum	12.64	330,000	7.1–8.1	—
Vicia faba	11.4	328,000	7.1	150,000
Acacia alata	11.63	—	7.9	—
Acacia longifolia	—	—	7.6	—
Non-legumes				
Beta vulgaris	13.9	~250,000		
Cannabis sativa	12.8	309,000		
Cucurbita maxima	12.1	340,000		

After Derbyshire et al., 1976 [24]

especially during the early stages of seed development (though they may only be sub-units of larger proteins), whereas in others (e.g. *Acacia* spp) 11S proteins seem to be absent.

There is evidence that in many cases the 7–8S fraction ("vicilin") is heterogeneous, consisting of more than one protein. Conglutin β of *Lupinus*, for example, is separable into several components differing in their polypeptide sub-units [53] and *P. sativum* vicilin is thought to contain two major proteins [24]. On the other hand, legumin seems to be more homogeneous; for example glycinin, the legumin from *Glycine*, appears on the basis of its immunochemical properties to consist of only one protein [53].

Both the 7–8S and the 11–12S proteins split into sub-units which can be separated by electrophoresis. The 11–12S protein of several species yields sub-units (acidic and basic due to relatively high levels of acidic and basic amino acids) falling into two groups of molecular weight — between 20,000–24,000, and 30,000–37,000. Legumin may contain, say, three different types of sub-unit, and different numbers of each type. The 11S protein of *V. sativa* for example, is made up of six sub-units of a component with MW 24,300, four sub-units of a 37,600 MW component and two sub-units of a 32,000 MW component, giving a legumin molecule of a total MW of about 360,200, a characteristic order of size for many seed legumins. *V. faba* legumin, on the other hand, has six sub-units of a 36,200 component, and six sub-units of components with MWs between 20,100 and 23,800 (total 328,000) [24]. The sub-units of various 11–12S proteins have many common properties suggesting that during evolution there has been conservation of parts of these molecules. Between 3 and 5 sub-units are apparently present in any one 7–8S globulin but less is known about their properties; this is because of the heterogeneity of this protein. Five different sub-unit types have been described for the 7–8S proteins, ranging in molecular weight from 23,000 to 56,000. Total molecular weights of vicilin are generally lower than those of legumin, often between 140,000 and 200,000 [24]. Interestingly, the sub-units of the 7–8S protein of both *V. faba* and *P. sativum* show qualitative changes during seed development [14,77]; this is another indication of the heterogeneity of this protein.

The amino acid compositions of some legume storage proteins are shown in Table 2.4. On the whole, the legumin and vicilin of different species are similar in their composition. An important common characteristic is the comparatively high level of the amides asparagine and glutamine, and of arginine, A feature of the 7–8S protein is that it lacks or has a very low content of the sulphur-containing amino acid, cysteine. In many cases these globulins have been shown to be glycoproteins, i.e. they contain carbohydrate moieties. Besides those shown in Table 2.4 we may note that both legumin and vicilin of *P. sativum* contain glucose, mannose and glucosamine, from about 0.5% to 1.35% in total (see [53]).

2.3.2. Storage Proteins of Gramineae

Most investigations on these proteins have, not surprisingly, been carried out on the major cultivated members of this group, i.e. the cereals. The four types

Table 2.4. Composition of some 11 S and 7S globulins

Amino acid	Phaseolus vulgaris		Phaseolus aureus	Vicia faba	
	11S[a, c]	7S[b, e]	7S[b, d]	11S[a, c]	7S[b, c, f]
Asp	9.5	12.4	13.4	12.3	11.9
Thr	4.9	3.4	3.0	3.7	2.9
Ser	7.3	6.7	7.3	7.4	5.1
Glu	13.1	15.1	19.9	19.9	17.6
Pro	5.1	2.9	2.9	5.4	nr
Gly	8.0	2.7	5.4	7.7	2.5
Ala	6.9	3.0	5.6	6.3	3.1
Val	7.0	5.2	6.6	5.1	4.3
$^1/_2$ Cys	0.6	0.3	nr	0.0	0.3
Met	1.5	0.7	0.3	0.3	0.4
Ile	4.9	5.6	4.5	4.3	5.2
Leu	8.7	9.1	9.4	8.5	9.3
Tyr	2.9	3.5	1.9	2.1	3.8
Phe	3.6	6.6	6.1	3.2	6.8
Lys	7.8	5.6	6.0	4.2	8.1
His	3.0	2.6	2.0	2.4	2.4
Arg	4.8	5.0	5.5	8.0	7.8
Trp	0.7	0.8	nd	nd	nr
Neutral sugars	<1.0%	4.46%	1.8%	0.1%	0.5%
Mannose		3.2	1.0%		
Glucose		—	trace		
Galactose		trace	trace		
Glucosamine		0.99%	0.2%	<0.2%	<0.2%
Xylose		0.35%	—		

nd=not determined; nr=not recorded
[a] amino acid as mol percent
[b] amino acid as percent of protein (wt/wt)
[c] From Derbyshire et al., 1976 [24]
[d] From Ericson and Chrispeels, 1973 [27]
[e] From Pusztai and Watt, 1970 [60]
[f] From Bailey and Boulter, 1972 [12]

of protein described by Osborne are found in these grains. In barley, for example, the prolamin (=hordein), albumin, globulin and glutelin (=hordenin) fractions comprise respectively 4.0%, 0.3%, 1.95% and 4.5% of the grain weight. The levels of prolamins are thus frequently relatively high, and because they are deficient in some of the amino acids essential to the animal diet, cereal protein is of poor nutritional quality. Glutelin and prolamin predominate in wheat, but in oats 80% of the protein is globulin and is thus of higher nutritional quality.

The glutelin fraction of cereal protein is more heterogeneous than the other storage proteins and consists of several different proteins. Wheat gliadin, for example, is separable into four major fractions α, β, γ and ω but, in all, the gliadin may have as many as 46 components revealed by gel electrophoresis and isoelectric focusing [3a]. Various difficulties are encountered in defining the glutelins of different cereals. Wheat glutelin is relatively easy to separate and investigate but other cereals present problems connected partially with the effects of previous alcoholic extraction (to remove prolamins) which may denature other proteins [6].

The amino acid composition of barley proteins is shown in Table 2.5. Because of the differences in extraction techniques and of other variables, rigorous comparisons between cereal proteins are difficult to make but certain general points can nevertheless be stated. Characteristically, the prolamins are high in proline and amides (glutamine, asparagine), and have low relative contents of tryptophan, lysine and methionine. The globulins of certain cereals (e.g. barley) are similar in amino acid composition (e.g. high arginine) to other seed globulins, even to those of legumes. In fact, globulins with a sedimentation value of about 11S have been found in barley, wheat and rice, prompting the suggestion that a legumin type of protein occurs in these monocots [24].

When we look at the amino acid compositions of the proteins from starchy endosperm and aleurone layers (Sect. 2.1.2) we can get some indication as to how the different kinds of protein are distributed within the kernel (Table 2.6). In wheat, the aleurone layer protein is clearly different from that in the remainder of the endosperm, being extremely rich in arginine: this could reflect a high globulin level. Rice aleurone grains, however, are rich in albumins [69].

We have referred above to the nutritional quality of cereal protein. Much energy has been expended in attempting to modify the amino acid composition using various breeding techniques. Protein of Z. mays, for example, is noto-

Table 2.5. Amino acid composition of barley proteins

Amino acid	Amino acid composition (g per 16 g N)[a]			
	Albumin	Globulin	Glutelin (=hordenin)	Prolamin (=hordein)
Asp	12.2	8.5	7.1	1.8
Thr	4.6	3.3	4.2	2.6
Ser	4.9	4.7	5.0	3.8
Glu	12.9	11.9	19.8	39.6
Pro	5.5	3.6	8.7	20.1
Gly	5.7	9.2	4.5	1.5
Ala	7.3	0.7	6.7	2.2
Val	7.8	5.5	6.6	4.7
Cys	2.1	3.6	1.2	2.1
Met	2.4	1.5	1.9	1.3
Ile	6.2	3.3	5.2	5.4
Leu	8.6	6.8	8.7	6.9
Phe	5.1	2.8	3.6	3.0
Lys	6.7	5.3	4.0	0.7
His	2.5	1.8	2.5	1.3
Arg	6.5	11.0	6.0	3.0
Trp	1.5	0.8	1.3	0.8
Amide N	5.9	5.1	10.3	23.0

[a] 16 g N approximates to 100 g protein. After Beevers, 1976 [15]

Table 2.6. Amino acid composition of protein from starchy endosperm and aleurone grains of wheat cv. Manitoba

Amino acid	Aleurone cell contents	Isolated endo-sperm protein
	Grams amino acid/16 g total nitrogen	
Asp	7.30	3.04
Thr	2.86	2.45
Ser	4.08	4.8
Glu	14.95	35.78
Pro	3.55	13.41
Ala	4.68	3.15
Val	4.95	3.88
Met	1.36	1.70
Ile	2.84	3.74
Leu	5.46	6.89
Tyr	2.69	3.52
Phe	3.68	5.47
Lys	4.17	1.50
Hist	3.60	2.13
Arg	10.49	3.18
$^1/_2$ Cys	3.29	0
Total	83.63	94.64

From Stevens et al., 1963 [67] recalculated by V. Gaba

riously deficient in lysine (this is because of the predominance of prolamin), but in the mutants opaque-2 and floury-2 this characteristic is significantly ameliorated. In these mutants the zein (a prolamin) content is lower but the glutelin fraction with a higher lysine content (cf. Table 2.5) is raised [41].

Like those of legumes and other groups each different type of storage protein in cereals consists of several components which can be revealed by ultracentrifugation and electrophoresis, especially after treatment with urea which breaks the sulphur bridges between polypeptides. In wheat, electrophoretic separation of reduced gliadin (a prolamin) yields three components having molecular weights of 36,500 (the major component), 44,200 and 11,400; intact gliadin has five sub-units with molecular weights ranging from 25,600 to 78,100 [17]. Additional components are revealed after isoelectric focusing. Glutenin, on the other hand, consists of 15 polypeptides of molecular weights between 11,600 and 133,000. Our knowledge of the structure of cereal proteins is not as advanced, however, as that concerning the legume proteins.

2.3.3. Protein Bodies

Sub-cellular bodies consisting of protein were first isolated from certain oil seeds by Hartig in 1855 [34]. He called these protein bodies aleurone grains using the Greek aleuron, meaning meal or flour. Since then a large number of species has been examined and in almost all cases storage protein is found to be located in discrete bodies, between 1 and 20 μ diameter. Aleurone grains have been described in oil-seeds such as *Arachis, Ricinus, Gossypium, Capsella, Sinapis, Yucca, Cannabis, Glycine,* in the cereals *Zea, Hordeum, Oryza* and developing *Triticum* kernels, and in the protein/starch legumes such as *Vicia, Phaseolus* and *Pisum* (Figs. 2.6, 2.7, 2.8).

Protein bodies are generally distributed throughout the protein-storing tissue though certain cells might be richer than others. In cereals, these bodies are found in most cells of the endosperm but they cannot all be called aleurone grains since this term is reserved for the protein in the peripheral endosperm cells, the aleurone layer. Unfortunately, then, somewhat different meanings are implied by the term aleurone grain and it might indeed be preferable to restrict its usage, as has recently been suggested, to those protein bodies in a recognized aleurone layer [53]. We shall use the term in this restricted sense.

Protein bodies are oval to circular in section and are bounded by a lipoprotein unit membrane [25, 78]. Inclusions frequently occur but not in all cereals [55] and even when present, not all protein bodies contain them. Two such inclusions have been described—crystalloids and globoids (Fig. 2.6). The crystalline inclusions are themselves proteinaceous, sometimes identifiable as a single protein type. In *Cannabis,* for example, the crystalloid is separable from the rest of the protein body and has been shown, by electrophoretic and other methods, to be the storage protein edestin (an 11 S globulin) [66]. The globoids on the other hand, as seen in a variety of species (*Gossypium, Capsella, Arachis*), are non-crystalline, globular structures. They are the sites of deposition of phytin, the potassium, magnesium and calcium salts of phytic acid, and in some species contain virtually all the seed's reserves of this compound [25] (see below). Globoids are virtually absent from the protein bodies of cereal starchy endosperm but do occur in the aleurone grains (see Fig. 2.6 D and E). Lipid and RNA have also been reported to be present in barley aleurone grains [40] and soybean protein bodies [70]. Barley aleurone grains also contain carbohydrate but in a unit distinct from the globoid, called the protein-carbohydrate body (Fig. 2.6 D, E) Globoids can also contain protein, however, in some cases identifiable as enzymes. Phosphatase is present in *Gossypium* globoids [47]; and in *Hordeum* the fine structure around the protein body (which has been assumed by some to represent the globoid) apparently contains phytase.

Various hydrolases are in fact associated with the protein bodies, though we should realize that some of the enzymes may be contaminants from the cytoplasm. Acid proteinases have been found in protein bodies of *Gossypium* [47, 79], *V. faba* [54], *Hordeum* [55] and *Cannabis* [66]. In the last, the proteinase is active against the crystalloid moiety (edestin) and has, therefore, been called edestinase [66]. The occurrence of these various enzymes for the hydrolysis of the native protein therefore suggests the possible lysosomal nature of protein bodies.

The characteristics of the protein bodies may vary even in one seed and their distribution can also be unequal. In *Yucca,* protein bodies of the perisperm (the major storage tissue) have a different ultrastructure from those in the embryo [39]. The starchy endosperm of *Z. mays* has large (about 2 μ diameter) and small (about 0.2 μ) bodies embedded in a protein matrix but many cells lack them and contain only the matrix. It is interesting that the matrix protein is entirely glutelin while the protein bodies are rich in the prolamin, zein [23]. The aleurone layer, of course, has an abundance of relatively large protein bodies, but the subaleurone layer in this species, and in *Oryza,* generally contains only smaller protein granules. In *Triticum,* protein bodies occur in the starchy endosperm of developing kernels only [20a]. The accumulating starch grains probably disrupt these protein bodies thus dispersing the protein among the starch grains. Well-defined aleurone grains are of course retained within the aleurone layer.

Since, as we have already seen, more than one type of storage protein exists within a seed it is a sensible question to ask if individual protein bodies contain only one or more than one protein. In *V. faba* [18] and other legumes, protein bodies of only one kind can be recognized ultrastructurally though the seed contains albumin, legumin and vicilin. There is, nevertheless, evidence from immunofluorescence studies that some protein bodies in *Vicia* contain only vicilin or even only albumin; most bodies seem however to consist of both legumin and vicilin [30]. The heterogeneity in size of cereal protein bodies might reflect different protein compositions. The smallest protein granules in the starchy endosperm of wheat are rich in gliadin and those of *Z. mays* are thought to be composed mainly of zein [75]. It is interesting that the opaque-2 mutant of *Zea* which has a lower zein content than the normal strain also has fewer smaller protein granules in the starchy endosperm. During kernel development smaller granules composed of glutelin are present but these become disrupted as maturation proceeds. Glutelin seems to be the major, and globulins the minor constituent of protein bodies in rice endosperm.

2.3.4. Lectins

Some of the protein in several species of seeds act as lectins. Lectins are proteins (usually glycoproteins) which bind to animal cell surfaces sometimes causing agglutination [21]. Their function in plants is obscure, though since they mostly occur in seeds they have been suggested to be simply storage proteins. About 2.5% of the protein in jack bean (*Canavalia ensiformis*) is the lectin concanavalin A and 1.5% of soybean seed protein is an agglutinin [21]. If they are indeed storage proteins their binding abilities to carbohydrate (say of an animal cell membrane) would seem to be a superfluous property.

2.4. Other Nitrogenous Seed Reserves

Many non-protein nitrogenous substances such as alkaloids, free amino acids and amides occur in seeds. Well-known alkaloids are theobromine in *Theobroma cacao,* caffeine in *Coffea* and strychnine in *Strychnos nux-vomica.* Some of the free amino acids are the ones which are also present in proteins (including the amides glutamine and asparagine) but in addition some non-protein amino acids are found in certain species.

It is not clear that all of the nitrogenous substances mentioned above function to support early seedling growth. But in many cases, because the compound exists at such relatively high levels which are depleted during seedling growth, it does seem that they act as reserves of nitrogen and carbon. Many of the non-protein amino acids come into this category. About 8.3% of the dry weight of seeds of *Dioclea megacarpa* is canavanine which contains approximately 30% nitrogen by weight. This amino acid is metabolized after germination, the nitrogen presumably being transferred by transamination to furnish the amino acids for protein synthesis. Examples of other non-protein amino acids which because of their high concentration in seeds would seem to act as food reserves are hydroxytryptophan (14% dry weight in *Griffonia*) and dihydroxy-phenylalanine (L-dopa) which forms 6% of *Mucuna* seeds [16].

2.5. Phytin

Phytin, the insoluble mixed potassium, magnesium and calcium salt of myo-inositol hexaphosphoric acid (phytic acid) is the major storage form of phosphate and macronutrient mineral elements in seeds. It is invariably present within a globoid in protein bodies [38, 45, 47, 56, 68, 69, 71].

Phytic Acid

In cereal grains it is associated with the aleurone grains in the aleurone layer and is more or less absent from the protein bodies of the starchy endosperm. In dicot seeds, globoids appear in the protein bodies of the cotyle-dons and/or endosperm and in gymnosperms in the protein bodies of the mega-gametophyte. Not all protein bodies or aleurone grains contain globoids, e.g. in cotton seeds only about 15% of the bodies have such inclusions.

Phytin content of whole seeds and protein bodies varies with species and cultivar. Examples of the chemical composition of the aleurone grains of rice and protein bodies (and isolated globoids) of cotton seed are shown in Table 2.7. Note, in particular, the higher phosphorus but lower protein content of the cereal aleurone grain in comparison with the dicot protein body. The concentration of the associated macronutrient elements can also be seen to differ. Qualitative as well as quantitative differences occur between the macronutrient compositions of different seeds, e.g. in *C. maxima* the globoid is rich in phosphate, potassium

Table 2.7. Chemical composition of globoids and protein bodies from cotton seed and aleurone grains from rice

| Composition | Cotton seed | | Rice |
	Globoid (weight %)	Protein body (weight %)	Aleurone grain (weight %)
Protein	4.2	57.72	11.7
Carbohydrate	1.35	9.01	7.93
Inorganic phosphorus	0.36	0.14	0.06
Phospholipid phosphorus	0.07	0.06	0.004
Organic phosphorus[a]	13.85	2.25	11.4
Inositol	13.2	nd	9.41
Moisture	9.7	10.07	14.9
K^+	6.4	2.4	9.45
Ca^{2+}	1.3	0.3	0.42
Mg^{2+}	1.7	0.5	8.3

[a] Mainly phytin; nd=not determined; Based on Lui and Altschul, 1967 [47] and Tanaka et al., 1973 [69]

and magnesium, with calcium being more or less absent. This does not appear to be related to the availability of macronutrients to the mother plant during seed development but is internally regulated [45]. Globoids of *Protea compacta* seeds, in contrast, have a high calcium content and low potassium and magnesium [71]. There are reports that in some seeds, manganese, copper and sodium are also present in the globoid in association with phytin.

2.6. Carbohydrates

The principal, most widespread storage carbohydrate of seeds is starch. Less frequently, other polysaccharides ("hemicelluloses," "amyloids," galactomannans) are the major carbohydrate store. Various sugars occur as quantitatively relatively minor reserve components in many species, though in a few cases they are the major carbohydrate.

2.6.1. Starch

This polysaccharide is the chief organic reserve of cereals where it accompanies much lower proportions of fat and protein. Appreciable levels of starch are also found in seeds which are rich in fat and/or protein (Table 2.1). Starch is not a single polysaccharide but consists of two polymers of D-glucose, one linear and the other branched. The linear polymer—amylose—is formed from glucose units joined only by α-1,4 glucosidic linkages. The other polymer—amylopectin—also has α-1,4 linkages but occasionally branching occurs where α-1,6

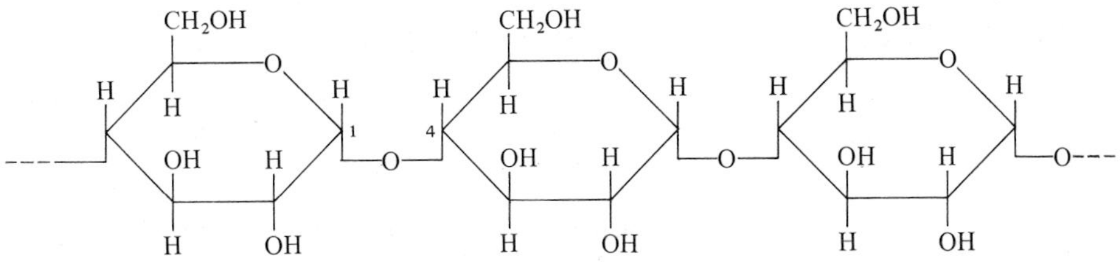

Amylose

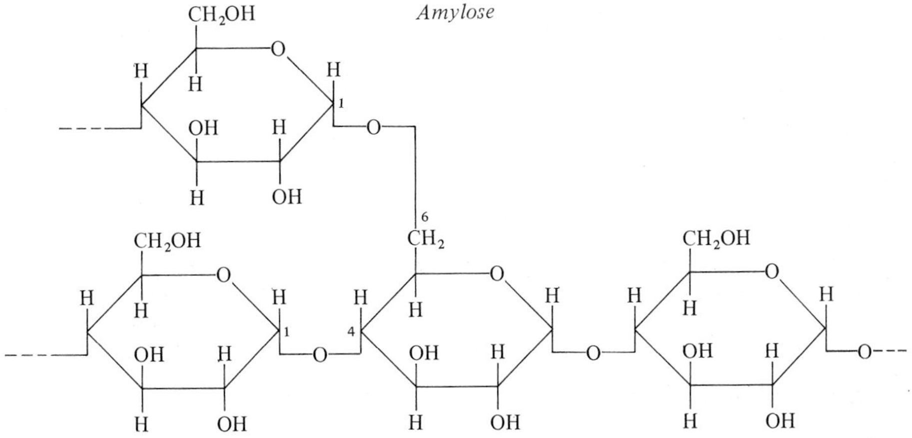

Amylopectin

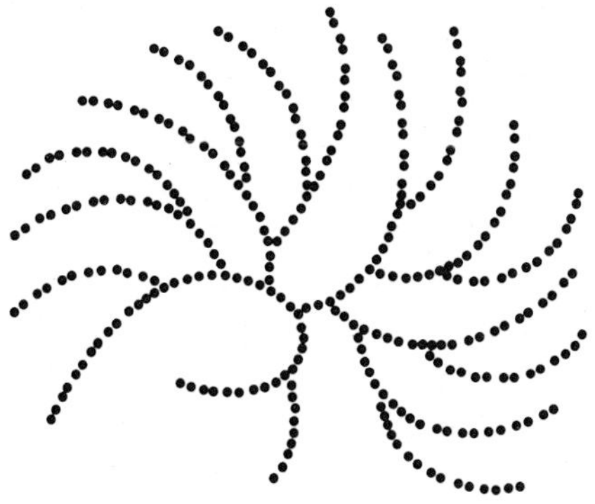

Fig. 2.9. Chemical structure of starch. Below amylopectin molecule

linking takes place (Fig. 2.9). In wheat amylopectin there is a branch for every 20–25 glucose residues.

Molecular weight determinations on the two starch components are difficult, among the problems being the degradation that can occur when the polysaccharides are isolated. Molecules of quite a wide range of size probably exist in seeds so that published figures are likely only to be averages. As one example though, wheat amylose and amylopectin have been reported to have, respectively, molecular weights of 14×10^3 and 4×10^6 [59].

The proportions of amylose and amylopectin vary among seeds of different species and even among different strains of the same species. Rice starch may consist of between 15–37% amylose depending upon the variety. Certain varieties of wheat, maize and barley have about 25% amylose, but in the "waxy" mutants of cereals (excepting wheat which has no much mutants) amylose levels are much lower and the starch is instead rich in amylopectin. Conversely, in wrinkled peas amylose forms 66% or more of the starch compared with about 35% in smooth peas and other legumes [11].

2.6.2. Starch Grains (see also Sect. 3.2.2)

Starch is laid down in discrete sub-cellular bodies, the starch grains. These have a characteristic appearance for each species (Fig. 2.10) and may be spherical, angular or elliptical (for details the reader should consult Seidermann [63]). Interestingly, grain shape depends on the amylose content, the less angular, rounded grains having relatively higher amylose levels [2a]. The grains also display a great range of size among the different species (between 2–100 μ diameter) and even in any one seed. Starch grains of wheat are of two types, lens-shaped and spherical (Fig. 2.8 A). The lenticular grains are in general larger, the largest about 28–33 μ across, and the smallest, spherical ones range between 2–8 μ, but there is in fact a continuum of size from one type to the other [51]. In barley, however, the grains seem to be separable into two distinct groups—large round grains, about 25 μ in diameter, and the much more numerous, smaller grains measuring up to 5 μ across (Fig. 2.8 B, C). Although the smaller grains account for about 90% of the total number they actually contain only approximately 10% of the total starch by weight [58]. Amylose and amylopectin are often present together in the same grain though not in fixed proportions, for some grains may be richer in one of the polymers. Moreover, the composition of the grains can vary with the location in the seed. In *Sorghum,* for example, the starch grains in the central cap area consist of amylopectin, staining red with iodine, whereas those in other regions of the kernel have a core of amylose which becomes blue when treated with iodine [44]. Very many of the starch grains in the waxy kernels of cereal mutants contain only amylopectin.

Starch grains of many species seem to form around a central or eccentric point, the hilum, around which "shells" of polysaccharide are deposited. The linear or branched molecules are arranged radially in these shells, and in some regions are hydrogen-bonded to give a crystalline or micellar structure [10]. These shells probably reflect diurnal periodicity in the synthesis and laying down of starch since they are absent from starch grains of seeds developing in continuous light under experimental conditions.

When isolated starch grains are examined various proteins are found associated with them. Some of the protein is of course carried over during the extraction procedure but this can be removed by repeated washing. Starch grains of both waxy and non-waxy corn (Z. *mays*) for example, have about 0.5% protein much of which is enzymic. One enzyme, ADPG-glucosyl transferase (Sect. 3.2.3.) is present in grains of non-waxy kernels but interestingly the grains of the waxy types of cereal kernels (e.g. in corn and rice), which are high in amylopectin, contain none of this enzyme. The enzyme, as well

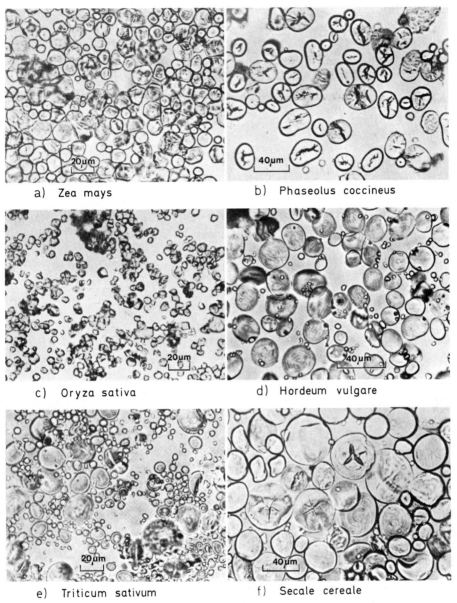

a) Zea mays

b) Phaseolus coccineus

c) Oryza sativa

d) Hordeum vulgare

e) Triticum sativum

f) Secale cereale

Fig. 2.10. Starch grains of various seeds. From Seidemann, 1966 [63]

as ADPG pyrophosphorylase, is also absent from the *Zea* mutants which have predominantly sucrose in their endosperm.

2.6.3. Sugars

Rarely sugars are the predominant storage carbohydrate. This is the case, however, in *Acer saccharum* which is reported to contain 6.4% sucrose, 5.21% of other sugars and no starch [2]. Di- and oligosaccharides are common as minor reserves (i.e. minor by weight but nonetheless physiologically important) in many seeds such as in the embryos of cereals. For example, about 20% of the dry weight of defatted wheat embryos is soluble sugar made up of sucrose and raffinose (58.5 and 41.5% respectively of the total sugars). Some mutants of *Z. mays* such as the brittle-1 and shrunken-2 strains bear little starch but instead have considerable quantities of sucrose [11]. Many legume seeds (e.g. *P. sativum*) have sucrose, raffinose and stachyose in their embryonic axis (see Sect. 6.3.1).

2.6.4. Other Polysaccharides

Many seeds contain reserves classed as hemicelluloses and amyloids. The very hard endosperm or perisperm of certain species is due to hemicelluloses laid down in the cell walls (there are no hemicellulose storage bodies or grains) which consequently are extremely thick. Examples are the ivory nut (*Phytelephas macrocarpa*), the date (*Phoenix dactylifera*) and coffee bean (*C. arabica*). Hemicelluloses also impregnate the cell walls of cotyledons in *Lupinus* and other species. Many of these hemicelluloses are in fact mannans with small amounts of sugar (glucose, galactose, arabinose) present as side-chains on the main linear polymer of mannose residues. The perisperm of *Coffea* consists of about 5% mannan which itself has 2% galactose [76]. In *P. macrocarpa* two kinds of mannans, A and B, together make up 72% of the endosperm. Mannan A, the major component, contains only traces of galactose while mannan B has about 6% of glucose, galactose and arabinose. The composition of the date endosperm mannans is similar [52]. It is interesting that mannan is a major component of endosperm cell walls in lettuce (*Lactuca sativa*) which is not normally considered to be a significant storage tissue. The endosperm is, however, enzymically degraded during seedling growth and it has been suggested that the tissue acts as a source of reserve carbohydrate [31].

Increasing side-chain substitution of the mannose chain with galactose gives rise to a family of mucilages (Fig. 2.11). Such galactomannans are found as major storage components in the endosperm of seeds of the Leguminosae (see Ta-

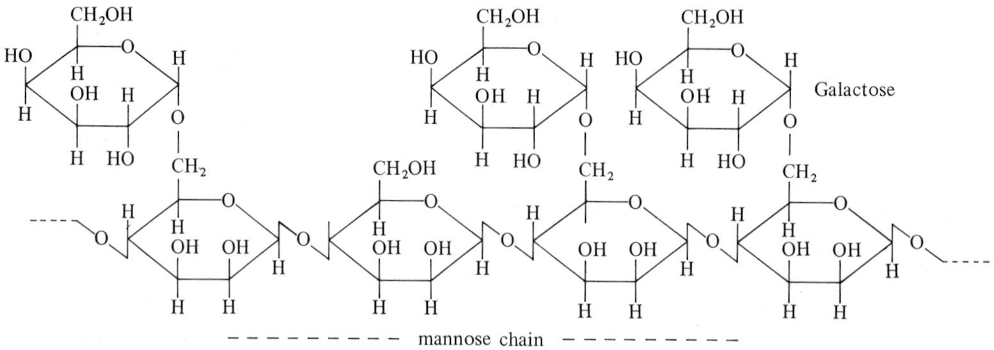

Fig. 2.11. Structure of galactomannan

Table 2.8. Some leguminous seeds containing galactomannans

Species	Percent galactomannan (dry wt)
Ceratonia siliqua (carob)	19
Cyamopsis tetragonolobus (guar)	20.5
Gleditsia triacanthes (honey locust)	18.5
Medicago sativa (lucerne)	9

From data of McCleary and Matheson, 1974 [49]

ble 2.8), Palmae, Annonaceae, Rubiaceae and Convolvulaceae [8, 50]. The degree of galactosylation varies with the species, a feature which may be useful taxonomically. In *Gleditsia ferox*, the endosperm galactomannan consists of 21% galactose whereas that in *Trifolium repens,* also in the Leguminosae, has 49% galactose [61].

Glucomannans occur as reserve materials in certain seeds such as bluebell (*Endymion non-scriptus*) while the mucilages of other species are or contain polyuronides. Xyloglucans (with galactose side chains) are found as "amyloids" in seeds of tamarind (*Tamarindus*) [65], *Annona* [43] and several other species. These polysaccharides again are deposited in the endosperm cell walls.

2.7. Oils or Fats ("Lipids")

Lipids are a chemically heterogeneous group of substances which have in common their solubility in organic solvents such as petroleum ether, hexane or chloroform/methanol. Various sterols, phospholipids and glycerides fall into this group. The predominant storage lipids of seeds are the neutral fats, or oils if they are liquid at "normal" temperatures. These are esters of glycerol and long-chain monocarboxylic acids, so forming the triglycerides or triacylglycerols (for molecular structure see below). Other storage lipids such as glycolipids and phospholipids are present in some seeds, often in significant proportions. In *Briza spicata,* for example, 78% of the total lipids are glycolipids [64]. A very large number of species have oil-storing seeds — more than those which store predominantly protein or carbohydrate. Many of our agriculturally important seeds are oil-seeds and there is a regular search for others which might be similarly exploited. The interested reader is referred to the papers of Earle and his colleagues [13, 26, 42] who have surveyed the oil and protein contents of about 4000 species.

The oil content of a few species shown in Table 2.9 illustrates the range found in seeds. Variations in the amount of oil in any one species are due to seasonal and geographical factors. Indeed, not only can the percentage of oil vary but so can the fatty acid composition, as we shall see later. Some families have members covering a wide range from low to high oil content. In the Leguminosae for example, are *Acacia* spp with only about 2% oil,

Table 2.9. Oil content and location in a variety of seeds

Species	Major storage organ	Oil composition (percent seed dry wt)
Queensland nut: *Macadamia ternifolia*	Cotyledons	75–79
Brazil nut: *Bertholletia excelsa*	Radicle/hypocotyl	65–68
Hazel: *Coryllus avellana*	Cotyledons	60–68
Opium poppy: *Papaver somniferum*	Endosperm	40–55
Almond: *Prunus amygdalus*	Cotyledons	40–55
Oil palm: *Elaeis guineensis*	Endosperm	50
Castor bean: *Ricinus communis*	Endosperm	35–57
Sunflower: *Helianthus annuus*	Cotyledons	32–46
Spurge: *Euphorbia* spp	Endosperm	23–49
Japanese red pine: *Pinus densiflora*	Megagametophyte	35
Soybean: *Glycine max*	Cotyledons	17–22
Tomato: *Lycopersicon esculentum*	Endosperm	15
Broad bean: *Vicia faba*	Cotyledons	8
Maize or corn: *Zea mays*	Embryo	4.7

Arachis hypogaea with up to 56% and many of intermediate values. On the other hand, most members of the Papaveraceae, Sapindaceae and Cucurbitaceae have high levels of oil in their seeds. Generally, seeds rich in oil tend also to be high in protein but not in starch. Interestingly, oil-rich seeds are smaller on average than those with abundant carbohydrate and protein. In the Leguminosae for example, which possesses both kinds of seed, out of a total of 678 screened species the average weight of oil/protein seeds is 109 g/1000 seeds but the weight of protein/starch seeds is 256 g/1000 seeds [42].

The neutral fats or oils are, as we stated above, esters of glycerol and fatty acids:

$$
\begin{array}{l}
\mathrm{CH_2O \cdot OC \cdot R^1} \\
\quad | \\
\mathrm{R^2CO \cdot O \cdot CH} \\
\quad | \\
\mathrm{CH_2O \cdot OC \cdot R^3}
\end{array}
$$

In the triglyceride, the carbon chains represented by R^1, R^2 and R^3 may be identical but usually they are not. Saturated and unsaturated fatty acids are found in the triglycerides. The saturated acids $[CH_3(CH_2)n \, COOH]$ all contain an even number of carbon atoms, n usually being between 4 and 24. Palmitic acid (n=14) is the most common saturated fatty acid of seed oils; some others are caproic (n=4), caprylic (n=6) capric (n=8) lauric (n=10) and myristic acid (n=12) all of which occur in the Palmae, and arachidic (n=18), behenic (n=20) and lignoceric acid (n=22), present in some leguminous seeds. The predominant fatty acids are, however, the unsaturated ones and two of these—oleic and linoleic—are estimated to account for over 60% by weight of all the oils in oil-seed crops. With only a few exceptions, the unsaturated

Fig. 2.12. Variation in fatty acid content as a function of temperature during seed development of rapeseed cv. Nugget. ○----○: oleic acid, ■—■: linolenic acid, △—△: erucic acid, ●—●: linoleic acid. After Canvin, 1965 [22]

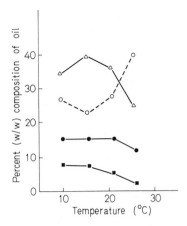

fatty acids of seed oils have 18 carbon atoms and contain 1, 2 or 3 unsaturated centres; thus the structure of an unsaturated fatty acid is sometimes indicated, by abbreviation, as 18:1, 18:2, 18:3 (saturated acids would of course be, say, 18:0 or 14:0). The formulae of the most common unsaturated fatty acids, oleic and linoleic, are respectively $CH_3(CH_2)_7CH:CH(CH_2)_7COOH$ (i.e. 18:1) and $CH_3(CH_2)_3(CH_2CH:CH)_2 (CH_2)_7COOH$ (i.e. 18:2).

Table 2.10 shows the fatty acid composition of some seed oils. It can be seen that the unsaturated acids are the most abundant; fats high in unsaturated acid are liquid, i.e. oils, at normal environmental temperatures. We have already mentioned, however, that considerable variation in the fatty acid content in seeds of any one species can be caused by genetic and environmental factors. This is illustrated by rapeseed (*B. napus*). Considering the genetic control first, in the cultivar "Regina" the major (about 40%) fatty acid is erucic acid (22:1) followed by oleic and linoleic acids, but in the cultivar "Oro" only a trace of erucic acid is present and oleic acid predominates [9]. Changes in fatty acid proportions can also be induced by altering the temperature during seed development (Fig. 2.12); this is an example of the important effect of the environment on oil composition. Another interesting example of how variation in fatty acid content can occur is provided by sunflower (*H. annuus*). Here, the position of the developing "seeds" on the inflorescence is influential such that at different locations on the head they show small but significant differences in the proportions of linoleic, palmitic and oleic acids in their oil [80].

2.7.1. Oil Bodies

The triglyceride reserves are laid down in the form of discrete sub-cellular organelles—the oil bodies. A variety of other names has been used for these storage organelles—spherosome, oleosome and lipid-containing vesicle. The mixed nomenclature reflects the uncertainty concerning the cellular origin of these

Table 2.10. Fatty acid composition of some seed oils

Species	Family	Percentage fatty acid														
		8:0	10:0	12:0	14:0	16:0	18:0	20:0	22:0	24:0	16:1	18:1	18:2	18:3	22:1[a]	Others
Arachis hypogaea	Leguminosae					8.6	3.6	←———	5.9	———→		54.5	27.4			
Glycine max	Leguminosae					7.0	5.5	←———	0.3	———→		23.0	60.0	3.0		
Helianthus annuus	Compositae					←—— 14.1 ——→						17.3	68.6			
Ulmus americana	Ulmaceae	5.3	61.3	5.9	4.6	2.9						11.0	9.0			
Ricinus communis	Euphorbiaceae						0.3					7.0	4.0			Ricinoleic 88[b]
Elaeis guineensis	Palmae	3.0	3.0	52.0	15.0	7.5	2.5					16.0	1.0			
Brassica napus	Cruciferae				0.4	1.5	0.4	0.5	2.0	1.8		14.0	24.0	2.0	55.0	
Zea mays	Gramineae				1.4	10.2	3.0				1.5	49.6	34.3			
Bertholletia excelsa	Lecythidaceae				1.9	14.3	2.7				58.3	22.8				

[a] Erucic acid: $CH_3(CH_2)_7 \ CH:CH \ (CH_2)_{11} \ COOH$

[b] Ricinoleic acid: $CH_3(CH_2)_4 \ CH_2CHOH \ CH_2 \ CH:CH(CH_2)_7 \ COOH$

See Butt and Beevers, 1966 [19] and Hitchcock and Nichols, 1971 [37] for detailed analyses of seed lipids. Only the major fatty acids are listed in this table. Other fatty acids such as ethylenic acids, acetylenic acids and substituted fatty acids (hydroxy-, polyhydroxy-, keto- and expoxy acids) occur in seed fats (for further details, see Hitchcock, 1975 [36]) bodies (see Chap. 3). Oil bodies range in size between 0.2–6 μ diameter according to the species. In those seeds with high oil deposits the bodies pack the cell, filling almost all the space left unoccupied by the other organelles (Fig. 2.7). Where oil is not the principal stored material these bodies are of course less abundant; in cereal aleurone cells, for example, they surround the aleurone grains (Fig. 2.6E) and also lie inside the plasmalemma in a narrow layer.

Connected with the controversial views on the origin of the oil bodies is the question as to whether they are bounded by membranes. Evidence for and against the existence of some kind of peripheral membrane is presented in Section 3.3.8. We should note, in addition, favourable evidence for an oil body membrane raised by the freeze-fracture studies on barley aleurone cells [20b] and by biochemical analyses of peanut and walnut oil bodies which reveal phospholipid and protein characteristic of membranes. Finally, we should be aware that in addition to the possible existence of membrane protein, enzyme protein may also occur in oil bodies. Enzymes for fatty acid biosynthesis have been reported in oil bodies of developing castor bean [35] and for triglyceride hydrolysis (acid lipases) in the mature endosperm of this species: the latter enzymes may not occur in oil bodies of other species (Chap. 6).

Some Articles of General Interest

1. Corner, E.J.H.: The Seeds of Dicotyledons. Cambridge: Univ. Press, 1976, 2 vols
2. Crocker, W., Barton, L.V.: Physiology of Seeds. Waltham, Mass.: Chronica Botanica, 1957
2a. Greenwood, C.T.: Starch. In: Advances in Cereal Science and Technology. Pomeranz, Y. (ed.). St. Paul, Minn.: Am. Assoc. Cereal Chem. Inc. 1976, pp. 119–157
3. Inglett, G.E. (ed.): Symp. Seed Proteins. Westport, Conn.: AVI Publ. 1972
3a. Kasarda, D., Bernardin, J.E., Nimmo, C.C.: Wheat proteins. In: Advances in Cereal Science and Technology. Pomeranz, Y. (ed.). St. Paul, Minn.: Am. Assoc. Cereal Chem. Inc. 1976, pp. 158–236
4. Martin, A.C., Barkley, W.D.: Seed Identification Manual. Berkeley and Los Angeles: Univ. Calif. Press, 1961
5. Pomeranz, Y. (ed.): Wheat—Chemistry and Technology. St. Paul, Minn.: Am. Assoc. Cereal Chem. Inc., 1971
6. Sylvester-Bradley, R., Folkes, B.F.: Cereal grains: their Protein Components and Nutritional Quality. Sci. Prog. (Oxford) **63**, 241–263 (1976)
7. Vaughan, J.G.: The Structure and Utilization of Oil Seeds. London: Chapman and Hall, 1970

References

8. Anderson, E.: Ind. Eng. Chem. **41**, 2887–2888 (1949)
9. Appelqvist, L.-Å.: In: Recent Advances in the Chemistry and Biochemistry of Plant Lipids. Galliard, T., Mercer, E.I. (eds.). London: Academic Press, 1975, Proc. Phytochem. Soc. **12**, 247–286
10. Badenhuizen, N.P.: Chemistry and Biology of the Starch Granule. Vienna: Springer, 1959
11. Badenhuizen, N.P.: The Biogenesis of Starch Granules in Higher Plants. New York: Appleton-Century-Crofts, 1969
12. Bailey, C.J., Boulter, D.: Phytochemistry **11**, 59–64 (1972)
13. Barclay, A.S., Earle, F.R.: Econ. Botany **28**, 178–236 (1974)
14. Basha, S.M.: Cited in Millerd, A.: Ann. Rev. Pl. Physiol. **26**, 53–72 (1975)
15. Beevers, L.: Nitrogen Metabolism in Plants. London: Ed. Arnold, 1976
16. Bell, E.A.: FEBS Lett. **64**, 29–35 (1976)

17. Bietz, J.A., Wall, J.: Cereal Chem. **49**, 416–423 (1972)
18. Briarty, L.G., Coult, D.A., Boulter, D.: J. Exp. Botany **20**, 358–372 (1969)
19. Butt, V.S., Beevers, H.: In: Plant Physiology. Steward, F.C. (ed.). New York: Academic Press, 1966, Vol IVB, pp. 265–414
20a. Buttrose, M.S.: Australian J. Biol. Sci. **16**, 305–317 (1966)
20b. Buttrose, M.S.: Planta (Berl.) **96**, 13–26 (1971)
21. Callow, J.A.: Curr. Adv. Plant Sci. **7**, 181–193 (1975)
22. Canvin, D.T.: Can. J. Botany **43**, 63–69 (1965)
23. Christianson, D.D., Khoo, U., Nielsen, H., Wall, J.S.: Pl. Physiol. **53**, 851–855 (1974)
24. Derbyshire, E., Wright, D.J., Boulter, D.: Phytochemistry **15**, 3–24 (1976)
25. Dieckert, J.W., Dieckert, M.C.: In: Symp. Seed Proteins. Inglett, G.E. (ed.). Westport, Conn.: AVI Publ. 1972, pp. 52–85
26. Earle, F.R., Jones, Q.: Econ. Botany **16**, 221–250 (1962)
27. Ericson, M.C., Chrispeels, M.J.: Pl. Physiol. **52**, 98–104 (1973)
28. Evers, A.D.: Ann. Botany (London) **34**, 547–555 (1970)
29. Evers, A.D.: Die Stärke **23**, 157–162 (1971)
30. Graham, T.A., Gunning, B.E.S.: Nature (London) **228**, 81–82 (1970)
31. Halmer, P., Bewley, J.D., Thorpe, T.A.: Planta (Berl.) **130**, 189–196 (1976)
32. Hamly, D.H.: Botan. Gaz. **93**, 345–375 (1932)
33. Harris, N., Chrispeels, M.J.: Pl. Physiol. **56**, 292–299 (1975)
34. Hartig, T.: Botan. Z. **13**, 881–882 (1855)
35. Harwood, J.L., Stumpf, P.K.: Lipids **7**, 8–19 (1972)
36. Hitchcock, C.: In: Recent Advances in the Chemistry and Biochemistry of Plant Lipids. Galliard, T., Mercer, E.I. (eds.). London: Academic Press, 1975, Proc. Phytochem. Soc. **12**, 1–19
37. Hitchcock, C., Nichols, B.W.: Plant Lipid Biochemistry. London: Academic Press, 1971
38. Hofsten, A.V.: Physiol. Plantarum **29**, 76–81 (1973)
39. Horner, H.T., Arnott, H.J.: Botan. Gaz. **127**, 48–64 (1966)
40. Jacobsen, J.V., Knox, R.B., Pyliotis, N.A.: Planta (Berl.) **101**, 189–209 (1971)
41. Jiminez, J.R.: In: Proc. High-Lysine Corn Conference. Mertz, E.T., Nelson, O.E. (eds.). Washington, D.C.: Corn Ind. Res. Found. 1966, pp. 74–79
42. Jones, Q., Earle, F.R.: Econ. Botany **20**, 127–155 (1966)
43. Kooiman, P.: Phytochemistry **6**, 1665–1673 (1967)
44. Lampe, L.: Botan. Gaz. **91**, 337–376 (1931)
45. Lott, J.N.A.: Pl. Physiol. **55**, 913–916 (1975)
46. Lott, J.N.A., Larsen, P.L., Darley, J.J.: Can. J. Botany **49**, 1777–1786 (1971)
47. Lui, N.S.T., Altschul, A.M.: Arch. Biochem. Biophys. **121**, 678–684 (1967)
48. Martin, A.C.: Am. Midland Naturalist **36**, 513–660 (1946)
49. McCleary, B.V., Matheson, N.K.: Phytochemistry **13**, 1747–1757 (1974)
50. McCleary, B.V., Matheson, N.K.: Phytochemistry **14**, 1187–1194 (1975)
51. McMasters, M.M., Hinton, J.J.C., Bradbury, D.: In: Wheat – Chemistry and Technology. Pomeranz, Y. (ed.). St. Paul, Minn.: Am. Assoc. Cereal Chem. Inc., pp. 51–113
52. Meier, H.: Biochim. Biophys. Acta **28**, 229–240 (1958)
53. Millerd, A.: Ann. Rev. Pl. Physiol. **26**, 53–72 (1975)
54. Morris, G.F.I., Thurman, D.A., Boulter, D.: Phytochemistry **9**, 1707–1714 (1970)
55. Ory, R.L.: In: Symp. Seed Proteins. Inglett, G.E. (ed.). Westport, Conn.: AVI Publ. 1972, pp. 86–98
56. Ory, R.L., Henningsen, K.W.: Pl. Physiol. **44**, 1488–1498 (1969)
57. Osborne, T.B.: The Vegetable Proteins, 2nd ed. London: Longmans Green, 1924
58. Palmer, G.H.: J. Inst. Brew. **78**, 326–332 (1972)
59. Potter, A.L., Hassid, W.Z.: J. Am. Chem. Soc. **70**, 3774–3780 (1948)
60. Pusztai, A., Watt, W.B.: Biochim. Biophys. Acta **207**, 413–431 (1970)
61. Reid, J.S.G., Meier, H.: Z. Pflanzenphysiol. **62**, 89–92 (1970)
62. Rest, J., Vaughan, J.G.: Planta (Berl.) **105**, 245–262 (1970)
63. Seidemann, J.: Stärke-Atlas. Berlin-Hamburg: Paul Parey, 1966
64. Smith, C.R., Jr., Wolff, I.A.: Lipids **1**, 123–127 (1966)

65. Srivistava, H.C., Singh, P.P.: Carbohyd. Res. **4**, 326–342 (1967)
66. St. Angelo, A., Yatsu, L.Y., Altschul, A.M.: Arch. Biochem. Biophys. **124**, 199–205 (1968)
67. Stevens, D.J., McDermott, E.E., Pace, J.: J. Sci. Food Agr. **14**, 284–287 (1963)
68. Suvorov, V.I., Buzulukova, N.P., Sobolev, A.M., Sveshnikova, L.N.: Fiziol. Rast. **17**, 1223–1231 (1970)
69. Tanaka, K., Yoshida, T., Asada, K., Kabai, Z.: Arch. Biochem. Biophys. **155**, 136–143 (1973)
70. Tombs, M.P.: Pl. Physiol. **42**, 797–813 (1967)
71. Van Staden, J., Comins, N.R.: Planta (Berl.) **130**, 219–222 (1976)
72. Vigil, E.L.: J. Cell Biol. **46**, 435–454 (1970)
73. Werker, E., Vaughan, J.G.: Planta (Berl.) **116**, 243–255 (1974)
74. Winton, A.L., Winton, K.B.: The Structure and Composition of Foods. New York: J. Wiley and Sons, 1932, Vol. I
75. Wolf, M.J., Khoo, U., Seckinger, H.L.: Science **157**, 556–557 (1967)
76. Wolfram, M.L., Laver, M.L., Patin, D.L.: J. Org. Chem. **26**, 4533–4535 (1961)
77. Wright, D.J., Boulter, D.: Planta (Berl.) **105**, 60–65 (1972)
78. Yatsu, L.Y.: J. Cell Biol. **25**, 193–199 (1965)
79. Yatsu, L.Y., Jacks, T.J.: Arch. Biochem. Biophys. **124**, 466–471 (1968)
80. Zimmerman, D.C., Fick, G.N.: J. Am. Oil Chem. Soc. **50**, 273–275 (1973)

Chapter 3. The Legacy of Seed Maturation

3.1. General Developmental Pattern

It is not our intention to present a detailed account of the morphological and anatomical aspects of seed development; for this the reader is directed towards the relevant references in the bibliography [3, 14, 16]. It is necessary, nevertheless, to review briefly the developmental processes in the Gymnosperms and Angiosperms so that we can later set the physiological and biochemical events in their appropriate context.

The Gymnosperm egg nucleus is situated within the female gametophyte and fertilization is effected by a motile gamete released from the pollen tube. Subsequent development of the zygote of conifers into the embryo involves the production of several free nuclei around which cell walls are then laid down to form the proembryo. A sequence of cell divisions results in the formation of embryonal and suspensor cells. This cell complex may develop to form a single embryo and an attached, elongated suspensor, but often there is cell separation to form four separate embryos side by side (polyembryony). One of these eventually secures ascendancy and the others degenerate. As the young embryo commences development in the seed, the central portion of the female (mega)gametophyte (which is haploid) breaks down to form the "corrosion cavity," into which the embryo is pushed by elongation of the suspensor. Lipid, starch and protein reserves are deposited into the persistent parts of the gametophyte, subsequently to be utilized following seed germination. Thus the female gametophyte in Gymnosperms serves the dual function of bearing the gametes and nourishing the embryo.

The distinguishing feature of the fertilization process in Angiosperms is the involvement of two male nuclei. One non-motile male nucleus released from the pollen tube fuses with the egg nucleus to restore diploidy in the zygote, and the other fuses with two polar nuclei to yield a triploid nucleus. Development of the zygote does not involve a free-nuclear stage; the first division results in an axial (distal or apical) cell and a basal cell. In the dicots the basal cell undergoes limited cell division to produce the suspensor, and may even contribute to the embryo: the axial cell divides and differentiates to form the embryo. In the monocots the basal cell rarely undergoes further division but forms the terminal (or haustorial) cell of the suspensor: the embryo and the other few cells of a limited suspensor are formed from the axial cell. As the name implies the dicots incorporate two cotyledons into the mature structure of the embryo, whereas the monocots possess only one. The single cotyledon of the Gramineae is reduced to the absorptive scutellum which, in the mature grain, lies adjacent to the nutritive endosperm. Other special features of mature Graminaceous embryos include a tissue which covers the root (the coleorhiza) and one which is around the plumule and covers the first foliage leaf (the coleoptile) (Fig. 2.5).

The true endosperm is a tissue unique to the Angiosperms and arises by division of the triple fusion nucleus; hence its cells are characteristically triploid though in a few species it can be diploid, tetraploid, pentaploid or polyploid. There are 2 major types of endosperm development: (1) nuclear, where the endosperm nucleus undergoes a series of free-nuclear divisions prior to cellularization (e.g. wheat, apple, charlock); (2) cellular, where there is no free nuclear phase (e.g. *Impatiens* and *Lobelia*). Both of these types occur in the monocots and dicots: a rarer form of development, the helobial type (which is intermediate between the other two since free nuclear division is preceded and followed by cell formation), is exclusive to some monocots. During growth of the endosperm some nutrients are drawn in from adjacent tissues (in some species the endosperm develops haustoria to penetrate these tissues) and others are synthesized *in situ* using transported material. The effect is often to surround the developing embryo with an available food source upon which it can draw during its maturation and subsequent germination/growth stages. In cereals there is limited utilization of the endosperm reserves during embryo maturation. The remnants of the depleted endosperm cells form the intermediate layer which lies between the starchy endosperm and the scutellum of the mature embryo (Fig. 2.5). In non-endospermic dicot seeds the endosperm reserves are depleted during embryo maturation and then reorganized in the cotyledons which act as storage organs for the germinated embryos (e.g. pea, peanut and many other legumes, cucumber, and certain other oil seeds, Chap. 2). In the mature lettuce seed the endosperm persists as a 2–3-cell-thick structure surrounding the embryo (Fig. 2.3 B) and acts as a mechanical barrier to impede radicle emergence. In contrast, the endosperm remains as a permanent storage tissue in endospermic dicot seeds (e.g. castor bean, Fig. 2.4 C) and in some species this persistent storage tissue may even have an aleurone layer (e.g. fenugreek, crimson clover and lucerne — see Chaps. 2 and 6). The nucellus develops into the perisperm in a relatively small number of species (e.g. *Yucca, Coffea*); in most cases however, development is limited and no recognizable perisperm occurs in the majority of seeds.

The integuments bounding the ovule undergo conspicuous changes during seed maturation becoming organized into a seed coat (testa). Seed coat structure is variable between different species and has considerable taxonomic value. Seed coats are reviewed briefly in Chapter 2 and the importance of the coat in the maintenance of seed dormancy will be discussed in Volume 2 of this book.

A large section of this Chapter is devoted to the production and storage of reserves in the seed, for it is upon such reserves that the embryo will eventually depend, not so much for germination, but for its successful subsequent growth and development into a seedling. Work on the physiology and biochemistry of seed maturation has been limited almost exclusively to a few commercially important annuals. For convenience we shall consider nutrient assimilation in the cereals, which have a persistent endosperm, then those dicots in which the endosperm is resorbed during maturation, and finally we will outline what little is known about the endospermic dicot seeds.

Other changes associated with the metabolism of the embryo or the axis during the maturation period will receive more limited attention, for they are poorly understood. Attempts have been made to culture very young embryos and follow their development in the test tube [e.g. 125a]. This approach has so far not met with general success since many embryos, once excised, do not develop further but either precociously germinate or, alternatively, die. The major drawback so far appears to be the lack of clearly-defined culture conditions to allow the young embryos to continue their development. Perhaps this is understandable because we know little about the factors which are provided by the mother plant to elicit the normal maturation process of the attached embryo. Once this hurdle is overcome though, we can anticipate many exciting discoveries in the field of embryogenesis, particularly in relation to the regulatory mechanisms involved in the control of ordered differentiation and development.

But one special aspect of the events in the developing embryo, apart from reserve synthesis, which will be considered concerns nucleic acid and protein metabolism. We will devote some attention to the synthesis of mRNA species which may remain untranslated in the embryo but which are conserved in the mature, dry seed perhaps to be used in germination and growth. In short, we will consider the possibility that these "messages" are laid down to prepare the embryo for subsequent germination.

Finally, we discuss the hormonal changes that occur in developing seeds and the role that these substances may play in seed development itself and in preparation for germination.

3.2. Filling of the Grain (Kernel) in Cereals

3.2.1. Source of Assimilates for Starch Formation

Carbohydrate (starch) reserves in the grains of temperate cereals are largely derived from the products of the photosynthesis taking place during grain development (or filling). Carbohydrate stored in the stems prior to anthesis (splitting of the anthers) may contribute no more than 5–10% of the final grain weight. The major source of assimilate for wheat and barley grains, which in both species are arranged in a spike inflorescence, is the current photosynthesis carried out in the flag leaf, stem, and ear, these regions making quantitatively different contributions. There can be little doubt that the contribution by the flag leaf is high, whereas that by the ear itself is variable and is dependent upon the size of the awn. In oats (where the grains are in a panicle inflorescence) the penultimate leaf appears to be of equal importance to the flag leaf in supplying assimilates to the filling grain [95]. Studies are required on a larger number of cereals to determine if the contributions by the penultimate and flag leaves are related to inflorescence type.

In a comparison between two awnless wheat varieties (Jufy I and Atle) and two awned barley varieties (Proctor and Plumage Archer), Thorne [179] showed that 17–30% of the wheat grain weight was attributable to fixation of CO_2 by the ear, but more than this was lost by grain respiration. Consequently, the flag leaf contributed 110–120% of the total carbon assimilated in the grain. In barley, the CO_2 fixation into assimilates in the grain was approximately 50% derived from the flag leaf and 50% from the ear. The differences in contribution by the ears of these two species was attributed to differences in the sizes of their awns. Similar results have now been obtained in experiments using awned and awnless varieties of the same species [39, 55]. It is interesting to note that assimilates from the flag leaf of wheat are distributed preferentially to the lower central spikelets, and setting of the grains in the upper florets (and hence grain yield per ear) may be reduced by the rapid development of grains from the first flowers to reach anthesis [148]. On the other hand, photosynthates from the awn are predominantly supplied to the grain which it subtends [194]. It would seem, therefore, that selection of varieties with a large and photosynthetically persistent awn could result in fuller and more uniformly developed cereal grains. This may not always be so, however, for although in dry conditions awned varieties of wheat give higher yields than unawned ones this does not occur in wetter climates where yields of the former varieties may even be reduced [112]. Metabolic activity of the developing grain itself is also an important factor to be considered (see later).

The transported form of photosynthetically fixed carbon is sucrose, and to understand how this is transported into the developing grain we have first to consider the anatomy of the vascular connection between the grain and the mother plant (Fig. 3.1A). The vascular tissue which supplies the unfertilized ovule with nutrients is obliterated by the developmental processes which follow fertilization and there is no direct vascular connection between the rachilla (small stalk upon which the grain is borne) and the inside of the grain itself. In wheat, there is a discontinuity of the vascular system running from the mother plant to the crease (or longitudinal furrow) of the grain in the nodal region where the glumes, lemma, palea and grain are attached to the rachilla. Part of this discontinuity is filled by specialized transfer cells which might regulate the transfer of solutes to the grain [206]. Vascular tissue runs in the crease of the grain (Fig. 3.1A, B and C) and this has phloem connections from the node to the apex, but the xylem is discontinuous beyond the mid-point of the grain. Between the vascular tissue and the inside of the grain lies the funiculus-chalazal (or placento-chalazal) region and the nucellar projection, and photosynthetic assimilates transported to the grain in the phloem must pass through these and into the endosperm.

There is still considerable debate as to whether sucrose can be transported from the phloem into the developing endosperm in an unmodified form, or whether it must first be hydrolysed to glucose and fructose and then resynthesized into the disaccharide. The funiculus-chalazal region of Zea mays shows increased invertase activity at about the time when starch synthesis commences in the endosperm [163], and this is associated with an increase in free hexoses in this region. One current hypothesis [155] favours the cleavage of sucrose during

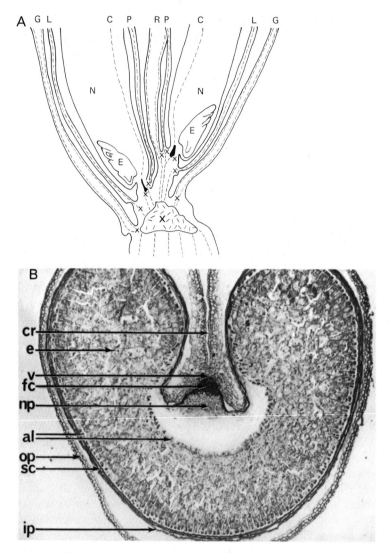

Fig. 3.1. (A) Sagittal section of a wheat spikelet showing the relation between the caryopses (grains) and vascular tissues of the mother plant. *E*: embryo (note lack of vascular connections between this and the mother plant or the rest of the grain), *N*: starchy endosperm, *C*: vascular tissue running inside the crease, *X*: regions where transfer cells are associated with the vascular tissue, *R*: rachilla, *P*: paleas, *L*: lemmas, *G*: glumes. The distribution of vascular tissues is indicated by broken lines. After Zee and O'Brien, 1971 [206]. (B) Cross section through a developing wheat kernel 14 days after anthesis at midpoint between base and apex. *cr*: crease, *e*: starchy endosperm, *v*: vascular bundles, *fc*: funiculus-chalazal region, *np*: nucellar projection *al*: aleurone layer, *op*: outer pericarp, *sc*: seed coat, *ip*: inner pericarp. From Sakri and Shannon, 1975 [155].
(C) Cross section through the crease region of a nearly full-sized wheat grain containing approximately 40% moisture, taken 3 weeks after anthesis. *OE*: outer epidermis of pericarp, *IE*: inner epidermis of pericarp, *I*: outer and inner integuments, *NE*: nucellar epidermis, *C*: crease, *VB*: vascular bundle, *CH*: chalaza, *NP*: nucellar projection, *A*: aleurone layer, *E*: endosperm. From Frazier and Appalanaidu, 1965 [60]

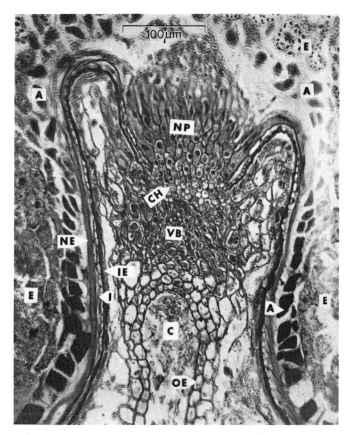

Fig. 3.1 C

phloem unloading by a cell wall-bound invertase and the monosaccharides diffuse, via the apoplast, through the funiculus-chalazal region, the nucellar projection, and into the free space of the endosperm. The restoration of sucrose then occurs after uptake into the endosperm cells [161]. Although starch deposited in the grain is synthesized from sucrose transported from photosynthetic organs, the accumulation of starch in wheat grains during its linear phase of synthesis is regulated not so much by the supply of sucrose to the grain from the mother plant as by processes operating within the grain itself [87–89]. The mode of regulation might be effected by resistances within the grain to the movement of sucrose from the vascular tissue to the endosperm and, in maize at least, control of starch formation might be mediated through the activity of invertase controlling the supply of incoming precursors. Eventually there is the onset of the declining phase of starch accumulation, and kernels begin to ripen. In wheat, this decline is apparently not attributable to a reduction or restriction in the supply of assimilate to the ripening grain, but to a fall in the capacity of the grains to utilize the assimilate to synthesize starch [89]. In summary then, it seems that the supply of assimilate to the developing grain is a less important determinant of grain growth than are factors operating within the grain itself.

3.2.2. Development of the Starchy Endosperm and Aleurone Layer

Development of the endosperm and aleurone layer of the English spring wheat varieties Svenno and Caesar have been followed from anthesis to maturity [57]. Anthesis is the accepted starting point for such studies, since it is assumed that pollination and fertilization quickly follow this event, even though such processes are not themselves detectable by the experimenter. The pattern of development is shown in Figure 3.2. From day 0 to day 2 after anthesis free division of the triploid nucleus occurs and cell walls begin to form, but of course the endosperm is barely discernible. At day 4 the endosperm has enlarged somewhat, due to cell division, and an outer meristematic layer is evident. This layer undergoes tangential divisions to produce cells to the inside, and radial divisions to extend its surface area as the inner mass of endosperm cells increases. From day 6 to day 14, events take place which will determine the final shape of the mature wheat grain. Division ceases in the thick-walled region where the vascular and funiculus-chalazal areas will eventually be situated and there is differential expansion on the flanks to cause them to turn down and form the crease between them. This meristematic, outer layer of the endosperm ceases to divide at about day 14, and is destined for a different function from the bulk of the endosperm: it becomes the aleurone layer. (Throughout this Chapter and the rest of the book we will refer to this outer layer of endosperm as the aleurone layer or tissue and the rest of the endosperm as the starchy endosperm or simply, endosperm.) All further increase in size of the endosperm after this time is by cell expansion.

After the phase of meristematic activity ceases, the aleurone layer and endosperm cells become modified for their future role. In wheat (cv. Heron) there is some starch present in the aleurone layer about 10 days after anthesis, persisting until about the 21st day when it is hydrolysed [127]. At day 14 the cytoplasm has vacuoles containing electron-dense bodies: these could be the phytin cores of the future aleurone grains (see Chaps. 2 and 6). After four weeks the thin cell walls of the aleurone layer start to thicken and there is proliferation of mitochondria and endoplasmic reticulum. Lipid bodies are in evidence around the aleurone grains at the 5th week of development and by six weeks the aleurone layer is easily distinguishable from the underlying (endosperm) regions. Some micrographs of the ultrastructure of mature aleurone cells of barley and wheat are shown in Chapters 2 and 7.

Starch deposition begins in the region of the furrow (the region nearest the sucrose source) after cell division in the endosperm ceases. The young endosperm cells have prominent nuclei, numerous mitochondria, endoplasmic

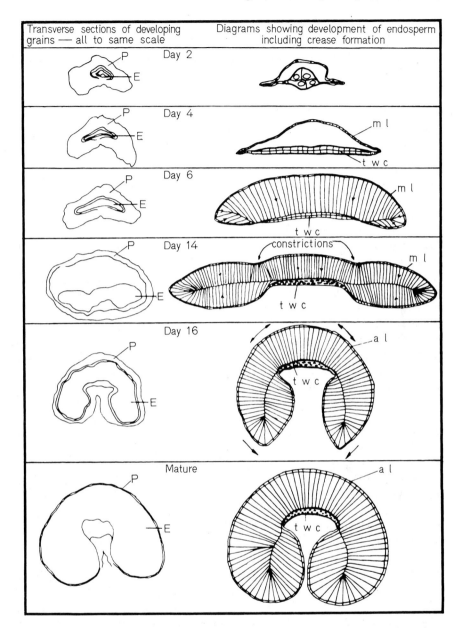

Fig. 3.2. The development of the wheat endosperm. *P*: pericarp, *E*: endosperm, *ml*: meriste-matic layer, *twc*: thick-walled cells, *al*: aleurone layer. The constrictions indicated on the Day 14 sample are now recognised as artefacts of fixation. By Day 16 enough storage material is present to resist distortion. After Evers, 1970 [57]

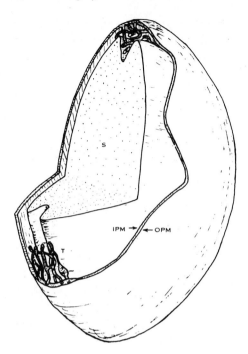

Fig. 3.3. A three-dimensional diagram illustrating the structure of the developing wheat endosperm amyloplast. *S*: starch, *OPM*: outer plastid membrane, *IPM*: inner plastid membrane, *T*: tubular invaginations of IPM. From Buttrose, 1963 [37]

reticulum, golgi bodies (dictyosomes) and proplastids [36]. It is within the proplastids that the starch granules form, one granule per proplastid in barley, wheat and rye and many per proplastid in oats and rice. Proplastids are bounded by a double membrane, the inner membrane in barley, oats and wheat (but not maize) being invaginated to form tubular processes (tubuli) into the plastid stroma (Fig. 3.3). These tubuli when present are detectable during starch formation but are few, or absent, in the mature starch grain (amyloplast). It has been suggested that they are involved in starch synthesis as sites where enzymes bind, or they aid the transport of precursors into the granule. While there is no evidence to substantiate these suggestions, it is known that some enzymes of starch synthesis are bound to starch grains. However, even maize starch grains, which do not have tubuli, have enzymes bound to them (see later). The first sign of starch synthesis is the appearance of granules which initially occupy only a small volume of the proplastid. In barley only one small granule is visible (Fig. 3.4A) but in rice and oats several can be seen (Fig. 3.4B) each of which develops into a mature granule within the grain (Fig. 3.4C): there may be over 100 granules per mature grain in oats. No new proplastids arise in the endosperm of barley later than two weeks after anthesis. Small granules appear in the developing starch grains after this time in the narrow layer of stroma between the large granule and the plastid membrane (Fig. 3.4D). This eventually separates from the parent plastid by constriction of the plastid membrane. Since the role of the endosperm cells is to act as sites of storage of nutrients, to the exclusion of the cytoplasm and the active metabolic components therein, the presence of small starch grains (3 μ diameter and smaller) between the larger ones (up to 35 μ diameter) leads to a more efficient packing of the cell.

Characteristic of starch granules are concentric rings or shells (Fig. 3.4E). Each ring probably represents one day's "growth" of starch. The biochemical basis of ring formation, which is due to a sudden discontinuity in the packing density of the starch, is completely unknown but physiologically it may be related in some way to changes in the ambient environmental conditions.

A developing kernel of Z. *mays* is composed of cells of varying physiological ages, and there appears to be a major gradient in cell development from cells located in the basal endosperm region to those in the central crown region and a minor gradient from the periphery towards the centre of the endosperm [162]. As a consequence, at any one time during the development of the kernel there is variability between cells as to their starch content. For example, in the lower third of the endosperm of maize kernels 20–30 days after anthesis some cells are completely devoid of starch, others have tiny amyloplasts and still others are essentially filled with granules.

3.2.3. The Synthesis of Starch

Whether it is transported unmodified into the endosperm, or resynthesized there from its component monosaccharides, sucrose is the starting point for starch formation. It is converted to fructose and uridine diphosphoglucose (UDPG) by sucrose synthetase (sucrose-UDP glucosyl transferase):

$$\text{Sucrose} + \text{UDP} \rightleftharpoons \text{Fructose} + \text{UDPG}$$

Fructose is converted to glucose-1-phosphate (G-1-P), as is the UDPG, the latter by UDPG pyrophosphorylase:

$$\text{UDPG} + \text{PPi} \rightleftharpoons \text{G-1-P} + \text{UTP}$$

This G-1-P must now be combined to form another sugar nucleotide, ADPG, by ADPG pyrophosphorylase:

$$\text{G-1-P} + \text{ATP} \rightleftharpoons \text{ADPG} + \text{PPi}$$

The glucose from the ADPG is then added on to a small glucan primer to increase its chain length by one glucose unit: this is repeated until the starch

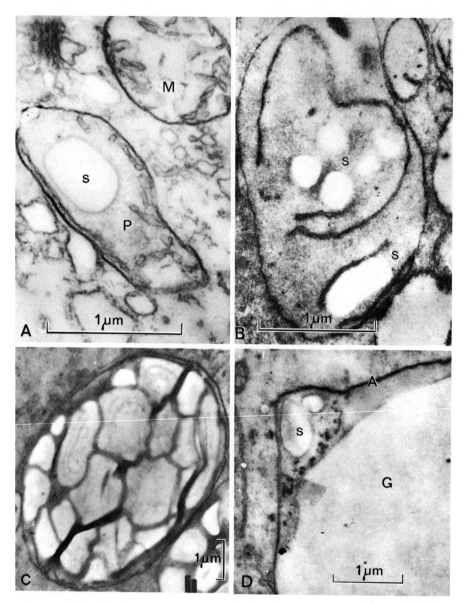

Fig. 3.4A–E. Development and structure of starch grains in cereal endosperms. (A) Very young barley endosperm cell showing proplastid (*P*) with single starch granule (*s*) and mitochondrion (*M*). (B) Proplastid from young oat endosperm with several starch granules (*s*). (C) Maturing compound starch grain from oat endosperm. (D) Development in barley of small starch granules (*s*) between a large granule (*G*) and the amyloplast membrane (*A*) at 2 weeks after anthesis. (E) Waxy maize starch grain, acid-treated to show concentric rings. From Buttrose, 1960 [36]

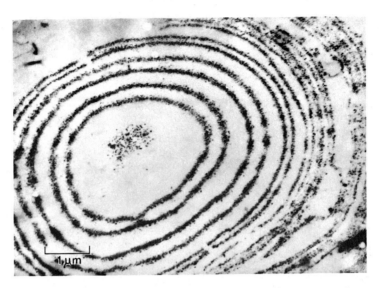

Fig. 3.4 E

molecule is completed. The enzyme involved is ADPG-starch synthetase (ADPG-starch glucosyltransferase; ADPglucose α-glucan 4α-glucosyltransferase):

$$\text{ADPG} + \text{Primer} \ (G_n) \ \rightarrow \ \text{Glucosyl primer} \ (G_{n+1}) + \text{ADP}$$

Another complement of enzymes capable of catalyzing starch synthesis are the phosphorylases, four of which are present at different stages of development of the maize endosperm [182]. These enzymes utilize G-1-P as starting substrate:

$$\text{G-1-P} + \text{Primer} \ (G_n) \ \rightleftharpoons \ \text{Glucosyl primer} \ (G_{n+1}) + \text{Pi}$$

The role of these enzymes in starch synthesis in the developing endosperm is debatable. It has been suggested that in maize and barley [26, 183] phosphorylase and/or UDPG-starch synthetase could play a role in the early production of a starch primer—a small glucose oligomer—thus initiating starch synthesis, which is then completed by the more efficient ADPG-primed starch synthetase system. The phosphorylase enzymes of maize can initiate primer-free starch synthesis, or use maltose as primer, but the significance of this process is unclear because these enzymes (and the UDPG-starch synthetase) arise in the developing endosperm when the endosperm already contains low levels of starch. One possible explanation for this concurrent presence of priming enzymes and developing starch grains in the maize endosperm is that the enzymes are priming starch synthesis in some cells, while more developed cells (see earlier) are already synthesizing starch. This explanation might be unnecessary if a more recent observation proves to be correct, i.e. that ADPG-starch synthetase is present from the earliest stages of starch formation and is capable of initiating starch synthesis in the absence of a primer [138].

There can be little doubt as to the importance of ADPG in starch synthesis, for shrunken-2 (sh_2) mutants of maize which contain only 10–12% ADPG pyrophosphorylase activity synthesize only 25–30% as much starch as normal maize [46, 181]. The question as to whether UDPG can also act as a glucose donor in a reaction catalyzed by a UDPG-starch synthetase, as suggested earlier, is still being debated. No mutants deficient for UDPG pyrophosphorylase have been isolated yet. Maize ADPG starch synthetase is a soluble enzyme, but there are also starch granule-bound nucleoside diphosphate glucose starch synthetases, which contain at least 10% of the starch-synthesizing capacity [180]. One of these bound synthetases might use UDPG as substrate. The waxy mutant of maize produces no amylose (but amylopectin instead), and has no granule-bound starch synthetase. On the basis of this observation it has been postulated that the soluble (ADPG primed) starch synthetase forms amylopectin and the bound (ADPG or UDPG primed) synthetase produces amylose. Since 75% of normal maize starch is amylopectin and only 25% is amylose, and since the enzymes for starch production from ADPG are the most active, this idea does not seem unreasonable. In the developing barley endosperm, both UDPG and ADPG seem capable of acting as glucosyl donors [25]. While the ADPG appears to be the more active donor, the synthetase enzyme, which utilizes it, is associated exclusively with the amyloplast fraction and the UDPG starch synthetase is mainly soluble, with a little bound activity. In rice endosperm the soluble starch synthetase only utilizes ADPG [178] and is the major form in waxy grains.

In addition to the linear (α-1,4)-linked glucose residues of amylose, starch also contains the branched amylopectin; indeed, in certain cereals such as maize, amylopectin is the major component. The synthesis of amylopectin is less well understood but branching (i.e. formation of α-1,6-D-glucosidic linkages) is known to be brought about by the branching enzyme, also called the Q-enzyme. Amylose and amylopectin of the starch grains of normal cereals are synthesized concurrently so some mechanism must exist to protect the developing amylose molecules from the branching enzyme. One suggestion is that starch synthetase binds to the developing amylose chain to protect it from attack by the branching enzyme [24], although somehow, for amylopectin synthesis both enzymes must work cooperatively. A more recent hypothesis invokes the participation of phospholipids in the regulation of amylopectin synthesis [187]. Phospholipids can inhibit the activities of starch hydrolases (α- and β-amylase, starch phosphorylase: see Chap. 6) and the branching enzyme, but have a relatively minor effect on starch synthetase activity and the anabolic activity of starch phosphorylase. So, if phospholipids were to bind to certain amylose chains they would protect them from breakdown by the starch hydrolases (although this is probably not important in developing grains because they are present only in insignificant quantities) and also from the branching enzyme, but not from further elongation by the synthetases. On the other hand, any amylose chains without attached phospholipids would be converted to amylopectin. While a role for phospholipids in the control of starch synthesis still requires verification, the proposed hypothesis is exciting and worthy of further investigation.

To conclude this section, the pathway for starch synthesis is summarized below:

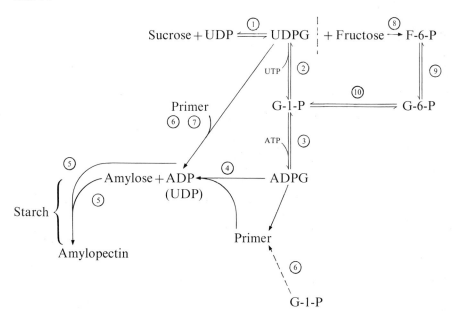

Enzymes: 1: Sucrose synthetase   6: Starch phosphorylase
          2: UDPG pyrophosphorylase   7: UDPG starch synthetase
          3: ADPG pyrophosphorylase   8: Fructokinase
          4: ADPG starch synthetase   9: Hexose phosphate isomerase
          5: Branching (Q-) enzyme   10: Phosphoglucomutase

Formation of G-1-P by the route of sucrose hydrolysis (by invertase), conversion of glucose and fructose to G-6-P, and then to G-1-P is a possible alternative pathway to ① and ②, but seemingly unimportant (except perhaps at the earliest stages of starch synthesis in some cereals).

3.2.4. Protein and RNA Synthesis in the Developing Endosperm

The process of endosperm development clearly involves the synthesis of many proteins, not only those enzymes involved in general cell metabolism and the synthesis of reserves, but also those proteins which are themselves laid down as storage material. Biosynthetic activity of the maize kernel increases from about the 8th day after anthesis up to about the 28th day (Fig. 3.5), after which it declines. The initial increase may, in part, be a reflection of increased enzyme synthesis within individual cells of the maize endosperm and in part an increase in the number of cells commencing protein synthesis as they approach

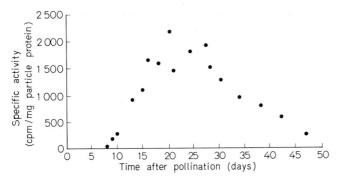

Fig. 3.5. The amino acid-incorporating activity of unwashed particles from maize kernels at different stages of development. A polysome-rich fraction (unwashed particle) was extracted from maize kernels at different stages of development and its capacity to catalyse protein synthesis in vitro tested. Each point represents the protein-synthesizing capacity of whole kernels at the time of extraction. After Rabson et al., 1961 [146]

a state of sufficient maturity. After 25–35 days about 80% of the protein-synthesizing capacity resides within the endosperm which, at this time, reaches maximum size and starts drying. The embryo, which is still developing, accounts for the rest of the protein-synthesizing activity, but at later stages of development the contribution by the endosperm falls dramatically and the embryo then accounts for 75% of the observed protein synthesis. Eventually, by day 50, the starchy endosperm and embryo can no longer synthesize proteins, the former because the actively metabolizing components of the cells have been crushed by the expanding starch and protein reserves and the latter possibly because the increasing water stress imposed by the drying of the grain has restricted the capacity of the ribosomes to initiate protein synthesis on the messenger RNA. In the dry embryo the ribosomes and messenger RNA are conserved separately within the cells (see later, and Chap. 5).

The above data on the protein-synthesizing capacity of maize was obtained using a waxy single cross WF9 × Bear 38. Other changes within the endosperm and embryo have been followed using the inbred line WF9 [85], in which the time course of grain development is a few days slower. Here, the massive synthesis of the reserves in the endosperm begins after about two weeks (Fig. 3.6A) at which time the embryo also starts to grow. Fresh weight (Fig. 3.6B) increases up to about 30 days and then drying commences. Synthesis of protein in maize endosperms appears to be biphasic (Fig. 3.6C). There is an increase between days 15–25, followed by a lag, and then a larger and more rapid increase after day 37. The free amino acid content of the endosperm increases up to the time when the second burst of protein synthesis takes place and then declines rapidly (Fig. 3.6D). The initial increase in protein synthesis probably reflects an increase in enzymes and structural proteins for endosperm growth and metabolism; the second increase, which occurs when protein-synthesizing capacity and the amino acid pool is declining, is a reflection of a terminal event in the endosperm, the synthesis of reserve protein.

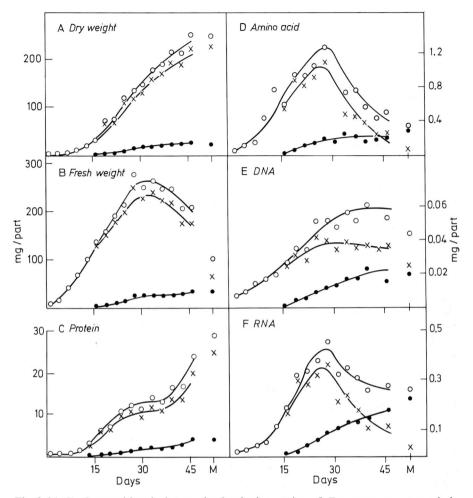

Fig. 3.6A–F. Compositional changes in developing grains of *Zea mays*. ○——○: whole grains; ×——×: endosperm; ●——●: embryo; *M*: mature grains. After Ingle et al., 1965 [85]

Synthesis of storage protein as a late event during endosperm maturation is also characteristic of rice and wheat, and while in neither is protein synthesis biphasic [44, 90], there is an initial increase and then a decline in the free amino acid pool of the endosperm. In rice, the amount of storage protein synthesized appears to be related (at least partially) to the size of the amino acid pool, protein-rich kernels having a larger pool at early stages of development. A major proportion of the storage protein is sequestered within protein bodies in the endosperm. In wheat, these bodies appear to arise at about the same time as the starch grains, as cores of protein lying within large sacs delimited by membranes [37, 129]. Outside these bodies, associated with the periphery, clusters of ribosomes can be observed. These may be involved in the synthesis of the protein body storage contents, and recent studies on *Zea mays* have shown that membrane-bound polysomes, some isolated from the outer membranes of protein bodies[1], can synthesize a protein which is probably zein [34, 99a]. Some storage protein (e.g. glutelin) found in the cytoplasm unassociated with the protein bodies might be synthesized by an independent, soluble system [128] although the experimental evidence is suspect because precautions to preclude contributions by contaminating bacteria were inadequate. Thus, whether two separate protein-synthesizing systems do indeed exist in the developing endosperm is still unresolved. It would be useful to have more information on when the free and sequestered proteins are being synthesized during maturation, if their synthesis is dependent upon the immediate prior synthesis of new RNA (particularly messenger RNA), their sites of biosynthesis and the mechanism of their deposition. The once-accepted contention that developing protein bodies themselves independently synthesize their own proteins within has now been discredited.

Predictably, DNA content of the embryo continues to increase up to the time of maturation (Fig. 3.6E). DNA in the endosperm increases up to the time when cell division ceases (ca. day 20) and then gradually declines as the dying cells become filled with storage materials. The RNA content of the embryo increases throughout the developmental period and reflects the growth and increasing metabolism of this region (Fig. 3.6F). RNA components necessary for protein synthesis during subsequent germination are stored in the dry embryo. In the endosperm, however, RNA levels increase during the early stages when the non-storage proteins are mainly synthesized, and decrease during the time when the major storage protein synthesis occurs. The decrease is probably due to an increase in endospermal ribonucleases. It is possible that protein synthesis occurring during the latter stages of maturation involves RNA synthesized earlier, and after use this is then degraded. It would be of interest to know if the messenger RNA for storage proteins is synthesized at an early developmental stage and then gradually depleted as the endosperm matures. Could the durability of this messenger RNA in some way determine how often, and for how long it is used, and play a role in determining the amount of proteins synthesized? In the wheat endosperm, RNA content increases up to about the 20th day of development (when approximately 50% of the total protein has been synthesized, but less than 30% of the storage protein) and then remains at a steady level [91]. Turnover of RNA during the latter stages of endosperm development has not been studied, so we do not know whether or not the early synthesized RNA is used in later protein synthesis.

An understanding of the mode and sequence of production of the components and factors affecting protein synthesis could be of considerable use to plant breeders and geneticists. Proteins of some seeds form an important part of the human diet in many countries, but are often deficient in certain amino acids, and if these are not provided from alternative food sources serious deficiency symptoms ensue. In normal maize, zein (the major protein body constituent) is poor in lysine and tryptophan, and is the protein which is synthesized late during maturation. Glutelins (the non-sequestered proteins) are synthesized

---

[1] Attachment of membrane-bound polysomes to the protein bodies might occur during extraction.

earlier and have a higher lysine content. Likewise in wheat, proteins rich in lysine decline progressively with endosperm development. An understanding of the mode of synthesis of these proteins could perhaps eventually lead to manipulation of their deposition and an alteration in their balance to one more suitable for the human diet. A number of studies have been initiated, particularly on those maize mutants which show differences in protein content [4], e.g. the opaque-2 mutant (o_2) which is poor in zein but richer in glutelins. In this mutant the ribonuclease content of the grain is very high late in development and this may act to destroy new, or even latent messenger RNA in the endosperm cells before it can be used for zein synthesis, whereas the glutelin message, already attached to ribosomes might initially be more resistant to enzyme attack. While the o_2 gene itself is probably not directly responsible for such enzyme changes, this suggestion does present a simple (and maybe oversimplified) explanation. Other explanations are possible, and these have been adequately considered elsewhere [4]. In other cereals, e.g. wheat, qualitative changes in the pattern of protein synthesis to yield lysine-poor proteins at later stages of development might be accompanied by selective degradation of lysine-rich proteins.

3.2.5. Storage of Phosphate

Phosphate is stored in the mature seed as phytic acid, which is the hexaphosphoric ester of myo-inositol—see Chapter 2. This combines with potassium, calcium and magnesium to form the salt called phytin. Accumulation of phytin in cereals is almost exclusively in the aleurone grains of the aleurone layer where it is deposited as a discrete globoid in association with protein. The precise mechanism of phytin deposition and the enzymes involved therein have been little studied and will receive no more attention here. The importance of phytin as a phosphate reserve is discussed in Chapter 6.

3.3. Establishment of Cotyledon Reserves in Dicots

3.3.1. Non-Endospermic Legumes

Most work on carbohydrate and protein reserve deposition in dicots has been carried out using commercially important non-endospermic legumes, particularly the garden pea (*Pisum sativum*), field pea (*Pisum arvense*) and broad bean, field bean or horse bean (*Vicia faba*), with fewer studies on the french bean or bushbean (*Phaseolus vulgaris*).

The pattern of events accompanying seed maturation in peas is outlined in Figure 3.7.

Firstly, let us consider the source(s) of the reserves laid down in these seeds. The pea pod (formed from the ovary wall) first increases in length and width, and then in wall thickness, attaining maximum fresh weight before the seeds contained therein commence their deposition of storage reserves [58]. It

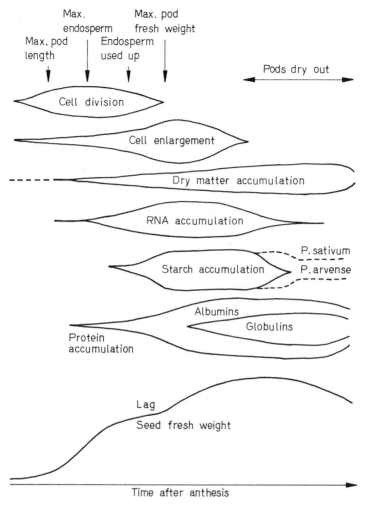

Fig. 3.7. The cardinal events in seed maturation of *Pisum* spp. The common time scale is the growth curve for seed fresh weight. All events depicted refer to rates within the developing cotyledons, divergent lines denoting periods of increasing rate, convergent lines times of decreasing rate. Differences between *P. sativum* and *P. arvense* are indicated. After Pate, 1975 [13]

has been found that the leaf and pod (i.e. pericarp) at the lowest reproductive node of the field pea are responsible for providing approximately two-thirds of the carbon required by seeds ripening at that node. Figure 3.8 summarizes the daily record of the provision of carbon to these seeds for each of the 5–28 days after anthesis, a period when they accumulate some 85–90% of their final carbon content (mainly as sucrose). The leaflets serve as major contributors throughout seed development; the stipules are important during early growth but their contribution declines as the pod becomes the predominant source. Eighty-five per cent of the carbon from the leaflets and stipules is furnished

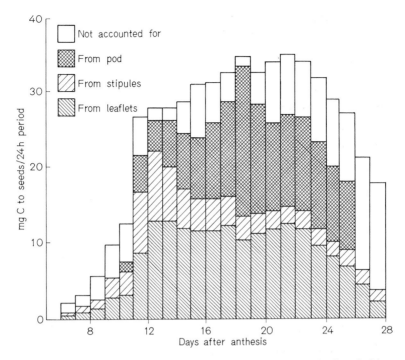

Fig. 3.8. A day-by-day record of the provision of carbon to seeds borne at the first fruiting node of field pea by their pod and by the stipules and leaflets of their subtending leaf. The estimates include photosynthetically fixed carbon, carbon available through mobilization of dry matter and, in the case of pods, carbon reassimilated from respiring seeds. These data for transfer are matched against the daily requirement of the seeds for carbon, the latter being calculated by summation of daily losses of carbon by the seeds in respiration and daily gains of carbon by the seeds as dry matter. Consequently, the unhatched areas of the histograms (labelled "not accounted for") represent carbon required by the seeds but not derived from organs at the fruiting node. After Flinn and Pate, 1970 [59]

from photosynthetically fixed carbon; the remainder is available from mobilization of dry matter. Recent studies on the diurnal changes in the leaf-phloem-fruit translocatory system of *Lupinus albus* have given us an interesting insight into the daily conditions under which photosynthates might be moved to the fruit. As seen in Figure 3.9 there are diurnal changes in sugar levels in the leaf laminae, phloem sap and fruits (seeds and pod) which are more or less in phase with each other, all exhibiting a minimum at dawn and a maximum around noon. Noticeably there is a close synchronization of the morning sugar rise in the leaf, phloem and fruit and an abrupt fall in sugar levels in leaf and phloem over the period from midnight to dawn. Phloem amino acids appear to fluctuate in phase with those of the translocated sucrose, although the relative amino acid composition does not vary significantly over the daily cycle. These results are suggestive of a synchronized leaf-phloem-fruit translocatory system, although the experimenters themselves outline some of the difficulties in interpre-

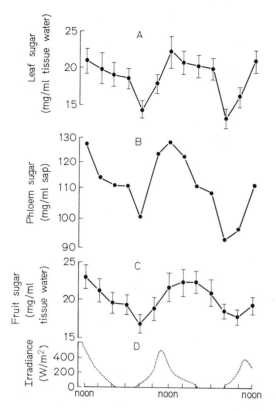

Fig. 3.9. Synchronization of the diurnal changes in leaf laminae, phloem and fruit of *Lupinus albus*. A: sugar level in leaf laminae representing the photosynthetic catchment of the fruits, B: sucrose level in fruit phloem sap, C: free sugar level in fruits, D: solar irradiance in the days of study of *Lupinus albus* over a two-day period commencing when the oldest fruit was 40 days old. After Sharkey and Pate, 1976 [166]

tation of this and related work [166], and warn us at this stage against making sweeping conclusions.

Carbon provided by the pod is derived by fixation of CO_2 from the internal atmosphere and from mobilization of dried matter accumulated earlier in the life of the pod. The bulk of this contribution, however, is attributed to the recycling of respired carbon from the developing seeds, although the amount of CO_2 fixed by pod photosynthesis would appear to be much lower than the large amount of CO_2 respired from the seed. The suggestion has been made that dark fixation by the seed itself of its own released CO_2 could enhance its recycling [79], especially since the light intensity within the pod is perpetually low. Phosphoenol pyruvate (PEP) carboxylase (a key enzyme in dark-fixation of CO_2) is high in the pod wall and testa some ten days after anthesis of *P. sativum,* and then declines in the former region by day 15 prior to reserve deposition in the cotyledons. But in the testa the enzyme reaches a maximum at day 15 and decreases slowly thereafter as the seed matures up to the 69th day. PEP carboxylase in the seed testa might, therefore, play a role in recycling respired CO_2 for the synthesis of reserves. This enzyme is formed in the cotyledons late in development (after 35 days) and might be stored there during maturation for use during subsequent germination/growth.

It is worth noting that in contrast to some cereals where net photosynthesis in the reproductive organs is high, here in peas (and also in soybean [145] and french bean [43]) their photosynthetic contribution is low compared to that of the leaves. In some oil-bearing seeds, e.g. rape, where the leaves are wholly or partially shaded by the developing pods, the photosynthetic contribution by the pod may be much more significant.

Considering dicots as a whole, amino acids supplied to developing seeds can be derived from a variety of sources including, as shown for pea, the photosynthesizing leaves. The onset of fruit and seed development in annual plants is frequently accompanied by the correlated senescence of the older leaves on the plant. One feature of leaf senescence is the hydrolysis of proteins; the liberated amino acids may be translocated to the developing seeds and fruits which are known to act as metabolic sinks.

Translocation of nutrients to the developing cotyledons of legumes occurs from the phloem of the mother plant through the vascular bundles running in the pod, which in turn are connected to the integuments (incipient testa) of the developing seed [139]. There is no direct vascular connection between the cotyledons or axis and the testa, so the embryo absorbs the nutrients released from the integument. Little is known about the transfer of nutrients from integuments to embryo but absorption may be aided in some species (e.g. *Vicia, Lathyrus* and *Pisum*) by specialized cells with increased surface area (transfer cells) on the outer epidermis of the embryo taking in materials released by similar cells on the inner epidermis of the integument [69]. Alternatively the nutrients released by the outer layers might be transported to the developing embryo via the suspensor, which in some species (e.g. *Phaseolus*) is well-developed and also has transfer cells. (In others, e.g. *Vicia* and *Lathyrus,* the suspensor is poorly developed.)

The levels of protein and amino acids in all seed and pod tissues have been determined during seed formation in the field pea [58]. Table 3.1 presents an estimation of the dry weight and total nitrogen content of the mature embryos of the six seeds of a single fruit and the relationship of these to the available dry weight and nitrogen from the pod, endosperm and seed coats. From these data it can be seen that (maximally) only 19% of the final dry weight and 23% of the total nitrogen of the seeds can be furnished from metabolites stored in the pod, endosperm and seed coats. So, although there is a build-up of nutrient tissue in the ovarian and maternal tissue of the seed itself, a continuous flow from the mother plant must take place during seed development. The principle amino acids derived from photosynthesis in garden-pea leaves are serine, glycine and alanine, and these, along with the amides asparagine and glutamine make up the bulk of amino acids transported in the phloem to the seed [106]. Important amino acids found in the stored proteins of the cotyledons include tyrosine, leucine, phenylalanine, histidine, lysine and arginine, which are not translocated into the cotyledons. Thus, the synthesis of these

Table 3.1. Estimates of the maximum possible transfer to embryos of pea of reserves in the fruit

Source	Period of availability (days after anthesis)	Amount available (mg)
(a) Expressed in terms of dry matter		
Decrease in dry weight of pods	32–57	159.1
Loss of dry matter from seed coat and endosperm of seeds	20–57	33
	Total	192.1
(b) Expressed in terms of nitrogen		
Loss of total nitrogen from pod	32–57	6.2
Nitrogen of endosperm of seeds	20–24	0.6
Loss of nitrogen from seed coats	24–48	2.7
	Total	9.5

The embryos of the six seeds of a fruit together contained 25.4 mg of nitrogen, and had a dry weight of 1070 mg. After Flinn and Pate, 1968 [58]

amino acids must occur within the developing seeds themselves: presumably the amino group of the amides from the phloem sap serves as the nitrogenous component for the amino acids, and the carbon skeletons are furnished by the translocated carbohydrates. A similar pattern of carbon and nitrogen availability has been reported for the broad bean [93], although both pods and seeds of this and some other legume species can also produce amino nitrogen by reduction of translocated nitrate [157].

Asparagine accounts for 50–70% of the nitrogen carried in the translocatory channels serving the fruit and seeds of L. albus [19]. This supply greatly exceeds the requirement of the developing seed for amide, either as a soluble reserve or as a building block for protein. An asparaginase arises in the seeds some 4–5 weeks after anthesis and its increasing activity coincides with the early stages of protein synthesis in the cotyledons. It is possible that this enzyme is responsible for the deamidation of the excess asparagine and that the amido group is used for the synthesis of those protein amino acids not provided to the seed in sufficient quantity in the translocation stream to the fruit. The carbon of asparagine might also be donated to protein, but principally to non-amino compounds of the soluble and insoluble components of the developing seed.

The sequence of changes in starch, protein and nucleic acid content of developing garden pea cotyledons is shown in Figure 3.10A–C. While peas and other legumes in general fit this pattern, the time taken for these events to occur is variable. For example, we note that garden pea seeds (cv. Greenfast) took only 26 days to mature on plants grown at 25°C, 80% relative humidity and a 16-h photoperiod (80 W/m^2) (Fig. 3.10), whereas those of the cultivar

Victory Freezer took 54 days to mature under undefined field conditions during September–October in Sydney, Australia [21]. Such differences in experimental practice make it difficult to integrate much of the published data: obviously more uniformity is needed if there is to be a combined effort by plant physiologists to unravel satisfactorily the characteristics of seed development.

The following pattern of events nevertheless emerges from the studies which have been carried out. Cotyledons of developing legume seeds pass through an initial, relatively brief, cell division phase, followed by an extended phase of growth by cell expansion (Fig. 3.7). During the early stages of cell expansion in non-endospermic legumes (e.g. peas and beans) the endosperm is occluded. (For a consideration of reserve formation in endospermic legumes see later.) In broad bean, field and garden pea seeds the majority of the stored reserves are laid down during the cell expansion phase (e.g. see Fig. 3.10), although both starch deposition and reserve protein synthesis in these seeds (and those of the french bean [137]) commence prior to the cessation of cell division.

Starch, which accounts for 34% (wrinkled-seeded varieties) to 45% (smooth-seeded varieties) of the dry weight of mature pea seeds, starts accumulating before protein—a feature common to several legumes. Bain and Mercer [21] observed that in their field-grown garden peas (cv. Victory Freezer) starch grains become visible in the stroma of some chloroplasts 10 days after anthesis and that most chloroplasts contain a single starch grain by day 19. Presumably, therefore, initial starch synthesis involves the utilization of the products of photosynthesis. The single starch grain enlarges within the chloroplast and by day 28 the lamellae and stroma are compressed against the limiting plastid membrane. For this phase of starch accumulation sucrose stored within the cotyledon cells (some of which has been obtained from the disintegrating endosperm) and translocated sucrose are used. The quantity and quality of light penetrating through the pod to the seed at later stages of development is probably insufficient to maintain photosynthesis. A few small chloroplasts do not form starch and their lamellar system remains intact, accounting for the "pea-green" colour of the mature seed. The conversion of sucrose to starch in peas is via the pathway outlined earlier for cereals (Sect. 3.2.3) [184]. A similar pattern for starch synthesis and chlorophyll formation and degradation is shown in Figure 3.10A for Greenfast peas.

Protein accounts for about 25% of the dry weight of the mature pea cotyledons, 60% of this being stored globulin (Chap. 2). As indicated in Figure 3.10B vicilin synthesis in the garden pea precedes that of legumin (such is also the case in *V. faba*), and synthesis of the albumin fraction commences even earlier [28]. The bulk of the synthesis of all these proteins occurs during the cell expansion phase of growth, when there is also a continuing increase in the DNA and RNA content of the cotyledons (Fig. 3.10C). Synthesis of RNA is not unexpected because the expanding cells are actively involved in the large-scale production of protein, both storage and metabolic. The continuing synthesis of DNA in cells which are not dividing is not so readily explained. It has been suggested [156] that polyploidization makes available extra copies of those cistrons involved in the synthesis of storage proteins during seed development. The evidence to support this hypothesis is far from unequivocal, how-

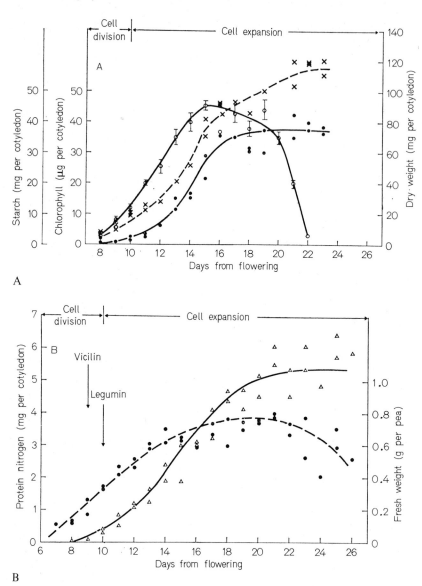

A

B

Fig. 3.10. (A) Changes in dry weight (×) and content of starch (●) and chlorophyll (○) during development of pea cotyledons. (B) Changes in fresh weight (●) and protein content (△) during development of pea cotyledons. *Vertical arrows* indicate the stages of development at which the storage proteins vicilin and legumin can first be detected. (C) Changes in DNA (○) and RNA (●) content during development of pea cotyledons. After Millerd and Spencer, 1974 [122]

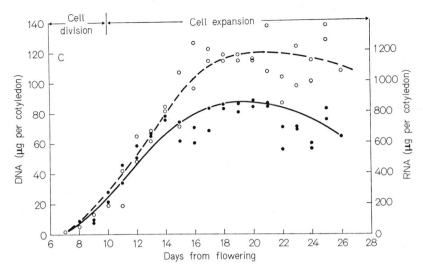

Fig. 3.10 C

ever, and we cannot rule out the possibility that the extra DNA merely has a storage function as a reserve of deoxyribonucleotides destined to help support the extensive cell division in the axis which occurs after germination of the mature seed [122]. The general pattern of RNA, DNA and protein synthesis is similar in the field pea [172], french [137], and broad bean [123].

While some of the reserve proteins of mature legumes appear to be deposited in small protein bodies interspersed in the cytoplasm amongst the larger starch grains (these bodies might arise by dilation of the ER cisternae or from the dictyosomes), the major storage proteins are laid down within protein bodies in the vacuole. Using fluorescent antibodies specific for broad bean legumin and vicilin, it has been shown that these storage proteins are present only in vacuolar protein bodies and that such bodies always contain both globulins [68]. Cells of immature cotyledons also possess other protein bodies lacking both legumin and vicilin, but it is not clear whether these exist in mature cotyledons. Since legume cotyledons accumulate albumins as well as globulins during development (in beans and peas to an approximate ratio of 1:1.4; albumin:globulin) it is conceivable that these other bodies persist as albumin-storing packages.

Globulin synthesis in legume cotyledons appears to follow a common pattern of events. In developing soybean [29], broad bean [30, 141], pea [2] and french bean [137] the initiation of synthesis of storage protein is accompanied by a proliferation of ribosomes associated with the membrane of the protein bodies or adjoining endoplasmic reticulum. Apparently there might also be invagination of the cytoplasm into the vacuole causing its division into segments, in each of which will be formed a protein body. That the associated ribosomes synthesize storage protein has never been proven, nor is it understood how the synthesized

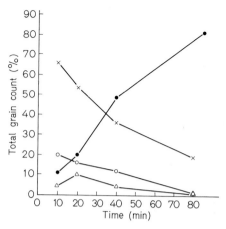

Fig. 3.11. Distribution of radioactive protein in the cotyledon cells of *Vicia faba* during development. Seeds were taken from 60 ± 4-day-developed pods and slices of cotyledon tissue incubated for 10 min in a medium containing ^{3}H-leucine. One slice was then cut into thin sections for study by electron microscopy following autoradiography (see Chap. 5.3 for details of this technique) to determine the location of radioactive amino acid incorporated into protein. Other slices were placed in non-radioactive leucine (the chase) for a further 10, 30 and 70 min and again the fate of the initially incorporated leucine (i.e. the de novo synthesized protein) followed. The distribution of radioactivity (as per cent total grain count) in various parts of the cell is illustrated in this figure. The percentage figures are derived from the following total grain counts: 10 min:127; 20 min:188; 40 min:430; 80 min:153. × : ER, ● : protein body, ○ : other organelles, △ : cytoplasm. After Bailey et al., 1970 [20]

proteins cross the tonoplast and/or membranes of the protein bodies. Nevertheless, there is some evidence from work on *V. faba* that proteins synthesized on the endoplasmic reticulum take about 80 min somehow to be transported into protein bodies, with no accumulation in the cytoplasm or in other organelles. See Figure 3.11 and its legend for relevant experimental details and result.

It is pertinent at this time to recall that the reserve proteins in legumes, legumin and vicilin, are glycosylated, albeit to a much lower extent than glycoproteins of animal origin. Glycosylation of the storage globulins occurs after peptide synthesis and while the glycosyl units added to legumin do not undergo modification, there appears to be a progressive shortening of the oligosaccharide chain during seed development [23]. The significance of the glycosylation process is unknown and has received little comment. In the light of the suggestion that mammalian secreted proteins are glycosylated to aid in their transfer through cell membranes, we might speculate that since storage proteins must cross the delimiting membrane of a protein body to be sequestered therein, glycosylation might aid this process of internal secretion.

The protein-synthesizing system of the developing cotyledon has itself been the subject for study in a number of laboratories. Cell-free protein-synthesizing systems have been obtained and polysomal (using native mRNA) and ribosomal (using synthetic messages, i.e. polyuridylic acid) activities determined at different stages of development. Attempts have also been made to link in vivo polysomal

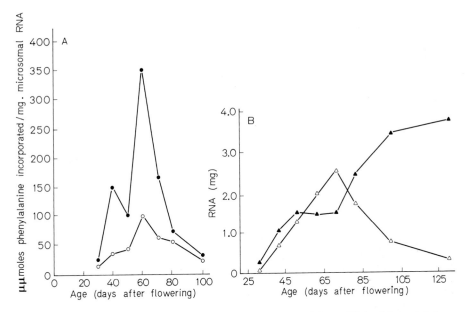

Fig. 3.12. (A) *In vitro* activity of ribosomal pellets extracted from *Vicia faba* cotyledons at different stages during development. ●——●: +added messenger RNA (poly U), ○——○: no added messenger RNA. After Payne et al., 1971 [140]. (B) The quantitites of free and membrane-bound ribosomes in cotyledons of *Vicia faba* during development. ▲——▲: free, △——△: membrane bound. After Payne and Boulter, 1969 [141]

activity to possible increases in messenger RNA availability during development, but without notable success so far. While cell-free systems using native messenger RNA and ribosomes (i.e. using extracted polysomes) can successfully incorporate amino acids into polypeptides, only in *P. vulgaris* has any link been made between such activity and the production of storage protein [176a].

In broad bean, polysomes from developing cotyledons show maximum in vitro activity if extracted and assayed at a time coincidental with the maximum storage protein production in vivo [140]. Furthermore, even the single ribosomes at this time are more active in producing polypeptides using synthetic messenger RNA (Fig. 3.12A). At the time of maximum in vitro activity, the quantities of membrane (probably ER)-bound ribosomes are maximal (Fig. 3.12B). Loss of in vitro activity coincides with a decrease in vivo in the rate of storage protein synthesis, and the quantities of free ribosomes increase at the expense of the membrane-bound ones. There is little or no loss of total RNA from the cotyledons during disruption of the membrane-bound polysomes and cessation of protein synthesis during maturation and drying: maturation likewise results in little loss in total RNA from pea cotyledons (Fig. 3.10C). We still do not know the reason for the increase and decline in in vivo and in vitro ribosomal activity. Increase in polysome activity could reflect increased availability of messenger RNA and a decrease in its destruction. The decline in ribosome activity is more difficult to explain, however, particularly since active ribosomes are conserved in the dry cotyledons (Fig. 3.12A) and are active (in vivo) on rehydration.

In peas, increased albumin and globulin synthesis in vivo more or less parallels an increase in polysomes and newly-synthesized RNA [28, 143]. Preparations from older cotyledons past the time of maximum protein synthesis exhibit a declining capacity for in vitro protein synthesis, fewer polysomes and an abundance of ribosomes. These ribosomes still incorporate amino acid into proteins in the presence of an artificial messenger. The decline in protein-synthesizing capacity with age could be due to one or more of a variety of factors: e.g. (1) decreased messenger RNA turnover (i.e. decreased synthesis, but continued breakdown); (2) reduced ribosomal activity; (3) decline in activity or availability of a supernatant factor involved in protein synthesis; or (4) decreased tRNA availability or the capacity to form amino acyl tRNA. There is no unequivocal evidence at present to allow us to conclude whether the switching on and off of synthesis of storage protein occurs at the level of transcription or translation. The biochemical mechanisms involved in messenger RNA synthesis and its association with ribosomes in an active protein-synthesizing complex are still incompletely understood in plants. Nevertheless, the developing dicot storage organ presents an interesting and potentially useful challenge for determining the mode of plant protein synthesis and its control. It has recently been observed that detached, immature pea cotyledons continue to grow for several days in liquid culture (containing only asparagine as the nitrogen source; see earlier) and synthesize chlorophyll, starch, DNA, RNA and the storage proteins, vicilin and legumin. Immature cotyledons, containing no detectable vicilin or legumin when transferred to liquid culture, can still synthesize the former protein, but synthesis of the latter is not initiated [124]. This ability of such isolated storage organs to undergo various developmental processes in defined, manipulatable culture conditions might prove to be very useful for future studies on the mechanism of reserve deposition.

3.3.2. Endospermic Legumes

Many legumes, such as fenugreek (*Trigonella foenum-graecum*) have endospermic seeds which store galactomannan in this tissue [150] (see Chaps. 2 and 6 for more details). Galactomannan begins to be laid down in the seeds of *Trigonella* at an early stage of development and continues to be formed until the seeds reach their full size and become yellow. Galactose and mannose residues seem to be laid down at the same time by a specific (undefined) mechanism, because no change in the proportion of galactose to mannose units can be detected at any time of development. It appears that the galactomannan is a cell wall polysaccharide whose formation outside the plasmalemma (presumably by secretion from within) continues until it completely fills the endosperm cells [151] and occludes the cytoplasm therein. Furthermore, this synthesis commences in the cells nearest the embryo and the last cells to be filled are those bordering the aleurone layer (see Fig. 6.4 for seed structure). The aleurone layer itself does not contain reserve galactomannan.

Low molecular weight carbohydrates (e.g. sucrose, raffinose and stachyose) are stored within the embryo. Early during seed development sucrose, raffinose, myo-inositol, glucose and galactinol are present, but all these decrease as stachyose levels increase until, in the dry seed, this is the only low molecular weight sugar present in significant proportions.

3.3.3. Lipid-Storing Seeds

Anatomical studies of developing lipid-storing dicot seeds have given us considerable insight into the cellular changes which occur during development from the zygote to the mature seed. Early development has been studied in the seeds of several members of the Cruciferae, the classical work on *Capsella bursa pastoris* (shepherd's purse) now having found its way into many botanical and biological textbooks. The reader is directed to the relevant references in the bibliography [3, 16] for detailed accounts of the developmental anatomy of dicot seeds. A much abbreviated account of the sequence of developmental events occurring within the seed of the oil-bearing Crucifer, *Sinapis arvensis* (charlock) is presented in Figure 3.13 — see figure legend for details.

3.3.4. Fat (Oil) Synthesis

Development of the reserves in fat-storing seeds has been approached experimentally from two main standpoints: (1) to elucidate the pathway of fatty acid biosynthesis and (2) to measure the accumulation of fat and dry matter and analyze it for fatty acid content.

3.3.5. Fatty Acid Synthesis

Several tissues have been used to study the mechanism of fatty acid synthesis, particularly the avocado mesocarp, spinach and lettuce chloroplasts, and etiolated barley seedlings, but some major advances have been made from studies of maturing oil-bearing seeds, e.g. castor bean (in which 90% of the fatty acid content is ricinoleic acid) and safflower, *Carthamus tinctorius* (76% linoleic and 1% oleic). Oleic acid plays a central role in plant fatty acid anabolism as the precursor of the major unsaturated fatty acids, e.g. it is the first detectable fatty acid formed when ^{14}C acetate is fed to the developing castor bean and is itself a precursor of ricinoleic acid.

Fatty acid biosynthesis utilizes acetyl CoA. Radioactive acetate is the common experimental substitute, but in the developing seed sucrose from the mother plant is the initial source of substrate. Biosynthesis is a multi-step process (Fig. 3.14) which firstly involves the formation of malonyl CoA by carboxylation of acetyl CoA with carbon dioxide. This malonyl CoA is then accepted by acyl carrier protein (ACP) which is part of a multienzyme complex called the ACP fatty acid synthetase complex. The malonyl CoA is then condensed with a further acetyl CoA molecule to yield acetoacetyl (4C) CoA, and CO_2 is released. Further 2C acetyl residues are sequentially condensed with acetoacetyl CoA on the ACP complex until a 16C fatty acid, palmitic acid (16:0)-ACP complex is formed. One further acetyl moiety is added to this by an elongation process to yield a stearic acid-ACP complex, from which the saturated 18C fatty acid (stearic acid 18:0) is released. Aerobic desaturation of stearic acid by stearyl desaturase to form an 18C fatty acid with one double bond is then possible: the product is oleic acid (18:1). Further desaturation to yield linoleic acid (18:2) occurs in safflower, oxidation of oleic acid to ricinoleic acid (12-hydroxy 18:1) occurs in castor bean, and elongation of oleic acid to erucic acid (22:1) in rape seed. See Figure 3.14 for a summary of these events. It is noteworthy that one of the most remarkable metabolic blocks that has been found in mammalian systems is the inability to desaturate oleic acid to yield linoleic acid; hence this fatty acid, which is an essential dietary component, must be obtained from plants.

3.3.6. Synthesis of Triglycerides (Triacylglycerols)

Fatty acids ultimately are esterified with glycerol to form the triglycerides. The reaction sequence by which this takes place in plants is quite well understood but surprisingly few studies of the process have been made on developing oil-storing seeds. There seems, nevertheless, no reason to doubt that the pathway outlined in Figure 3.15 operates in such seeds. In addition, recent evidence suggests that a variety of phospholipids (e.g. phosphatidyl choline) can be metabolized to diglycerides and can, incidentally, also act as acyl donors [200]. Developing seeds of *Crambe abyssinica* [71] readily incorporate radioactively-labelled glycerol phosphate principally into triglycerides but also into diglycerides and phospholipids. Similarly, developing flax seeds rapidly incorporate the label

---

▶

Fig. 3.13A–E. Stages during seed development in *Sinapis arvensis*. (A) Transverse section of the ovary at the level of an ovule containing a young embryo (*E*). Approx. 7 days after fertilization. (B) Young embryo at the octant stage of development lying in the free nuclear endosperm (*EN*) surrounded by the nucellus (*N*). Approx. 7 days after fertilization. (C) Heart-shaped proembryo (*PR*) now embedded in the endosperm, which is cellular. The nucellus is present on the inside of the inner integument (*IE*), which is comprised of a palisade layer (*P*) with cell wall thickenings, and a layer to the inside of large thin-walled cells. The outer integument (*OE*) consists of an epidermis and sub-epidermal layers of cells. Approx. 14 days after fertilization. (D) Oblique section of a developing ovule through the base of the cotyledons and shoot meristem showing embryo having grown and curved within the seed. Endosperm cells become absorbed as the embryo develops. The testa is more obvious. Approx. 21–28 days after fertilization. (E) Shoot meristem (*S*) and base of cotyledons (*C*) of the mature seed. Note the loss of endosperm, nucellus and reduction of the inner integument within the thick, well-developed testa (*T*). The outer integument becomes mucilaginous. Time to embryo maturity from fertilization approx. 28–35 days. From Edwards, 1968 [50]

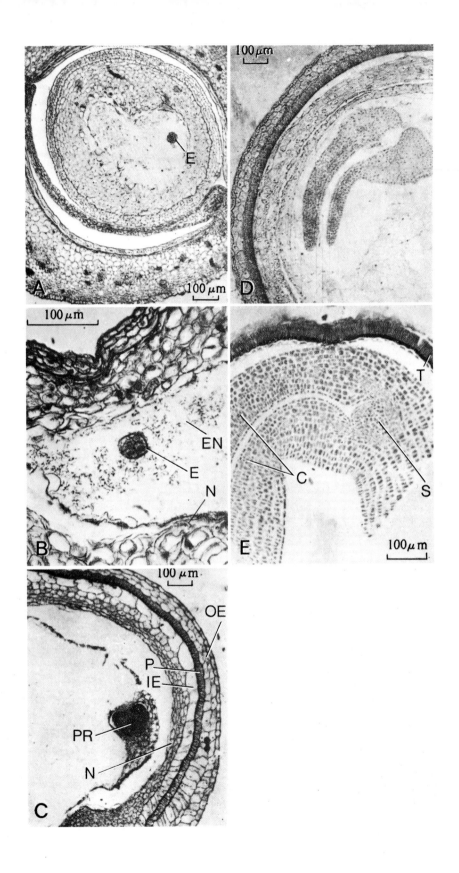

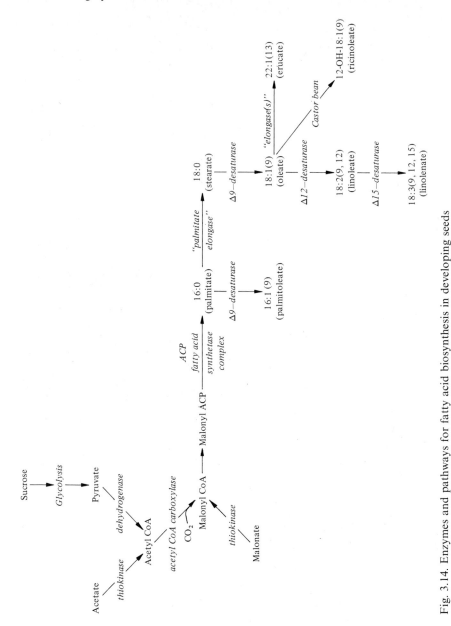

Fig. 3.14. Enzymes and pathways for fatty acid biosynthesis in developing seeds

from ^{14}C acetate (through fatty acids) into phospholipid, diglycerides, and ultimately triglycerides [49]. The fat fraction of a *Crambe* seed homogenate, consisting of only slightly contaminated oil bodies, also utilizes glycerol phosphate for triglyceride synthesis. The enzymes for this conversion therefore seem to be located in, or associated with the oil bodies themselves and we might justifiably infer that these are the sites of synthesis within the cell. Enzyme activities

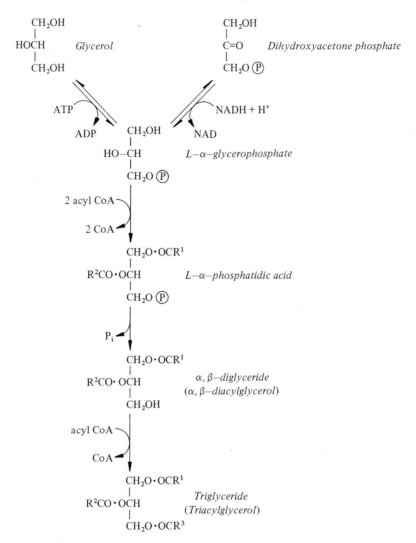

Fig. 3.15. Pathway for triglyceride biosynthesis

presumably change during the course of seed development, however, for seeds at different stages of development synthesize different amounts of the three products mentioned above. Young *Crambe* seeds when fed ^{14}C-acetate six days after flowering produce predominantly phospholipid but on nearing maturity a higher proportion of supplied label appears in triglycerides [71].

In the early stages of development of soybean seeds the lipid is virtually devoid of triglyceride and the main constituents are phospholipids and glycolipids. From the 9th day after flowering triglycerides are rapidly synthesized, coinciding with the onset of deposition of the lipid stores. Phospholipid synthesis continues throughout the period of deposition but at a low rate [144].

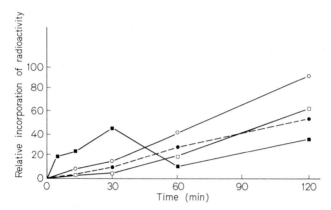

Fig. 3.16. The appearance of radioactivity in fractions extracted from developing *Crambe* seeds. Slices of developing *Crambe* seeds were fed with ^{14}C-glycerol and extracted at different times. The labelled fractions were then identified. ■——■: glycerol-3-phosphate, ○——○: phospholipid, ●---●: diglyceride, □——□: triglyceride. (Unidentified label in the lipid and aqueous fractions is omitted). After Gurr et al., 1974 [71]

Developing *Crambe* seeds also incorporate ^{14}C-glycerol firstly into glycerol phosphate; later the label is in di- and triglycerides and phospholipids (Fig. 3.16). The seeds therefore probably contain a glycerol kinase. It is not certain, however, that glycerol itself is normally the starting point for glycerol phosphate synthesis in seeds. Evidence from other plant tissues is that dihydroxyacetone phosphate is reduced to glycerol phosphate and that free glycerol may not be involved.

During seed development there occur qualitative changes in fatty acid content (see the next section). In reality, the changes are in the composition of the triglycerides which exhibit different proportions of fatty acids at the various stages of seed maturation. In soybean for example, much of the triglycerides in the early stages of development contain linolenic acid but closer to maturity the triglycerides come to contain relatively more linoleic and oleic acids. Interestingly, the positional arrangement of the fatty acids remains fairly constant, linolenic acid attaching at the α and β positions, linoleic at the β position and oleic acid at the α position of glycerol [153].

3.3.7. General Pattern of Oil Formation

In a number of seeds there appear to be distinct phases of development related to the time of oil formation:

1. An initial or lag phase. This is when there is no fatty acid synthesis. The length of this phase varies from species to species (and even in the same species under varying environmental conditions [168]). In castor bean (Table 3.2) it lasts 20–25 days after pollination during which time the embryo and seed coat develop [38, 201], in rape seed for the first 4 weeks of development [136], and for the first 10–15 days in safflower [113, 169]. The delay before the initiation of fatty acid synthesis is due to the absence from the developing tissue of

Table 3.2. Composition of castor bean seed at various stages of development

Age, days after pollination	Fresh wt, mg/seed	Dry wt, mg/seed	Moisture %	Oil	
				mg/seed	% dry wt
6–9	201	21	89.5	0.4	1.9
12–15	349.3	43.8	87.5	0.9	2.0
18–21	361.5	63.5	82.5	5.5	8.6
24–27	358.6	91.2	74.6	19.5	21.4
30–33	345.3	148.2	57.0	57.5	38.8
36–39	376.1	216.8	42.5	101.7	46.8
42–45	360	252.5	30.0	132.1	52.3
48–51	338.4	247.3	27.0	132.4	53.5
54–57	330.2	278.8	15.5	153.2	55.0
60	317.5	282.0	11.0	154.0	54.6
Commercial seed	300.2	278.6	6.6	140.6	49.6

After Canvin, 1963 [38]

Table 3.3. Weight of the individual fatty acids as mg/100 seeds at various stages of maturity of the castor bean

Age, days after pollination	Palmitic	Oleic	Linoleic	Linolenic	Ricinoleic
6–9	17	3	18	1.2	0
12–15	22	22	27	2.5	16
18–21	41	63	86	14.3	345
24–27	72	164	197	0	1532
30–33	98	276	397	0	4945
36–39	122	285	590	0	9173
42–45	132	462	726	0	11902
48–51	119	357	760	0	11992
54–57	138	490	674	0	14178
60	154	539	708	0	13983

After Canvin, 1963 [38]

the appropriate anabolic enzymes: presumably the tissue has to reach a certain state of maturity before they are able to be synthesized. During this phase of development in the rape seed there is some accumulation of starch and sucrose which will be used later for fatty acid synthesis [136], although the amount of carbohydrate stored would appear to be insufficient to provide for total fatty acid synthesis. In this seed, and in castor bean where there is little or no sucrose and starch accumulation, there is, presumably, sufficient mobilization of photosynthate into the developing seed at the appropriate time to provide the necessary amount of carbon skeletons for fatty acid synthesis.

　　2. Rapid synthesis of fatty acids. This lasts about 20 days in the rape and safflower embryo and in the castor bean endosperm [38, 113, 136] (Table 3.2). By the end of this phase over 90% of the reserves have been accumulated (Table 3.3). While there appears to be little controversy over the pathways

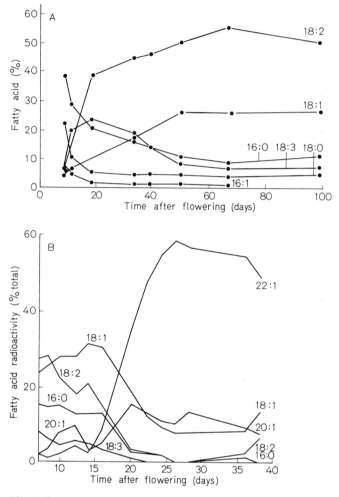

Fig. 3.17 (A) Changes in fatty acid composition of the triglycerides of soybeans during maturation. After Privett et al., 1973 [144]. (B) The influence of seed age on the incorporation of (2-^{14}C)-acetate into the seed-oil fatty acids of *Crambe abyssinica*. 16:0: palmitic, 16:1: palmitoleic, 18:1: oleic, 18:2: linoleic, 18:3: linolenic, 20:1: gadoleic, 22:1: erucic. After Appleby et al., 1974 [18]

by which saturated and subsequently unsaturated fatty acids are synthesized, there is uncertainty concerning the precise sites in the cell where these take place. Cells have been fractionated into sub-cellular organelle and soluble components, these fed with ^{14}C-acetate, and the products analyzed. Although the use of ^{14}C-acetate is "physiological" it has the disadvantage that for its conversion to malonyl CoA, acetyl CoA carboxylase is needed — an enzyme which might be absent, or limiting, or even inhibited in in vitro systems. Thus, while some experiments suggest that acetate is converted to oleic acid in the proplastids of castor bean endosperms [131, 208], others using malonate show that the fat fraction (this contains the oil bodies but it cannot utilize acetate in vitro) is the site for fatty acid biosynthesis [74]. It has been shown that the formation of oleic acid from sucrose, its natural precursor in the endosperm tissue, can take place in a particulate (possibly proplastid) fraction of the endosperm cell [202] but even in these experiments the possibility that the fat fraction could play an equally, if not more important, role in biosynthesis was not eliminated. Oil bodies isolated from developing castor bean endosperm do contain enzymes for triglyceride and fatty acid synthesis [74]. On the other hand, conversion of oleic acid to ricinoleic does not appear to occur in either the proplastid or fat fraction but depends on the activity of a mixed-function oxidase associated with the microsomal fraction of the cell [64]. In safflower seeds a similar site is also implicated since the enzyme responsible for linoleic acid synthesis, oleyl CoA desaturase, is associated with microsomes [188]. In contrast, the fat fraction of *Crambe* is the location of the conversion of oleic acid to erucic acid [18].

3. *The final phase.* This is when further triglyceride deposition is halted: it usually occurs at about the time the seeds begin to dehydrate (e.g. see Tables 3.2 and 3.3, days 40–60). It is possible that during this phase some of the enzymes associated with biosynthesis of storage triglycerides are destroyed and the capacity of the storing cells to resynthesize these enzymes is permanently suppressed. In this connection, we may note that during germination and growth of castor bean the endosperm never synthesizes ricinoleic acid [38], although this tissue produces many enzymes for its degradation (see Chap. 6). Similarly, maturing safflower seeds synthesize a high amount of linoleic acid and lesser amounts of palmitic, palmitoleic and oleic acid: germinated seeds can synthesize palmitic, stearic and oleic acids, but not linoleic acid. We do not know the control mechanism through which storage triglyceride synthesis is promoted and subsequently suppressed.

As a summary of some of the above events the sequence of fatty acid content in developing seeds of soybean and *C. abyssinica* is shown in Figure 3.17A and B. Starting about nine days after flowering in soybean the saturated fatty acids decrease as linoleic, then oleic acid accumulate (Fig. 3.17A). In *Crambe* the synthesis of erucic acid lags behind that of oleic and gadoleic acid, but from two weeks after flowering the biosynthetic pathway is geared almost exclusively to this acid (Fig. 3.17B). In *Sinapis alba* also the percentage of saturated fatty acid decreases as the seeds develop [45].

3.3.8. Development of Oil-Storing Bodies

The origin of the oil bodies is controversial. An early suggestion was that they develop from spherosomes [61] but this has been disputed on several grounds [175]. Firstly, spherosomes (which occur in non-fat-storing as well as fat-storing tissues) may be of a different size class from the oil bodies, and are said to fall in the range of 0.8–1.0 μ in diameter. The second point of dispute hinges on whether or not oil bodies possess a surrounding membrane. According to Frey-Wyssling et al. [61] they do, and on this account they resemble spherosomes, but Sorokin [175] denies the presence of a limiting membrane, which therefore distinguishes them from spherosomes. Spherosomes are "pinched off" terminal strands of endoplasmic reticulum; thus, if oil bodies are in fact "matured" spherosomes their origin is from the endoplasmic reticulum. In castor bean endosperm [159] and in peanut cotyledons [204] the membrane surrounding the oil body is held to be a half-unit membrane. It has therefore been suggested, at least in the case of castor bean, that newly-synthesized fats (triglycerides) accumulate between the "halves" of the unit spherosomal membrane, that the inner half-membrane eventually disappears, and thus an oil body having a limited half-unit membrane is formed [159]. On the other hand, no peripheral membrane is seen in oil bodies of developing seeds of *Crambe* [171], *Sinapis* [152] or, according to some workers, of castor bean [8] — evidence against an origin from the endoplasmic reticulum. Nonetheless, a well-defined interface with the cytoplasm can be observed which has been suggested to be a phospholipid monolayer [152].

Other evidence against the spherosomal nature and origin of oil bodies comes from studies on the enzymic properties of spherosomes. Several workers have noted that spherosomes from different sources, i.e. from oil-storing and other tissues, contain various hydrolytic enzymes such as acid phosphatases and non-specific esterases, whereas oil-bodies do not (e.g. in *Crambe* [171]); it is possible, of course, that enzymes become lost during development. There is therefore a strong possibility that spherosomes are not the precursors of oil bodies and we are not yet certain about how the latter do arise.

In castor bean endosperm, vacuolar inclusions containing particles are evident in the developing oil droplets; these particulate inclusions are revealed by electron microscopic examination of both the isolated fat (oil body) fraction and in oil bodies of tissue slices. Since they remain discrete within the droplet these particles are presumed to be hydrophilic, and are thought to be the enzyme complex involved in triglyceride deposition [75]. Perhaps oil accumulation ceases when these enzymes can no longer receive substrates from outside the oil body because of interference by the accumulated oil. In *Crambe* also, a particulate mass is associated with the developing oil bodies but in this case the particles surround the site of oil deposition [171].

It is thus too soon to make any firm conclusions concerning the development of oil reserves in these bodies and this is clearly an important area for further investigation.

3.3.9. Protein Synthesis in Fat-Storing Seeds

Many fat-storing seeds also synthesize considerable quantities of storage protein during development: note the similarity in the timing of fat formation in castor bean shown in Tables 3.2 and 3.3 to the synthesis of protein as shown in Table 3.4.

Whereas in the starch-storing legume seed *Vicia faba* it has been proposed that storage proteins are synthesized on the rough endoplasmic reticulum in close association with the protein-storing vacuoles or bodies, a different mecha-

Table 3.4. Protein content of the castor bean
seed during development

Age, days after pollination	Protein	
	% dry wt	mg/seed
6–9	15.0	3.2
12–15	15.5	6.8
18–21	15.3	9.7
24–27	15.4	14.1
30–33	14.0	20.6
36–39	14.5	31.4
42–45	13.8	34.8
48–51	14.8	36.6
54–57	16.0	44.6
60	16.3	46.0
Commercial seed	19.5	54.3

After Canvin, 1963 [38]

nism has been proposed for the mode of protein reserve deposition in some fat-storing seeds. This is illustrated in Figure 3.18. The hypothesis, as suggested on the basis of ultrastructural studies on shepherd's purse, cotton and peanut, is that storage proteins are made on the polysomes associated with the endoplasmic reticulum, concentrated into protein droplets in dictyosomes and then transported across the cytoplasm in membrane-bound vesicles to the protein-storing vacuoles. The vesicles then coalesce with the vacuole and empty their contents within. This interesting hypothesis has been based on a sequence of published electron microscope pictures [5] but until it is confirmed it must be recognized that electron microscopic studies of this type have severe limitations since it is impossible accurately to describe a dynamic pattern of events with a collection of two-dimensional pictures. Thus the summarizing scheme outlined in Figure 3.18 must be regarded as interesting but speculative.

3.4. Energy Supply in the Maturing Seed

In order for there to be cell proliferation and subsequent deposition of storage materials in the developing seed it is obvious that a plentiful supply of ATP energy must be available. The literature describes patterns of oxygen consumption by some developing seeds, but there has been only very limited interest in ATP levels. Thus, for example, it has been found that there is a marked increase in oxygen consumption by excised ovules of the opium poppy (*Papaver somniferum*) during, or immediately preceding pollination, division of the endosperm nucleus, cell wall formation in the endosperm, and cotyledon elongation [92]: the implication from these studies is that there is an increase in ATP from oxidative phosphorylation at key stages of development. Work on other

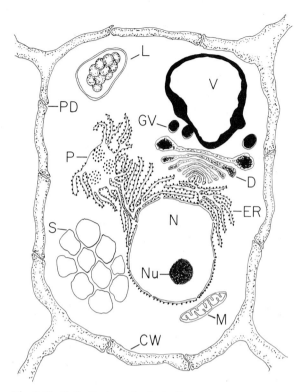

Fig. 3.18. Cell of a hypothetical oil seed producing storage proteins and sequestering them within a vacuole. *P*: polyribosomes, *D*: dictyosome, *V*: storage vacuole, *M*: mitochondrion, *CW*: cell wall, *GV*: membrane-bound vesicles. *L*: plastid, *S*: oil body, *PD*: plasmodesma *N*: nucleus, *Nu*: nucleolus, *ER*: endoplasmic reticulum. Kindly provided by J.C. and M.W. Dieckert. See also Dieckert and Dieckert, 1972 [5]: scheme devised from electron micrographs presented in this paper

seeds has likewise led us to the unsurprising conclusion that oxygen consumption increases to a maximum during times of greatest increases in dry weight, nitrogen content, etc.

A direct study on ATP content has been conducted on two varieties of maturing rape (*Brassica napus*) seeds [40] although, unfortunately, in this instance there were no simultaneous measurements made on oxygen consumption. Cell division in the developing rape seed ceases at about 30 days after anthesis, a time some 10 days prior to the attainment of maximum fresh weight (Fig. 3.19A). Following the normal pattern of development outlined earlier in the chapter, the major synthesis of reserves (in this case lipids and protein) occurs at the completion of the cell division phase (Fig. 3.19B and C). ATP production in the seed increases during cell division, presumably reflecting, in part, increased synthesis of mitochondria in the newly formed cells (Fig. 3.19D). An increase in available substrate for respiration, in the form of soluble carbohydrate, occurs for the first 25 days following anthesis (Fig. 3.19E). The subsequent decline in carbohydrate probably reflects its utiliza-

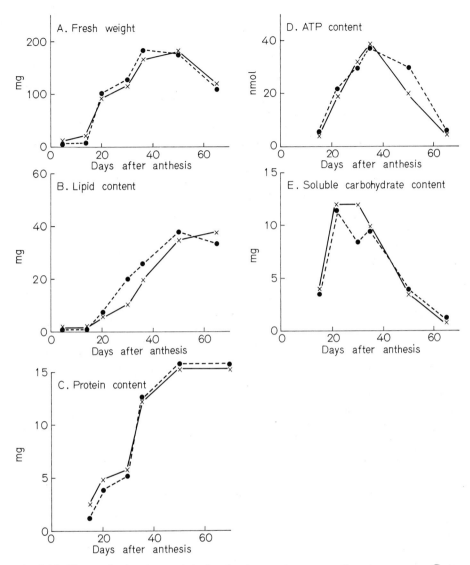

Fig. 3.19. Changes in the rape seed during development. ×——×: *Brassica napus* cv. Gorc-zanski, ●---●: *Brassica napus* cv. Victor. Based on Ching et al., 1974 [40]

tion for lipid biosynthesis, and this decline might in some way eventually limit respiration and thus account for the diminished ATP supply. The ATP level reaches a peak just after the commencement of reserve synthesis, with a decrease of 50% by day 50—the time when lipid and protein deposition ceases and drying commences (Fig. 3.19 D). It is obvious from these results that the levels of ATP in the seed tissues are high at a time when the synthesis of reserves is high, both events occurring when cell numbers in the rape seed are stabilized. The assumption which is made in these studies is that the location of ATP

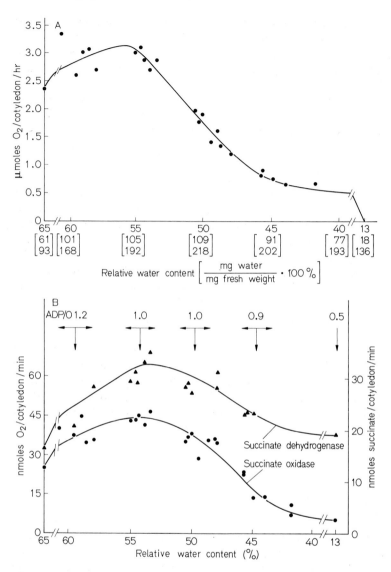

Fig. 3.20. (A) The respiration rate of pea cotyledons during the final phases of seed development plotted against the relative water content of the cotyledons. Figures in brackets represent water content and fresh weight (mg), respectively, per cotyledon. From these data the relative water content (*upper row*) was calculated. (B) The activity of the succinate oxidase system, and the succinate dehydrogenase activity of mitochondria from pea cotyledons during the final phases of seed development. ADP/O ratios are averages of data from several experiments (↔). *Vertical arrows* (↓) mark the average relative water content of the cotyledons used in these experiments. After Kollöffel, 1970 [96]

synthesis within the seed coincides with the major site of synthetic activity: this is not an unreasonable assumption to make even though it remains to be proven. The decline in ATP continues after completion of reserve synthesis (i.e. after day 50) until, in the mature dry seed, it reaches a negligible concentration.

Oxygen consumption by developing seeds expectedly diminishes as they dry out [167]. This loss of water is usually accompanied by a loss of integrity of the mitochondria and a decrease in the activity of some respiratory enzymes. Such changes are well-illustrated in the pea cotyledon during drying. As the cotyledon matures and dries its respiration rate (measured as oxygen consumption) rises slightly under stress and then falls (Fig. 3.20A). Changes in the activity of the citric acid cycle enzyme succinic dehydrogenase follow a similar pattern, as does the activity of succinic oxidase (the oxidase system whereby electrons are passed from the reduced FADH coenzyme of succinate dehydrogenase to oxygen via the electron transport chain) (Fig. 3.20B). Two other oxidase systems associated with the citric acid cycle (malate oxidase and α-ketoglutarate oxidase) also decline during drying, but malate dehydrogenase activity remains high in the dry seed. Respiratory control falls with increased drying, as does the ADP/O ratio (i.e. the amount of ADP used for ATP production per molecule of oxygen consumed by the terminal oxidase of the electron transport chain).

Studies with the electron microscope on the cotyledons of pea during the drying stage of development have shown us that there is disorganization of their fine structure, and in the dry seed mitochondria are poorly differentiated [21]. Similar observations have been made on several monocot and dicot embryos and cotyledons, and on the endosperm of castor bean. Reorganization of the respiratory system during imbibition following drying is discussed in Chapter 5.

3.5. The Fate of the Synthesizing Machinery

One of the most fascinating and yet least understood aspects of seed physiology is the extensive reversal in metabolism which ultimately takes place in the nutrient storage tissue after development has been completed, i.e. during germination and growth. Cells that have been factories for carbohydrate, lipid and protein production during embryogenesis become the sites of destruction and mobilization of these reserves during growth. Although in certain seeds there is limited turnover of some reserves following germination, in the main the synthesis of reserves ceases prior to seed maturation and the synthetic machinery is either completely sequestered or, more likely, wholly or partly destroyed. The fate of the components of the synthetic machinery is unknown, as is the "signal" to wind down the synthetic processes. Certainly, as we have seen earlier in this chapter, there occurs a time during development when there is maximum synthesis of reserves, which is probably a result of optimization of a number of factors — including maximal gene activation, template availability for enzyme production (and in the case of storage protein, template for reserve formation), activity of the cytoplasmic components involved in enzyme synthesis, energy availability, concentration of precursors, etc. A decrease in any of these factors could lead to reduced synthesis of reserves. It may be a very significant observation that depletion of reserve synthesis is associated with the onset of drying of the seed, and while the desiccation process per se has received only scant attention, it is likely to play an important role in the permanent switching-off of the synthetic events and perhaps also in the activation of processes for subsequent reserve mobilization.

During the formation of reserves in the cereal endosperm the actively metabolizing components of the cytoplasm are gradually occluded by the expanding starch grains. Thus, the machinery for continued protein synthesis (i.e. the cytoplasmic polysomes) is eventually destroyed. Some enzymes for reserve mobilization (e.g. α- and β-amylase) do appear to be present during the early phases of reserve formation, but their activity is obviously very low in relation to the activity of the synthetic enzymes, and prior to the completion of reserve synthesis they are inactivated. Enzymes for mobilization of cereal reserves are not synthesized in the starchy endosperm itself, but in the surrounding aleurone layer (Chaps. 6 and 7). Interestingly enough, the formation of "diastase" (amylase) following barley germination does not occur unless the grain has completed its maturation process including drying [31]. Thus, some unknown activation mechanism for the synthesis of α-amylase is apparently triggered during the drying phase of grain development. Furthermore, germination of the mature barley grain does not occur unless it has been dried [56].

Earlier in this chapter we saw that there is a decline in reserve synthesis which is coincidental with the onset of seed drying. We should also note that the GA-induced synthesis of α-amylase in the barley aleurone layer is dependent on the drying phenomenon [56]. If beneficial effects of dehydration on the control of enzyme synthesis were common in seeds and also if the components of the synthetic processes of maturation were destroyed during desiccation, then the enzymic activities of maturing and germinating seeds could be regulated simply by the interpolation of the physiological phenomenon of drought. The situation is probably more complex than this however, because some species of embryos and seeds dissected from the parent plant at several stages of development (without drying) can still germinate, e.g. isolated barley embryos [135], caryopses of *Avena fatua* [17], isolated axes of lima bean [94], *Phaseolus vulgaris* and *Phaseolus coccineus* [192]. Hydrolytic enzymes can also be produced, e.g. by cotton embryos [84] and isolated castor bean endosperms [109]. It is possible that at stages of seed development prior to complete maturity there is no requirement for activation of germination-related processes by drying, thus allowing for precocious germination of the dissected embryos, but that this requirement is gained only at complete maturity. Nevertheless, in these precociously germinating species there must be a control of some kind for the cessation of the events of maturation (e.g. storage-product synthesis) and for the switch-over to events related to germination and growth.

Protein synthesis has been observed to resume rapidly on imbibition of the mature, dried embryo, and certain aspects of this phenomenon will be dis-

cussed in detail in Chapter 5. Now we shall consider whether the RNA components of protein synthesis (viz messenger, transfer and ribosomal RNA)
formed in storage tissue during embryogenesis are conserved during drying
and are utilized during subsequent germination or growth of the seed.

3.5.1. Conservation of the Protein-Synthesizing Apparatus During Drying

Since a considerable body of work on this subject has been conducted on
the cotton seed this merits discussion by itself.

(i) Cotton

At maturity the cotton embryo consists of two thin, highly-convoluted cotyledons (storing lipid, protein and phytin) wrapped around the axis (Fig. 2.3A).
Following germination there arises in the cotyledons a proteinase (carboxypeptidase C) which is probably involved in the hydrolysis of storage proteins: this
enzyme is not present in the developing seed. Preliminary experiments on the
mature cotton seed showed that protein synthesis [47], including de novo synthesis of the proteinase [82], is not inhibited by actinomycin D (an inhibitor
of RNA synthesis) applied during the first three days after imbibition. It was
inferred from this observation that the production of this enzyme does not
require new RNA synthesis (including messenger RNA) during the early stages
following water uptake. Evidence was then sought to show that production
of the messenger RNA for the proteinase synthesized after germination in fact
occurred prior to seed maturation and was stored in the dry seed.

Immature embryos of cotton, when excised from the ovule during their
development, germinate — precociously. Mature embryos, before the desiccation
phase, weigh up to 125 mg (Fig. 3.21). Developing embryos, weighing as little
as 95 mg, germinate when they are removed from the ovule and placed on
agar gel; like the more mature (115 mg) embryos they also produce enzyme
(Table 3.5; minus actinomycin D). Dissected, developing embryos placed on
actinomycin D also germinate, but those at a stage of development where they
weigh less than 100 mg fail to produce the proteinase enzyme (Table 3.5). This
was taken to show that the enzyme messenger RNA (mRNA) is produced
within the embryo somewhere between the 95 and 100 mg stage of development,
some 15–20 days before the maturation process is completed (Fig. 3.21). It is
important to note that an assumption behind these experiments is that actinomycin D inhibits DNA-dependent mRNA synthesis in these embryos. This, in
fact, has never been shown directly and we must therefore regard the inferences
from these experiments with due caution.

If immature cotton embryos have the ability to germinate, demonstrated
by their behaviour when isolated, why do they not do so in the intact developing
seed? Vivipary, the germination of immature seeds within the fruit of the parent
plant, is rare in cotton and when it occurs it is lethal. Prevention of vivipary
might be explained by the fact that coincidental with the capacity of the embryo
to germinate precociously, and perhaps to produce messenger RNA for proteinase

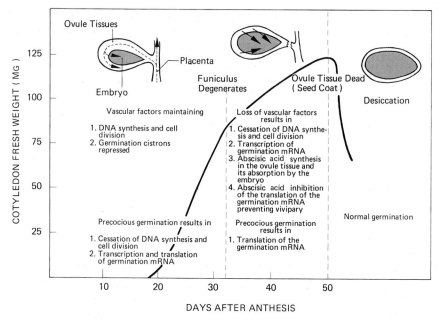

Fig. 3.21. Postulated scheme of developmental events in cotton cotyledon embryogenesis. After Ihle and Dure, 1972 [84]

Table 3.5. Development of proteinase enzyme activity during precocious germination of cotton embryos

Age	Enzyme units/cotyledon pair	
	− Actinomycin D	+ Actinomycin D
115 mg embryos, 2 days germinated	12.48	9.33
100 mg embryos, 4 days germinated	6.23	8.65
95 mg embryos, 6 days germinated	8.58	0

After Ihle and Dure, 1969 [82]

synthesis, there is also a rise in the concentration of abscisic acid in the ovule [83]. Apparently this abscisic acid diffuses into the developing embryo to inhibit germination and, in some unknown way, prevents translation of the transcribed message. At maturity, the ovule wall (integumentary) tissue dies during the formation of the sclerified seed coat and so an important source of abscisic acid is lost and normal germination of the mature seed can occur.

It has been suggested that on imbibition of the mature, dried seed the putative stored proteinase message has to be further processed before it can be translated. This might involve activation by post-transcriptional addition of adenylic acid residues on to the 3′OH end of the mRNA. Polyadenylation is inhibited by the chemical cordycepin and it is interesting that this also inhibits proteinase production in the cotyledons [191]. Obviously, the presence of this activation mechanism should not be a unique feature of the mature seed, unless the protein-ase message active in the precociously-germinated embryos dissected from the immature seed does not require similar processing. As yet this is undetermined.

A summarizing diagram outlining the postulated scheme of developmental events during cotton cotyledon embryogenesis is presented in Figure 3.21: more comprehensive accounts of the work in this field can be found in two recent reviews [7, 193]. Again, we emphasize the tentative nature of some of the evidence for this scheme.

Another enzyme which arises in germinated cotton seeds is isocitrate lyase, important in the conversion of the products of fat catabolism to hexose (see Chap. 6). This enzyme, like the proteinase, is absent from the maturing seed and is synthesized de novo following germination. One group of workers present evidence for its synthesis on conserved messages [84], but others make substantial claims that it is not [173].

While considerable attention has been paid to the conservation of the protein-ase messenger RNA in cotton seed, less has been given to the conservation of the other types of RNA required for protein synthesis. As far as transfer RNA is concerned, it appears that there are no significant differences between the levels of all cytoplasmic species at various developmental stages of the embryo; nor do any differences arise until several days after germination, when the chloroplasts mature [116]. Active ribosomes appear to be stored in the dry seed, and while synthesis of ribosomal RNA does occur during the early stages of germination of post 85-mg stage cotton embryos [193], it is not known whether this is involved in the translation process at any stage during early germination. Since germination and protein synthesis can both occur when the majority of RNA synthesis (including rRNA synthesis) is inhibited [195], it is possible that the synthesis of new ribosomes is not a pre-requisite for early translational events.

(ii) Other Seeds

Good evidence that germinating seeds can utilize ribosomes formed during the maturation phase has come from studies on *P. vulgaris* embryonic axes [190]. Here, embryo development requires 35 days for completion, and RNA synthesis can be observed up to day 24. After this time protein synthesis continues for several days, but there is no new RNA synthesis: while the pool size of some RNA precursors is reduced, the RNA content of the embryo remains constant throughout and subsequent to maturation. Most of this RNA is stored as ribosomes which become involved in protein synthesis during germination, prior to any new RNA synthesis.

The situation in peas appears to be somewhat similar to that in *Phaseolus* in that the RNA synthesized early in maturation is stable and is maintained (mainly in ribosomes as ribosomal RNA) during seed desiccation [143]. Many dried seeds contain active ribosomes (see Chapter 5 for a more complete consid-eration of protein and RNA synthesis during germination).

In some species, therefore, there is evidence for conservation of ribosomes and cytoplasmic RNAs. But in endosperm of castor bean and maize (Fig. 3.6F) we can see the opposite. Here, a dramatic decline occurs during maturation in ribosome and RNA content. Since the maize endosperm tissue dies following starch deposition it is not surprising that cellular components concerned with protein synthesis are lost. The castor bean endosperm is a persistent living tissue, however, responsible for the production of numerous enzymes for mobilization of reserves into the growing embryo (Chap. 6). Nonetheless, ribosomes seem to be lost from this tissue late in the maturation process, commencing at about the time of dehydration, re-synthesis occurring on imbibition of the mature seed (Fig. 5.13) [109, 110]. Whether a similar decline in RNA content also occurs in the maturing embryo is apparently unknown.

3.6. Hormones and Seed Development

Dormancy, germination and some early events in seedling development are thought in many cases to be regulated by various hormones. Two questions might therefore be asked: (1) What hormones are present in seeds? (2) Do these hormones arise during seed development? A third question, important in relation to seed development rather than to germination is: are any of the events of seed maturation, discussed above, influenced or controlled by endogenous hormones?

To answer the first question we need only recall that seeds were the first discovered higher plant sources of auxins, gibberellins and cytokinins. Indole acetic acid was identified in *Zea mays* kernels in 1946 [72], gibberellin-like activity in *Phaseolus* seeds in 1951 [125], and a purified cytokinin from *Z. mays* in 1963 [101]. Since these times it has become firmly established that both developing and mature seeds of very many species are rich in these three hormones and also in a fourth–abscisic acid. Indeed, it is not unlikely that all seeds contain these substances to some extent or another, especially during their development. In the following account we will first consider the occurrence and development of the individual hormones in seeds and then end with a brief discussion of the overall pattern of hormonal physiology in maturing seeds and the possible roles of these substances.

3.6.1. Gibberellins

It became clear in the early years after gibberellins were first discovered in higher plants that they are present in seeds of a wide range of species in concentrations much higher than are found elsewhere in the plant [10]. Values as high as 90 µg GA_3 equivalents per 100 g fresh weight have been found in mature seeds of certain Convolvulaceae (e.g. *Ipomoea batatas*). Table 3.6 lists just a few species in which the gibberellins have been positively identified. Besides these, numerous reports exist of seeds which yield "gibberellin-like" activity, i.e. substances which have biological activity as gibberellins together

Table 3.6. Some seeds in which gibberellins have been positively identified

Species and seed condition	Gibberellin	Concentration	Reference
Echinocystis macrocarpa (mature seed)	GA_1, GA_3, GA_4, GA_7	–	[54]
Coryllus avellana (mature seed)	GA_1 GA_9	1.02 nmol/embryo	[199]
Pyrus malus (mature seed)	GA_4, GA_7, GA_9	–	[170]
Cucumis sativus (mature seed)	GA_1, GA_3, GA_4, GA_7	–	[80]
Pyrus communis (immature seed)	GA_{45}	–	[27]
Pisum sativum (immature seed)	GA_{20}, GA_{29}	–	[52, 62]
Phaseolus multiflorus (immature seed)	GA_1 GA_5 GA_6 GA_8 GA_{17} GA_{19} GA_{20}	8.0 mg/kg fresh wt 1.5 mg/kg fresh wt 2.5 mg/kg fresh wt 30.0 mg/kg fresh wt 2.0 mg/kg fresh wt 0.5 mg/kg fresh wt 0.5 mg/kg fresh wt	[48] [48] [48] [48] [48] [48] [48]
Phaseolus vulgaris (immature seed)	GA_1, GA_4, GA_5, GA_6, GA_8, GA_{37}, GA_{38}, GA_8 glucoside	–	[203]
Phaseolus vulgaris (mature seed)	GA_1, GA_8, GA_8 glucoside, glucosyl esters of GA_1, GA_4, GA_3, GA_{38}	–	[203]
Calonyction aculeatum (immature seed)	GA_{17} GA_8, GA_{19}, GA_{20}, GA_{27}, GA_{29}, GA_{30}, GA_{31}, GA_{33}, GA_{34}	15 mg/kg fresh wt	[177]

with some of their chemical or physical properties but which have not been chemically characterized. With improvements in analytical techniques, new identifications of gibberellins in seeds are regularly appearing in the scientific literature. Structural formulae for five seed gibberellins are shown in Figure 3.22. Reference should be made to recent publications for further details [e.g. 9].

Seeds may also contain gibberellins conjugated to a sugar, usually glucose, to form glucosides and less commonly glucosyl esters. In addition to those shown in Table 3.6 we might also mention the 5 glucosides (of GA_3, GA_8, GA_{22}, GA_{26} and GA_{29}) that have been found in developing seeds of *Pharbitis nil* [205] and point to the evidence that free gibberellins in other seeds are conjugated during maturation (see below). It has been suggested that gibberellins might also become bound to proteins (and can be liberated by treatment with proteolytic enzymes) but there is still uncertainty about this possibility [126].

Several pieces of evidence show that the gibberellins are synthesized by the seeds themselves. Firstly, the build-up of gibberellin in the seed (e.g. of

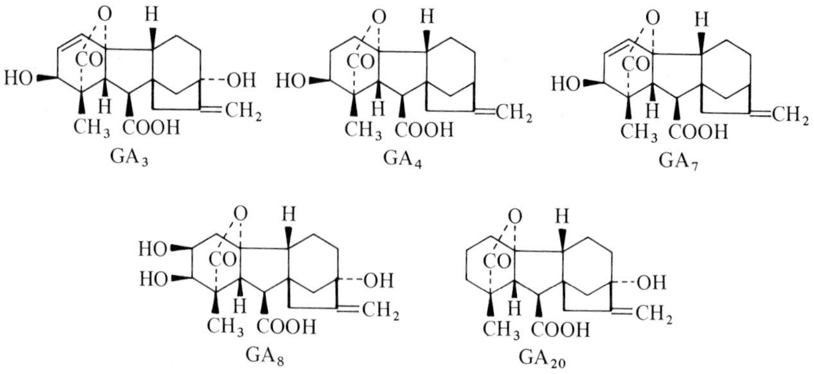

Fig. 3.22. Structures of five gibberellins commonly found in seeds

wheat, *Triticum aestivum*) does not coincide with increased export from other parts of the plant [197]. But more importantly, it has been shown directly that developing seeds are capable of gibberellin biosynthesis. Indeed, most of our knowledge about their biosynthesis in higher plants is derived from studies on immature seeds. The liquid endosperm of developing seeds of *Echinocystis macrocarpa* (wild cucumber), for example, converts mevalonic acid into gibberellin through various intermediates of the suspected biosynthetic pathway such as geranylgeranyl pyrophosphate, kaurene and kaurenol. Moreover, cell-free enzyme preparations from developing seeds (*Pisum sativum, E. macrocarpa* and *Cucurbita pepo*) also incorporate mevalonate, synthesize kaurene and convert hydroxykaurenoic acid to gibberellins [126]. Many of these conversions by cell-free systems are, expectedly, prevented by substances which inhibit gibberellin biosynthesis—the growth retardants AMO 1618, Phosfon D and CCC [99]. It is not possible here to discuss fully the biochemical mechanism of gibberellin synthesis. An outline is shown in Figure 3.23; more details can be found in recent literature [9, 196]. We might note, though, that good evidence for gibberellin interconversions in immature seeds comes from experiments in which the fate of fed, radioactively-labelled gibberellin has been traced [63].

The site of gibberellin biosynthesis in *Echinocystis* is the endosperm, the highest rates occurring while this nutritive tissue is still liquid before it becomes absorbed by the developing embryo. Once this has taken place the capacity to make gibberellins falls [67]. A similar situation occurs in some other seeds, for example in apricot, *Prunus armeniaca* [86], but in apple seeds gibberellins appear when digestion and absorption of the nucellus and primary endosperm begins. The amounts continue to increase even when this absorption has ended, at the time when embryo growth is accelerating [108]. A situation rather like this also holds in *P. sativum* seeds where there are two waves of gibberellin production, the first in the liquid endosperm and the second in the embryo (Fig. 3.35), though it has been claimed, surprisingly, that the liquid endosperm in this species has only a low capacity for kaurene biosynthesis [41].

A feature shown by all developing seeds is the drop in levels of free gibberellin as the seed matures. This is preceded by one (e.g. *T. aestivum* [197], *Echinocystis*

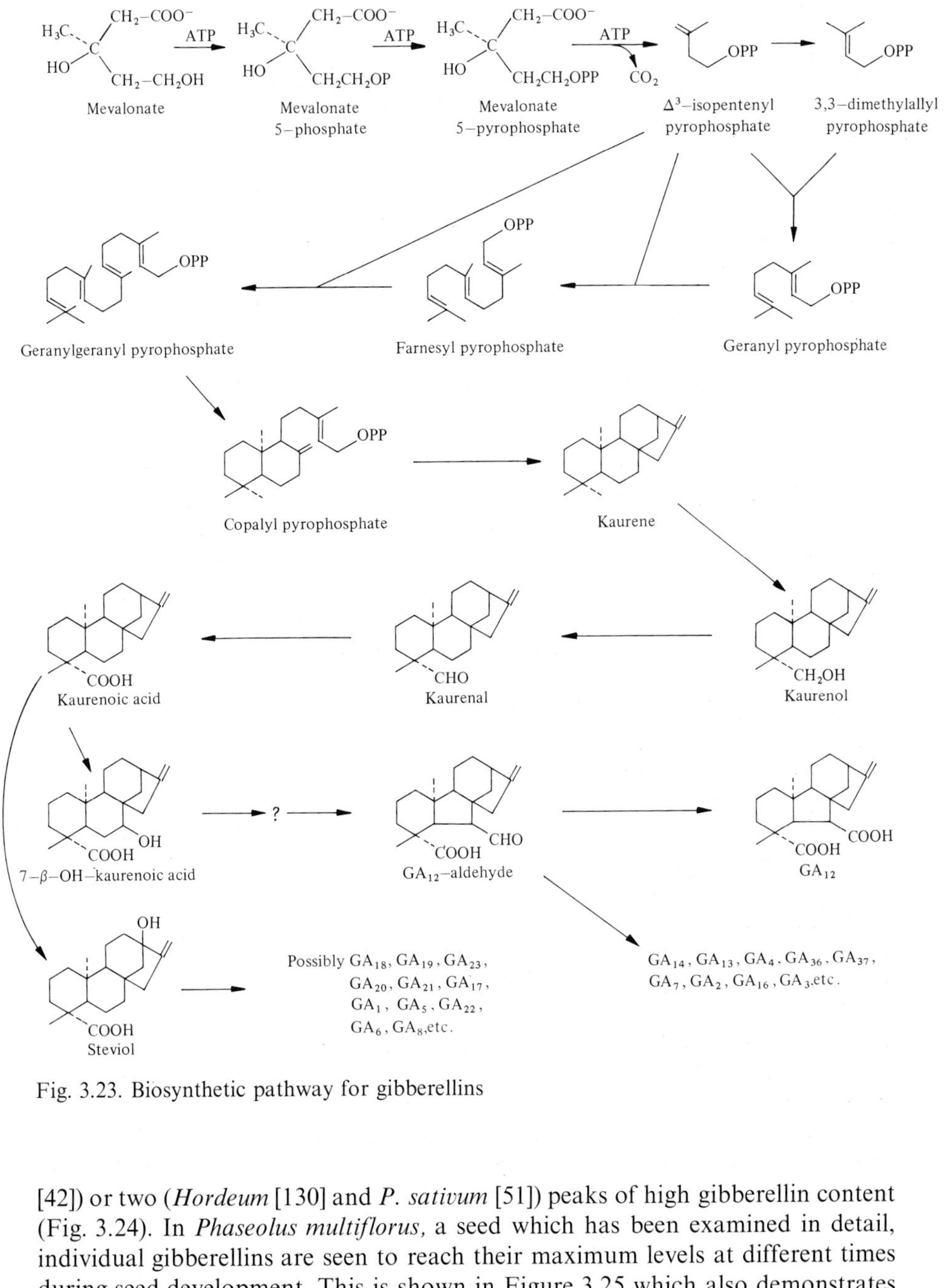

Fig. 3.23. Biosynthetic pathway for gibberellins

[42]) or two (*Hordeum* [130] and *P. sativum* [51]) peaks of high gibberellin content (Fig. 3.24). In *Phaseolus multiflorus,* a seed which has been examined in detail, individual gibberellins are seen to reach their maximum levels at different times during seed development. This is shown in Figure 3.25 which also demonstrates how seemingly different patterns of hormone content come from the expression of results on a fresh weight or per seed basis.

The characteristic drop in free gibberellin content towards the end of seed development has been shown in some cases (e.g. *Hordeum vulgare* [130], *P. vul-*

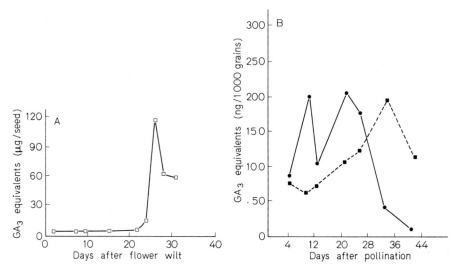

Fig. 3.24A and B. Changes in gibberellin content during seed development. (A) *Echinocystis macrocarpa*. After Corcoran and Phinney, 1962 [42]. (B) *Hordeum vulgare* cv. Brevia. After Mounla and Michael, 1973 [130]. ●——●: "free" gibberellin, ■---■: "bound" gibberellin

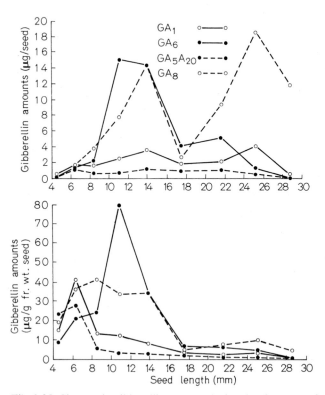

Fig. 3.25. Changes in gibberellin content during development of *Phaseolus multiflorus* seeds. After Durley et al., 1971 [48]

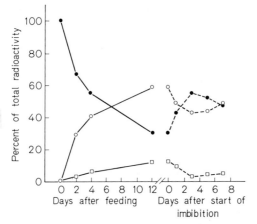

Fig. 3.26. Distribution of radioactive gibberellin during development and germination/ growth of *Pisum sativum* seeds. Developing seeds were fed with ^{3}H-GA$_1$ in vitro and allowed to develop to maturity. The distribution of radioactivity during development and subsequent germination was traced. ——: developing seeds, ---: germinating seeds/growing seedlings, ●——●: "free" GA (ethyl acetate fraction), ○——○: "bound" GA (aqueous fraction), □——□: tissue residue. After Barendse et al., 1968 [22]

garis [76]) to be associated with a rise in the conjugated gibberellin ("bound gibberellin") (Fig. 3.24 B). It seems likely, therefore, that as seed development terminates at least some of the gibberellins are sequestered in an inactive form probably in combination with sugars. Indeed, developing seeds of *Pharbitis nil* and *Pisum sativum* when supplied with radioactive (^{3}H) GA$_1$ apparently convert it into a water-soluble, presumably conjugated form (Fig. 3.26). Developing *Phaseolus multiflorus* seeds similarly convert radioactive gibberellin to hexa-pyranosyl and glycosyl esters [160]. A proportion of the sequestered gibberellin is liberated when pea seeds germinate (Fig. 3.26). Thus, germination might, in part, utilize gibberellins previously stored during maturation, though several workers have cast doubt upon this role for the conjugated gibberellins [63, 203]. Yamane and his co-workers, for example [203], showed in *P. vulgaris* seeds that there was little liberation of free radioactive gibberellin during germination but that, on the contrary, there was continued transfer of label into glucosides.

It does seem more likely that the gibberellins synthesized in developing seeds are much concerned with both seed and fruit growth. Work on *P. sativum* has demonstrated a fairly close relationship between the presence of high gibberellin levels in the seeds and the increases in both absolute and relative growth rates of seeds and ovary wall (i.e. the pod) (see Fig. 3.34). A similar picture has emerged for several other species [126].

3.6.2. Auxins

Turning now to the auxins we find that developing seeds are again a particularly rich source having, in general, much higher levels of the free hormone than

mature seeds. The auxins are indole-3-acetic acid (IAA) and its various derivatives (Fig. 3.27 and Table 3.7). From the Table we can see that in some seeds
IAA is combined with myo-inositol or arabinose as conjugated or bound auxin.
Combined forms of IAA are in fact present in many seeds, *Z. mays* being
one that has been studied in some detail. Here the bound auxins, amounting
to up to 100 mg/kg fresh weight, fall into two groups according to their molecular
weight and solubility [98]. One group, of high molecular weight, is lipid-soluble
and contains, among other compounds, IAA esterified with a β-1,4 glucan of
variable chain length [142]. The second group consists of low molecular weight,
water-soluble substances including IAA myo-inositols, and IAA myo-inositolga-
lactosides [185]. It is considered that one role of these substances may be to
furnish IAA during early seedling growth supplying, for example, the tip of
the cereal coleoptile.

During seed development auxins accumulate in the nutritive tissues, i.e.
in the endosperm and nucellus. This was shown firstly for the cereals (e.g.
rye [78]) and later for apple [107] and garden pea [51] among others. In the
garden pea the auxin produced in the endosperm was found to be 4-chloroindole
acetic acid, whereas its methyl ester occurred mainly in the embryo [65]. The
auxin extractable from the seed tissues is almost certainly produced there, presumably by the biosynthetic pathway which has been worked out from studies
on other parts of higher plants. These have shown that IAA is derived from
the amino acid tryptophan, by the route shown in Figure 3.28. Further details
can be found in recent reviews (e.g. [158]).

The auxins, like the gibberellins, show fluctuations in level during the course
of seed development. Many examples of this have been reported but only a
few will be mentioned here. In apple seeds there are two peaks in auxin content,
the first coinciding with the change from a free nuclear (i.e. coenocytic) endosperm to a cellular one (Sect. 3.1). Digestion of the endosperm then begins, ac-

Table 3.7. Some seeds containing auxins

Species	Auxin	Reference
Prunus cerasus (immature seeds)	IAA (11.2 ng/seed, maximum)	[81]
Fragaria (immature achenes)	IAA	[133]
Triticum aestivum (immature kernels)	IAA (0.098 ng/seed, maximum)	[197]
Pisum sativum (immature seeds)	CIAA	[66]
	MCIA	[51, 111]
Malus sylvestris (immature seeds)	IAA, Ethyl IAA	[189]
Zea mays	IAA	[72]
(immature kernels)	Ethyl IAA	[149]
(immature kernels)	IAA arabinoside	[165]
(immature kernels)	IAA myo-inositol	[132]
(immature kernels)	IAA myo-inositol arabinoside	[98]
(immature kernels)	Indole-3-pyruvic acid	[176]
(mature kernels)	IAA-glucan	[142]
(mature kernels)	IAA-myo-inositols	[53]

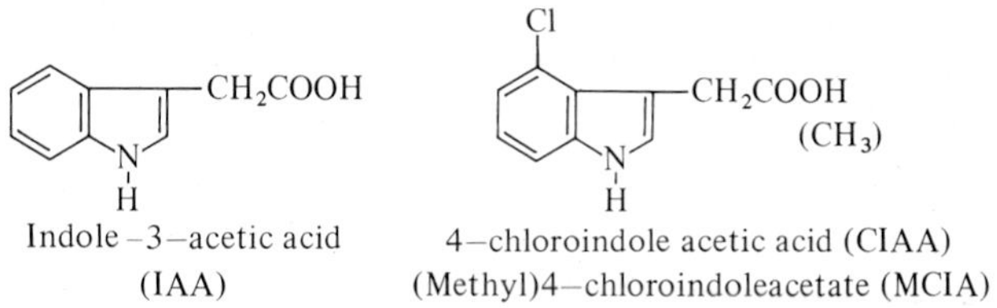
Cl
CH₂COOH
CH₂COOH
(CH₃)
N
H
N
H
Indole–3–acetic acid
(IAA)
4–chloroindole acetic acid (ClAA)
(Methyl)4–chloroindoleacetate (MClA)

Fig. 3.27. Structural formulae of auxins occurring in seeds

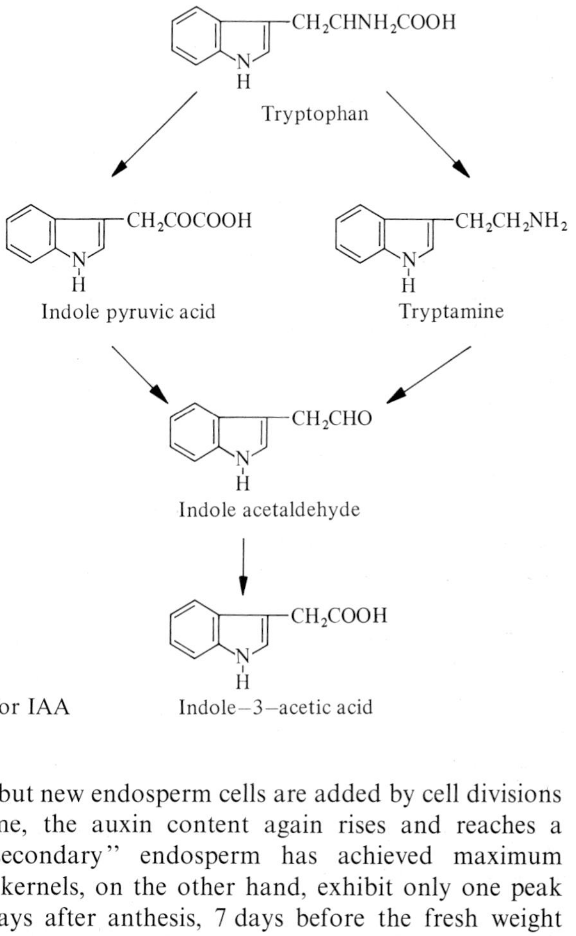
CH₂CHNH₂COOH
Tryptophan
CH₂COCOOH
Indole pyruvic acid
CH₂CH₂NH₂
Tryptamine
CH₂CHO
Indole acetaldehyde
CH₂COOH
Indole–3–acetic acid

Fig. 3.28. Biosynthetic pathway for IAA

companying embryo growth, but new endosperm cells are added by cell divisions at the periphery; at this time, the auxin content again rises and reaches a second peak when the "secondary" endosperm has achieved maximum size [107]. Developing wheat kernels, on the other hand, exhibit only one peak in auxin level at about 35 days after anthesis, 7 days before the fresh weight of the grain is at its maximum (Fig. 3.29). In *P. sativum* the auxin content of the endosperm climbs steeply to reach a high level in the 6-day-old endosperm; after the endosperm has been digested there is a small increase in the free auxin of the embryo at about 14 days which then falls to a very low value (Fig. 3.29). It seems then, in all the seeds which have been investigated, that the free auxin rapidly diminishes as the seed reaches full maturity. Conversion to other forms (esters and "bound" auxin) is known to occur in endosperm of maturing cereal kernels (e.g. maize [see Table 3.7] and rye [77]) and in seeds

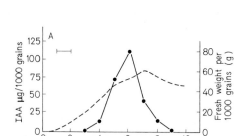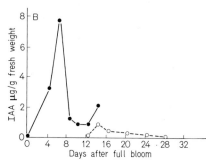

Fig. 3.29 A and B. Changes in auxin content during seed development. (A) *Triticum aestivum* cv. Kloka. *Horizontal line* shows the period of anthesis. ●——●: IAA in whole grain, ---: fresh wt. After Wheeler, 1972 [197]. (B) *Pisum sativum* cv. Alaska, ●——●: IAA in liquid endosperm, ○---○: IAA in embryo. After Eeuwens and Schwabe, 1975 [51]

of certain dicot species [189, 207]. Since it is not clear what role, if any, auxin plays in germination, the various changes in levels of this substance in immature seeds are at present best seen in relation to seed development itself (see Sect. 3.6.5). Liberation of IAA from bound or conjugated forms might, however, contribute to the control of early seedling growth, as has been mentioned above.

3.6.3. Cytokinins

Since zeatin was isolated from immature maize kernels a number of cytokinins have been identified or, on the basis of biological activity, have been shown to be present in seeds. Derivatives of zeatin occur in maize and yellow lupin (*Lupinus luteus*). Most cytokinins are adenine derivatives but one substance which can support cell division and therefore has cytokinin-like activity does not belong to this chemical group. It is diphenylurea from coconut liquid endosperm (Table 3.8 and Fig. 3.30) and it is this substance which is partially respons-

Table 3.8. Seeds containing identified cytokinins

Species	Cytokinin	Reference
Zea mays (immature kernels)	Zeatin (approx. 11 µg/kg fresh wt)	[101]
	Zeatin riboside	[102]
	Zeatin ribotide	[103]
Cucurbita pepo (immature seeds)	Zeatin	[70]
Cocos nucifera	Zeatin riboside	[104]
	Diphenylurea	[164]
Lupinus luteus (immature seeds)	Dihydrozeatin	[97]
Pyrus malus (immature seeds)	Zeatin	[105]
	Zeatin riboside	[105]
	Zeatin ribotide	[105]

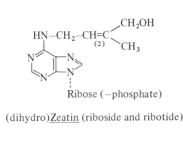

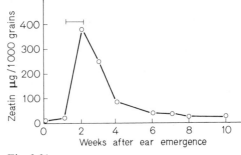

Fig. 3.31

N, N′–diphenylurea

Fig. 3.30

Fig. 3.30. Structural formulae of cytokinins found in seeds. The structure of zeatin is shown with substitutions to give the riboside or, if phosphorylated ribose is present, the ribotide. The bracketed number indicates the extra hydrogen present in dihydrozeatin

Fig. 3.31. Change in cytokinin content during grain development in *Triticum aestivum* cv. Kloka. *Horizontal bar* indicates the period of anthesis. After Wheeler, 1972 [197]

ible for the well-known growth-promoting properties of this tissue. In addition to the seeds shown in Table 3.8 which are known to contain chemically-identified cytokinins, a large number is known which yield semi-purified cytokinin-like compounds which have not been rigorously characterized. Examples are *P. sativum* [73], *Lactuca sativa* [186], *P. arvense* [35], *Leucodendron daphnoides, Protea compacta* [32], *Helianthus annuus* [121], and *T. aestivum* [197].

Since cytokinins are exported from the roots it has been suggested that these organs supply the developing seeds. There is evidence in wheat in favour of this hypothesis since here the maximum cytokinin levels transported through the stem precede the peak of cytokinin concentration in the developing kernels by just a few days [197]. On the other hand, *P. sativum* seeds, developing in the pod under in vitro culture, show approximately a 30-fold increase in cytokinin content over 18 days, thus establishing that the seeds themselves are capable of hormone synthesis [73].

Changing levels of cytokinin can be followed during seed development. Cytokinin in *Z. mays* kernels, for example, reaches a peak 11 days after pollination and declines considerably over another 10 days [120]. A similar pattern is found in *T. aestivum* (Fig. 3.31). One major and two minor peaks are seen in *Pisum arvense* (field pea) which coincide with the maximum volume of the endosperm and with the two periods of rapid growth of the whole seed and embryo [35]. The cytokinins of immature seeds are thought, therefore, to be involved in the phases of growth and development. Whether or not they also represent a store of hormone for later germination and growth is uncertain.

3.6.4. Abscisic Acid

Many seeds contain the inhibitor (+)-abscisic acid (ABA. Fig. 3.32) located in various seed parts according to the species. In addition to those few seeds which are listed in Table 3.9 several others yield identifiable ABA and many more have substances with similar biological activity but which have not been positively characterized. Abscisic acid esterified with glucose (β-D-glucopyrano-syl abscisate) is found in certain plants and has been reported in seeds (e.g. apple [33]).

The source of the seed's abscisic acid is not yet clear. Some, or all in certain seeds, may be derived from the mother plant but in at least one case (*T. sativum*) it has been demonstrated that the embryo and endosperm can synthesize this substance from supplied mevalonate [119]. It has been suggested also that as seeds suffer increasing water stress when they begin to dehydrate the ABA levels rise as they do in wilting leaves (e.g. [118]). This is not likely in all cases, however, since in wheat kernels for example, the ABA levels increase before any appreciable water loss occurs [147]. Synthesis of ABA is considered to follow the route for isoprenoid production from mevalonate (see [118] for further details). A suggested alternative is that ABA is not synthesized in its own right but that it arises from the carotenoid violaxanthin via the production of xanthoxin. This is a debatable point; it now seems unlikely [118], especially in seeds.

ABA is virtually undetectable in developing *P. sativum* seeds until about 16 days after pollination; thereafter the level steeply rises, reaches a peak at

Table 3.9. Some seeds containing abscisic acid

Species	Concentration (mg/kg fresh wt)	Reference
Pisum sativum (immature seeds)	1 (maximum)	[51]
Lactuca sativa (mature achenes)	0.70	[115]
Triticum aestivum (immature kernels)	0.12	[114]
Rosa arvensis (mature achenes)	0.16	[117]
Rosa canina (mature achenes)	0.53	[117]
Cocos nucifera (mature seed)	—	[117]
Persea gratissima (mature seed)	0.11	[117]
Pyrus malus (mature seed)	0.04	[154]
Corylus avellana (mature seed) Pericarp	0.37	[198]
Testa	5.03	
Embryo	0.02	
Fraxinus americana (mature seed)	0.45	[174]
Pericarp	0.73	
Fraxinus ornus (mature seed)	0.11	[174]
Pericarp	0.22	
Taxus baccata (mature seed)	0.71	[100]

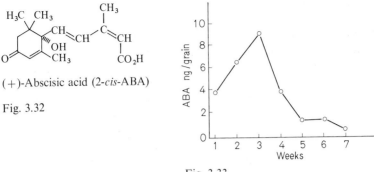

(+)-Abscisic acid (2-*cis*-ABA)

Fig. 3.32

Fig. 3.33

Fig. 3.32. Structural formula for naturally-occurring abscisic acid

Fig. 3.33. Change in abscisic acid content during grain development in *Triticum aestivum* cv. Arawa. Week 1 is 4 weeks after ear emergence and is 19 days after peak anthesis. After McWha, 1975 [114]

24 days, then disappears by full maturity (see Fig. 3.35). A similar pattern of change is also seen in developing wheat (*T. aestivum*) (Fig. 3.33). The metabolic fate of ABA during seed maturation is uncertain but breakdown to phaseic and dihydrophaseic acids seems likely [118]. Of course, in those mature seeds which possess appreciable amounts of ABA the compound is presumably carried over during maturation. Clarification is needed concerning exactly how much ABA survives the maturation process and what the previous pattern of ABA production is in those cases in which the substance is carried over into the dry seed. Any ABA which is consigned to the mature seed is probably very important in controlling germination since it is debatably the major chemical agent responsible for the initiation and maintenance of seed dormancy. Apart from this role, in the developing seed ABA might be responsible for the control of embryo growth or, more accurately, for the prevention of precocious germination, as we have seen earlier (Sect. 3.5.1). Germination of seeds while still on the mother plant (vivipary) may thus be arrested by the build-up of the inhibitor.

3.6.5. The Role of Hormones in the Developing Seed

In this account we have seen how patterns of appearance and disappearance of the various hormones occur during seed maturation. We have indicated, from time to time, what significance these substances might have but it may now be useful briefly to recapitulate these possibilities.

The physiological roles of these hormones fall into 4 groups: (1) They accumulate in preparation for their subsequent participation in the control of germination and early seedling growth. As examples, we can quote the probable involvement of ABA in dormancy and the possible release during germination and growth of gibberellin and auxin from "bound" and conjugated forms. (2) They control fruit growth and development. This is clearly the case in many

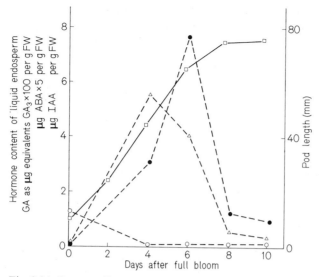

Fig. 3.34. Ovary wall growth and hormone content of seeds during early seed development in *Pisum sativum* cv. Alaska. □——□: pod length, ○---○: ABA, ●---●: auxin, △---△: gibberellin. After Eeuwens and Schwabe, 1975 [51]

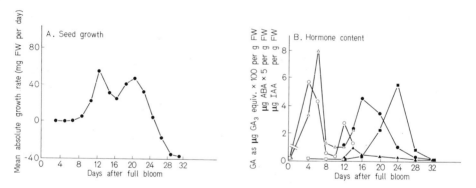

Fig. 3.35A and B. Seed development and hormone levels in *Pisum sativum* cv. Alaska. □——□: ABA in liquid endosperm, ○——○: GA in liquid endosperm, △——△: auxin in liquid endosperm, ■——■: ABA in embryo, ●——●: GA in embryo, ▲——▲: auxin in embryo. After Eeuwens and Schwabe, 1975 [51]

fleshy fruits [134] but is exemplified also by *P. sativum* in which the growth rate of the ovary wall (i.e. the wall of the pod) closely relates to changes in hormonal content of the seeds (Fig. 3.34). (3) Movement of dry matter (carbohydrates and nitrogenous substances) into the seed may be regulated by the hormones contained therein. In wheat, for example, the highest levels of auxin in the developing kernel are associated with the time of intensive accumulation of dry matter [197]. (4) Finally, the growth and development of the seeds themselves may be under hormonal control. The diauxic growth curve of developing seeds of the field pea (*P. arvense*) accompanies changes in the extractable cytokinin content [35]. Peaks in cytokinin level coincide with times of maximal development of endosperm and the two periods of rapid growth of the embryo. In *P. sativum* we can see a fairly close correlation between the phases of maximum growth of the seed and the levels of auxin and gibberellin in endosperm and embryo [51]; moreover, the decline in growth rate of the seed accompanies the build up of abscisic acid (Fig. 3.35). In this connection, we have already touched upon the possibility that abscisic acid might function to prevent viviparous germination. Such correlations as these do not, of course, prove that the hormones are causally involved with seed growth and development but they provide strong evidence fully consistent with this concept.

Some Articles of General Interest

1. Altschul, A.M., Yatsu, L.Y., Ory, R.L., Engleman, E.M.: Seed proteins. Ann. Rev. Pl. Physiol. **17**, 113–136 (1966)
2. Beevers, L.: Nitrogen Metabolism in Plants. London: Edward Arnold, 1976
3. Bhatnagar, S.P., Johri, B.M.: Development of angiosperm seeds. In: Seed Biology. Kozlowski, T.T. (ed.). New York: Academic Press, 1972, Vol I, pp. 77–149
4. Dalby, A., Cagampang, G.B., Davies, I. ab I., Murphy, J.J.: Biosynthesis of proteins in cereals. In: Symp. Seed Proteins. Inglett, G.E. (ed.). Westport, Conn.: AVI Publ. Co. Inc. 1972, pp. 39–51
5. Dieckert, J.W., Dieckert, M.C.: The deposition of vacuolar proteins in oilseeds. In: Symp. Seed Proteins. Inglett, G.E. (ed.). Westport, Conn.: AVI Publ. Co. Inc. 1972, pp. 52–85
6. Dure, L.S. III: Developmental regulation in cotton seed embryogenesis and germination. In: Developmental Regulation. Aspects of Differentiation. Coward, S.J. (ed.). New York: Academic Press, 1973, pp. 23–48
7. Dure, L.S. III: Seed formation. Ann. Rev. Pl. Physiol. **26**, 259–278 (1975)
8. Harwood, J.L.: Fatty acid biosynthesis. In: Recent Advances in the Chemistry and Biochemistry of Plant Lipids. Galliard, T., Mercer, E.I. (eds.). London: Academic Press, 1975. Proc. Phytochem. Soc. **12**, 247–286
9. Krishnamoorthy, H.N.: Gibberellins and Plant Growth. New Delhi: Wiley Eastern, 1975
10. Lang, A.: Effects of some internal and external conditions on seed germination. In: Encyclopedia of Plant Physiology. Ruhland, W. (ed.). Berlin: Springer, 1965, Vol XV/2, pp. 848–893
11. Manners, D.J.: The structure and metabolism of starch. Essays in Biochemistry. London: Academic Press, 1974, Vol X, 37–71
12. Millerd, A.: Biochemistry of legume seed proteins. Ann. Rev. Pl. Physiol. **26**, 53–72 (1975)
13. Pate, J.S.: Pea. In: Crop Physiology. Evans, L.T. (ed.). Cambridge: Univ. Press. 1975, pp. 191–224
14. Singh, H., Johri, B.M.: Development of gymnosperm seeds. In: Seed Biology. Kozlowski, T.T. (ed.). New York: Academic Press, 1972, Vol I, 21–75
15. Turner, J.F., Turner, D.H.: The regulation of carbohydrate metabolism. Ann. Rev. Pl. Physiol. **26**, 159–186 (1975)
16. Wardlaw, C.W.: Embryogenesis in Plants. London: Methuen and Co., 1955

References

17. Andrews, C.J., Simpson, G.M.: Can. J. Botany **47**, 1841–1849 (1969)
18. Appleby, R.S., Gurr, M.I., Nichols, B.W.: Europ. J. Biochem. **48**, 209–216 (1974)
19. Atkins, C.A., Pate, J.S., Sharkey, P.J.: Pl. Physiol. **56**, 807–812 (1975)
20. Bailey, C.J., Cobb, A., Boulter, D.: Planta (Berl.) **95**, 103–118 (1970)
21. Bain, J.M., Mercer, F.V.: Australian J. Biol. Sci. **19**, 49–67 (1966)
22. Barendse, G.W.M., Kende, H., Lang, A.: Pl. Physiol. **43**, 815–822 (1968)
23. Basha, S.M.M., Beevers, L.: Pl. Physiol. **57**, 93–97 (1976)

24. Baun, L.C., Palmiano, E.P., Perez, C.M., Juliano, B.O.: Pl. Physiol. **46**, 429–434 (1970)
25. Baxter, E.D., Duffus, C.M.: Planta (Berl.) **114**, 195–198 (1973)
26. Baxter, E.D., Duffus, C.M.: Phytochemistry **12**, 2321–2330 (1973)
27. Bearder, J.R., Dennis, F.G., MacMillan, J., Martin, B.C., Phinney, B.O.: Tetrahedron Lett. **9**, 669–670 (1975)
28. Beevers, L., Poulson, R.: Pl. Physiol. **49**, 476–481 (1972)
29. Bils, R.F., Howell, R.W.: Crop Sci. **3**, 304–308 (1963)
30. Briarty, L.G., Coult, D.A., Boulter, D.: J. Exp. Botany **20**, 358–372 (1969)
31. Brown, H.T., Morris, G.H.: J. Chem. Soc. **57**, 458–528 (1890)
32. Brown, N.A.C., Van Staden, J.: Physiol. Plantarum **28**, 388–392 (1973)
33. Bulard, C., Barthe, P., Garello, G., Le Page-Degivry, M.T.: C.R. Acad. Sci. Paris; Ser. D **278**, 2145–2148 (1974)
34. Burr, B., Burr, F.A.: Proc. Natl. Acad. Sci. **73**, 515–519 (1976)
35. Burrows, W.J., Carr, D.J.: Physiol. Plantarum **23**, 1064–1070 (1970)
36. Buttrose, M.S.: J. Ultrastruct. Res. **4**, 231–257 (1960)
37. Buttrose, M.S.: Australian J. Biol. Sci. **16**, 305–317 (1963)
38. Canvin, D.T.: Can. J. Biochem. Physiol. **41**, 1879–1885 (1963)
39. Carr, D.J., Wardlaw, I.F.: Australian J. Biol. Sci. **18**, 711–719 (1965)
40. Ching, T.M., Crane, J.M., Stamp, D.L.: Pl. Physiol. **54**, 748–751 (1974)
41. Coolbaugh, R.C., Moore, T.C.: Phytochemistry **10**, 2395–2400 (1971)
42. Corcoran, M.R., Phinney, B.O.: Physiol. Plantarum **15**, 252–262 (1962)
43. Crookston, R.K., O'Toole, J., Ozbun, J.L.: Crop Sci. **14**, 708–712 (1974)
44. Cruz, L.J., Cagampang, G.B., Juliano, B.O.: Pl. Physiol. **46**, 743–747 (1970)
45. Dasgupta, S.K., Friend, J.: J. Sci. Food Agr. **24**, 463–470 (1973)
46. Dickinson, D.B., Preiss, J.: Pl. Physiol. **44**, 1058–1062 (1969)
47. Dure, L.S. III, Waters, L.C.: Science **147**, 410–412 (1965)
48. Durley, R.C., MacMillan, J., Pryce, R.J.: Phytochemistry **10**, 1891–1908 (1971)
49. Dybing, C.D., Craig, B.M.: Lipids **5**, 422–429 (1970)
50. Edwards, M.M.: J. Exp. Botany **19**, 575–582 (1968)
51. Eeuwens, C.J., Schwabe, W.W.: J. Exp. Botany **26**, 1–14 (1975)
52. Eeuwens, C.J., Gaskin, P., MacMillan, J.: Planta (Berl.) **115**, 73–76 (1973)
53. Ehmann, A., Bandurski, R.S.: Carbohyd. Res. **36**, 1–12 (1974)
54. Elson, G.W., Jones, D.F., MacMillan, J., Suter, P.J.: Phytochemistry **3**, 93–101 (1964)
55. Evans, L.T., Rawson, H.M.: Australian J. Biol. Sci. **23**, 245–254 (1970)
56. Evans, M., Black, M., Chapman, J.: Nature (London) **258**, 144–145 (1975)
57. Evers, A.D.: Ann. Botany (London) **34**, 547–555 (1970)
58. Flinn, A.M., Pate, J.S.: Ann. Botany (London) **32**, 479–495 (1968)
59. Flinn, A.M., Pate, J.S.: J. Exp. Botany **21**, 71–82 (1970)
60. Frazier, J.C., Appalanaidu, B.: Am. J. Botany **52**, 193–198 (1965)
61. Frey-Wyssling, A., Grieshaber, E., Mühlethaler, K.: J. Ultrastruct. Res. **8**, 506–516 (1963)
62. Frydman, V., MacMillan, J.: Planta (Berl.) **115**, 11–15 (1973)
63. Frydman, V., MacMillan, J.: Planta (Berl.) **125**, 181–195 (1975)
64. Galliard, T., Stumpf, P.K.: J. Biol. Chem. **241**, 5806–5812 (1966)
65. Gandar, J.C., Nitsch, J.P.: In: Régulateurs Naturels de la Croissance Végétale. Nitsch, J.P. (ed.). Coll. Int. CNRS, No. 123. Paris. (1964), pp. 169–178
66. Gandar, J.C., Nitsch, J.P.: C.R. Acad. Sci. Paris Ser. D **265**, 1795–1798 (1967)
67. Graebe, J.E., Dennis, D.T., Upper, C.D., West, C.A.: J. Biol. Chem. **240**, 1847–1854 (1965)
68. Graham, T.A., Gunning, B.E.S.: Nature (London) **228**, 81–82 (1970)
69. Gunning, B.E.S., Pate, J.S.: In: Dynamic Aspects of Plant Ultrastructure. Robards, A.W. (ed.). London: McGraw-Hill, 1974, pp. 441–480
70. Gupta, G.R.P., Maheshwari, S.C.: Pl. Physiol. **45**, 14–18 (1970)
71. Gurr, M.I., Blades, J., Appleby, R.S., Smith, C.G., Robinson, M.P., Nichols, B.W.: Europ. J. Biochem. **43**, 281–290 (1974)
72. Haagensmit, A.J., Dandliker, W.B., Wittwer, S.H., Murneek, A.E.: Am. J. Botany **33**, 118–120 (1946)

73. Hahn, H., deZacks, R., Kende, H.: Naturwissenschaften **61**, 170–171 (1974)
74. Harwood, J.L., Stumpf, P.K.: Lipids **7**, 8–19 (1972)
75. Harwood, J.L., Sodja, A., Stumpf, P.K., Spurr, A.R.: Lipids **6**, 851–854 (1971)
76. Hashimoto, T., Rappaport, L.: Pl. Physiol. **41**, 623–628 (1966)
77. Hatcher, E.S.J.: Nature (London) **151**, 278–279 (1943)
78. Hatcher, E.S.J., Gregory, F.G.: Nature (London) **148**, 626–627 (1941)
79. Hedley, C.L., Harvey, D.M., Keely, R.J.: Nature (London) **258**, 352–354 (1975)
80. Hemphill, D.D., Baker, L.R., Sell, H.M.: Planta (Berl.) **103**, 241–248 (1972)
81. Hopping, M.E., Bukovac, M.J.: J. Am. Soc. Hort. Sci. **100**, 384–386 (1975)
82. Ihle, J.N., Dure, L.S. III: Biochem. Biophys. Res. Commun. **36**, 705–710 (1969)
83. Ihle, J.N., Dure, L.S. III: Biochem. Biophys. Res. Commun. **38**, 995–1001 (1970)
84. Ihle, J.N., Dure, L.S. III: J. Biol. Chem. **247**, 5048–5055 (1972)
85. Ingle, J., Bietz, D., Hageman, R.H.: Pl. Physiol. **40**, 835–839 (1965)
86. Jackson, D.I., Coombe, B.G.: Science **154**, 277–278 (1966)
87. Jenner, C.F.: In: Mechanisms of Regulation of Plant Growth. Bieleski, R.L., Ferguson, A.R., Cresswell, M.M. (eds). Wellington: Bull. Royal Sci., N.Z. 1974, Vol. XII, 901–908
88. Jenner, C.F., Rathjen, A.J.: Ann. Botany (London) **36**, 743–754 (1972)
89. Jenner, C.F., Rathjen, A.J.: Australian J. Pl. Physiol. **2**, 311–322 (1975)
90. Jennings, A.C., Morton, R.K.: Australian J. Biol. Sci. **16**, 318–331 (1963)
91. Jennings, A.C., Morton, R.K.: Australian J. Biol. Sci. **16**, 332–341 (1963)
92. Johri, M.M., Maheshwari, S.C.: Pl. Cell Physiol. (Tokyo) **6**, 61–72 (1965)
93. Kipps, A.E., Boulter, D.: New Phytologist **73**, 675–684 (1974)
94. Klein, S., Pollock, B.M.: Am. J. Botany **55**, 658–672 (1968)
95. Klinck, H.R., Sim, S.L.: Ann. Botany (London) **40**, 785–793 (1976)
96. Kollöffel, C.: Planta (Berl.) **91**, 321–328 (1970)
97. Koshimozu, K., Kusaki, T., Mitsui, T., Matsubara, S.: Tetrahedron Lett. **14**, 1317–1320 (1967)
98. Labarca, C.C., Nicholls, P.B., Bandurski, R.S.: Biochem. Biophys. Res. Commun. **20**, 641–646 (1965)
99. Lang, A.: Ann. Rev. Pl. Physiol. **18**, 537–570 (1970)
99a. Larkins, B.A., Bracker, C.E., Tsai, C.Y.: Pl. Physiol. **57**, 740–745 (1976)
100. Le Page-Degivry, M.-T., Bulard, C., Milborrow, B.V.: C.R. Acad. Sci. Paris Ser. D **269**, 2534–2536 (1969)
101. Letham, D.S.: Life Sci. **2**, 569–573 (1963)
102. Letham, D.S.: Life Sci. **5**, 551–554 (1966)
103. Letham, D.S.: Life Sci. **5**, 1999–2004 (1966)
104. Letham, D.S.: Physiol. Plantarum **32**, 66–70 (1974)
105. Letham, D.S., Williams, M.W.: Physiol. Plantarum **22**, 925–936 (1969)
106. Lewis, O.A.M., Pate, J.S.: J. Exp. Botany **24**, 596–606 (1973)
107. Luckwill, L.C.: J. Hort. Sci. **24**, 32–44 (1948)
108. Luckwill, L.C., Weaver, P., MacMillan, J.: J. Hort. Sci. **44**, 413–424 (1969)
109. Marrè, E.: In: Current Topics in Developmental Biology. Moscana, A.A., Monroy, A. (eds.). New York: Academic Press, 1965, Vol. II, 75–105
110. Marrè, E., Coccuci, S., Sturani, E.: Pl. Physiol. **40**, 1162–1170 (1965)
111. Marumo, S., Hattori, H., Abe, H., Munakata, K.: Nature (London) **219**, 959–960 (1968)
112. McKenzie, H.: Can. J. Pl. Sci. **52**, 81–87 (1972)
113. McMahon, V., Stumpf, P.K.: Pl. Physiol. **41**, 148–156 (1966)
114. McWha, J.: J. Exp. Botany **26**, 823–827 (1975)
115. McWha, J., Hillman, J.: Z. Pflanzenphysiol. **74**, 292–297 (1975)
116. Merrick, W.C., Dure, L.S. III: J. Biol. Chem. **247**, 7988–7999 (1972)
117. Milborrow, B.V.: Planta (Berl.) **76**, 93–113 (1967)
118. Milborrow, B.V.: Ann. Rev. Pl. Physiol. **25**, 259–307 (1974)
119. Milborrow, B.V., Robinson, D.R.: J. Exp. Botany **24**, 537–548 (1973)
120. Miller, C.O.: Ann. N.Y. Acad. Sci. **144**, 251–257 (1967)
121. Miller, C.O., Witham, F.H.: In: Régulateurs Naturels de la Croissance Végétale. Nitsch, J.P. (ed.). Paris: Coll. Int.

CNRS No. 123, 1964, pp. Erratum p. I–VI

122. Millerd, A., Spencer, D.: Australian J. Pl. Physiol. **1**, 331–341 (1974)

123. Millerd, A., Whitfield, P.R.: Pl. Physiol. **51**, 1005–1010 (1973)

124. Millerd, A., Spencer, D., Dudman, W.F., Stiller, M.: Australian J. Pl. Physiol. **2**, 51–59 (1975)

125. Mitchell, J.W., Skaggs, D.P., Anderson, W.P.: Science **114**, 159 (1951)

125a. Monnier, M.: Int. Assoc. Plant Tissue Culture Newsletter **19**, 6–9 (1976)

126. Moore, T.C., Ecklund, P.R.: In: Gibberellins and Plant Growth. Krishnamoorthy, H.N. (ed.). New Delhi: Wiley Eastern, 1975, pp. 145–182

127. Morrison, I.N., Kuo, J., O'Brien, T.P.: Planta (Berl.) **123**, 105–116 (1975)

128. Morton, R.K., Raison, J.K.: Biochem. J. **91**, 528–539 (1964)

129. Morton, R.K., Palk, B.A., Raison, J.K.: Biochem. J. **91**, 522–528 (1964)

130. Mounla, M.A.Kh., Michael, G.: Physiol. Plantarum **29**, 274–276 (1973)

131. Nakamura, Y., Yamada, M.: Pl. Cell Physiol. (Tokyo) **15**, 37–48 (1974)

132. Nicholls, P.B.: Planta (Berl.) **72**, 258–264 (1967)

133. Nitsch, J.P.: Pl. Physiol. **30**, 33–39 (1955)

134. Nitsch, J.P.: In: Plant Physiology. Steward, F.C. (ed.). New York: Academic Press, 1971, Vol. VI A, pp. 413–502

135. Norstog, K.: Phytomorphology **22**, 134–139 (1972)

136. Norton, G., Harris, J.F.: Planta (Berl.) **123**, 163–174 (1975)

137. Öpik, H.: J. Exp. Botany **19**, 64–76 (1968)

138. Ozbun, J.L., Hawker, J.S., Greenberg, E., Lammel, C., Preiss, J., Lee, E.Y.C.: Pl. Physiol. **51**, 1–5 (1973)

139. Pate, J.S., Sharkey, P.J., Lewis, O.A.M.: Planta (Berl.) **120**, 229–243 (1974)

140. Payne, E.S., Brownrigg, A., Yarwood, A., Boulter, D.: Phytochemistry **10**, 2299–2303 (1971)

141. Payne, P.I., Boulter, D.: Planta (Berl.) **84**, 263–271 (1969)

142. Piskornik, Z., Bandurski, R.S.: Pl. Physiol. **50**, 176–182 (1972)

143. Poulson, R., Beevers, L.: Biochim. Biophys. Acta **308**, 381–389 (1973)

144. Privett, O.S., Dougherty, K.A., Erdahl, W.L., Stolyhwo, A.: J. Am. Oil Chem. Soc. **50**, 516–520 (1973)

145. Quebedeaux, B., Cholet, R.: Pl. Physiol. **55**, 745–748 (1975)

146. Rabson, R., Mans, R.J., Novelli, G.D.: Arch. Biochem. Biophys. **93**, 555–562 (1961)

147. Radley, M.: J. Exp. Botany **27**, 1009–1021 (1976)

148. Rawson, H.M., Evans, L.T.: Australian J. Biol. Sci. **23**, 753–764 (1970)

149. Redemann, C.T., Wittwer, S.H., Sell, H.M.: Arch. Biochem. Biophys. **32**, 80–84 (1951)

150. Reid, J.S.G., Meier, H.: Phytochemistry **9**, 513–520 (1970)

151. Reid, J.S.G., Meier, H.: Caryologia **25**, suppl. 219–222 (1973)

152. Rest, J., Vaughan, J.G.: Planta (Berl.) **105**, 245–262 (1972)

153. Roehm, J.N., Privett, O.S.: Lipids **5**, 353–358 (1970)

154. Rudnicki, R.: Planta (Berl.) **86**, 63–68 (1969)

155. Sakri, F.A.K., Shannon, J.C.: Pl. Physiol. **55**, 881–889 (1975)

156. Scharpé, A., Parijis, R. van: J. Exp. Botany **24**, 216–222 (1973)

157. Schlesier, G., Muntz, K.: Biochem. Physiol. Pflanzen. **166**, 87–93 (1974)

158. Schneider, E., Wightman, F.: Ann. Rev. Pl. Physiol. **25**, 487–513 (1974)

159. Schwarzenbach, A.M.: Cytobiology **4**, 145–147 (1971)

160. Sembdner, G., Schneider, G., Weiland, J., Schreiber, K.: In: Plant Growth Regulators. S.C.I. Monograph No. 31, 1965, pp. 70–86

161. Shannon, J.C.: Pl. Physiol. **43**, 1215–1220 (1968)

162. Shannon, J.C.: Cereal Chem. **51**, 798–809 (1974)

163. Shannon, J.C., Dougherty, C.T.: Pl. Physiol. **49**, 203–207 (1972)

164. Shantz, E.M., Steward, F.C.: J. Am. Chem. Soc. **77**, 6351–6353 (1955)

165. Shantz, E.M., Steward, F.C.: Pl. Physiol. **32**, suppl. viii (1957)

166. Sharkey, P.J., Pate, J.S.: Planta (Berl.) **128**, 63–72 (1976)

167. Shirk, H.G.: Am. J. Botany **29**, 105–109 (1942)

168. Sims, R.P.A., McGregor, W.G., Plessers, A.G., Mes, J.C.: J. Am. Oil Chem. Soc. **38**, 273–276 (1961)

169. Sims, R.P.A., McGregor, W.G., Plessers, A.G., Mes, J.C.: J. Am. Oil Chem. Soc. **38**, 276–279 (1961)
170. Sinska, I., Lewak, S., Gaskin, P., Mac-Millan, J.: Planta (Berl.) **114**, 359–364 (1973)
171. Smith, C.G.: Planta (Berl.) **119**, 125–142 (1974)
172. Smith, D.L.: Ann. Botany (London) **37**, 795–804 (1973)
173. Smith, R.H., Schubert, A.M., Benedict, C.R.: Pl. Physiol. **54**, 197–200 (1974)
174. Sondheimer, E., Tzou, D.S., Galson, E.C.: Pl. Physiol. **43**, 1443–1447 (1968)
175. Sorokin, H.P.: Am. J. Botany **54**, 1008–1016 (1967)
176. Stowe, B.B., Thimann, K.V., Kefford, N.P.: Pl. Physiol. **31**, 162–165 (1956)
176a. Sun, S.M., Buchbinder, B.V., Hall, T.M.: Pl. Physiol. **56**, 780–785 (1975)
177. Takahashi, N., Murofushi, N., Yokota, T.: In: Proc. 7th Intern. Conference on Plant Growth Substances 1972, Canberra, Australia, Carr, D.J. (ed.), p. 175–180
178. Tanaka, Y., Akazawa, T.: Pl. Cell Physiol. (Tokyo) **12**, 493–505 (1971)
179. Thorne, G.N.: Ann. Botany (London) **29**, 317–329 (1965)
180. Tsai, C.-Y.: Biochem. Gen. **11**, 83–96 (1974)
181. Tsai, C.-Y., Nelson, O.E.: Science **151**, 341–343 (1966)
182. Tsai, C.-Y., Nelson, O.E.: Pl. Physiol. **44**, 159–167 (1969)
183. Tsai, C.-Y., Salamini, F., Nelson, O.E.: Pl. Physiol. **46**, 299–306 (1970)
184. Turner, J.F.: Australian J. Biol. Sci. **22**, 1145–1151 (1969)
185. Ueda, M., Ehman, A., Bandurski, R.S.: Pl. Physiol. **46**, 715–719 (1970)
186. Van Staden, J.: Physiol. Plantarum **28**, 222–227 (1973)
187. Vieweg, G.H., De Fekete, M.A.R.: Planta (Berl.) **129**, 155–159 (1976)
188. Vijay, I.K., Stumpf, P.K.: J. Biol. Chem. **247**, 360–366 (1972)
189. Raussendorf-Bargen, G. von: Planta (Berl.) **58**, 471–482 (1962)
190. Walbot, V.: Develop. Biol. **26**, 369–379 (1971)
191. Walbot, V., Capdevila, A., Dure, L.S. III: Biochem. Biophys. Res. Commun. **60**, 103–110 (1974)
192. Walbot, V., Clutter, M., Sussex, I.: Phytomorphology **22**, 59–68 (1972)
193. Walbot, V., Harris, B., Dure, L.S. III: In: Developmental Biology of Reproduction. Markert, C.L., Papaconstantinou, J. (eds.). New York: Academic Press, 1975, pp. 165–187
194. Walpole, P.R., Morgan, D.G.: Nature (London) **240**, 416–417 (1972)
195. Waters, L.C., Dure, L.S. III: J. Mol. Biol. **19**, 1–27 (1966)
196. West, C.A.: In: Biosynthesis and its Control in Plants. Milborrow, B.V. (ed.). London: Academic Press, 1973, Proc. Phytochem. Soc. Vol. IX, 143–169
197. Wheeler, A.W.: Ann. Appl. Biol. **72**, 327–334 (1972)
198. Williams, P.M., Ross, J.D., Bradbeer, J.W.: Planta (Berl.) **110**, 303–310 (1973)
199. Williams, P.M., Bradbeer, J.W., Gaskin, P., MacMillan, J.: Planta (Berl.) **117**, 101–108 (1974)
200. Wilson, R.F., Rinne, R.W.: Pl. Physiol. **57**, 556–559 (1976)
201. Yamada, M., Stumpf, P.K.: Biochem. Biophys. Res. Commun. **14**, 165–171 (1964)
202. Yamada, M., Usami, Q., Nakajima, K.: Pl. Cell Physiol. (Tokyo) **15**, 49–58 (1974)
203. Yamane, H., Murofushi, N., Takahashi, N.: Phytochemistry **14**, 1195–1200 (1975)
204. Yatsu, L.Y., Jacks, T.J.: Pl. Physiol. **49**, 937–943 (1972)
205. Yokota, T., Murofushi, N., Takahashi, N., Katsumi, H.: Phytochemistry **10**, 2943–2949 (1971)
206. Zee, S.-Y., O'Brien, T.P.: Australian J. Biol. Sci. **24**, 35–49 (1971)
207. Zenk, M.H.: In: Régulateurs Naturels de la Croissance Végétale. Nitsch, J.P. (ed.). Paris: Coll. Int. CNRS No. 123, 1964, pp. 241–250
208. Zilkey, B., Canvin, D.T.: Biochem. Biophys. Res. Commun. **34**, 646–653 (1969)

Chapter 4. Imbibition, Germination, and Growth

4.1. Uptake of Water

4.1.1. Some Theoretical Considerations

Water is essential for the rehydration of seeds as the initial step towards germination. The amount of water taken up by an imbibing seed depends upon a number of factors (e.g. size, hydratability of contents, etc.) but in absolute terms it is quite small and often may not exceed 2–3 times the dry weight of the seed. For establishment and subsequent growth of the seedling, a larger and sustained water supply is required.

Two major factors have to be considered when discussing water uptake by a seed. These are (1) the water relations of the seed; and (2) the relationship between the seed and its substrate, which in nature is of course the soil. In order to understand the water relations of a seed cell we must first consider the concept of water potential. Water potential (ψ) is an expression of the energy status of water. Net diffusion of water occurs down an energy gradient from high to low potential. The water potential in the cell (ψ cell) can be formulated thus:

$$\psi \text{ cell} = \psi_\pi + \psi_m + \psi_p$$

This means that ψ cell is affected by 3 factors: (1) ψ_π—this is the osmotic or solute concentration effect. The concentration of dissolved solutes in a cell will influence water uptake, the greater the concentration, the greater the attractive force to water; (2) ψ_m—the matric (or hydrational) potential is contributed by the ability of matrices (e.g. cell walls; protein bodies) to be hydrated and bind water; (3) ψ_p—the turgor (hydrostatic) pressure occurs because as water enters a cell the contents swell and exert a force upon each unit area of the cell wall. Turgor pressure is in fact the amount by which pressure inside the cell exceeds the atmospheric pressure outside.

The potential of pure water at atmospheric pressure is the highest value encountered and is by convention, zero. Values for ψ_π and ψ_m are negative, and for ψ_p positive. The sum of the three terms (i.e. the water potential) is a negative number, except in fully turgid cells where it becomes zero. Thus, increasing osmotic and matric effects within a cell will reduce the water potential to lower (more negative) values whereas increasing turgor pressure will increase the water potential to higher (less negative values). Water potential can be expressed in terms of pressure or energy. The *bar* is often used (1 bar = 10^3 dynes/cm^2 or 10^2 J/kg or 0.987 atmos).

The forces with which water is held in the soil, and the amount of work which has to be done by the seed to remove it, are obviously important in determining the extent to which a seed can become rehydrated. Water potential is an expression of the driving force for water transport, and a cell will only absorb water if the water potential of its surroundings is higher (less negative) than that of the cell. This is important, for soil also has its own water potential (ψ soil), which is the sum of ψ_π, ψ_m and ψ_p of the soil, although of these, as we shall see, only ψ_m appears to play a significant role. The flow of water through the soil and into the seed can be visualized most simply by an analogy to Ohm's law [39]

$$F = \frac{\psi_e - \psi_s}{i_1 + I + i_2}$$

where F=rate of flow from environment to seed; ψ_e=potential of the environment; ψ_s=potential of the seed; i_1=internal impedance of soil matrix or other matrix on which a seed is imbibed; I=external impedance=(degree of contact of seed with water supply)$^{-1}$; i_2=internal impedance of seed (including seed coat and air spaces).

Thus the difference in water potential between seed and soil ($\psi_e - \psi_s$) is only one of the factors on which the rate of flow of water from the environment to the seed depends. Initially upon imbibition the difference in water potential between the seed and soil is enormous, but as seed moisture content increases the water potential of the seed increases (becomes less negative) and that of the surrounding soil decreases as water is withdrawn. Since the seed cannot move towards the source of moisture, successful germination depends on the movement of sufficient moisture to the seed surface. This in turn depends on the water potential of the zones of soil immediately surrounding the seed and on the rate at which water moves through the soil, i.e. the hydraulic conductivity of the soil (the reciprocal of i_1 — the soil impedance). Impedance is important in soils which are in the process of drying, especially if they are coarsely textured, and in comparison the impedance to water flow exhibited by many seed coats (i_2) is very small. Another factor which is important is the area of contact between soil particles and the seed surface — the reciprocal of this value is the external impedance (I).

Other, more complex formulae have been derived to account for the variety of factors (including others not mentioned here) which play a role in the uptake of water into a seed and the movement of water through the soil environment. Discussion of the merits and limitations of these formulae and associated mathematical models is beyond the intended scope of this book, but the attention of the reader is directed towards appropriate sources of information [6, 13, 22, 23, 50, 52].

4.1.2. Soil Matric Potential and Seed-Soil Contact

When a seed is planted in soil its rate of imbibition and (assuming it has no inherent dormancy mechanism) subsequent germination may be determined

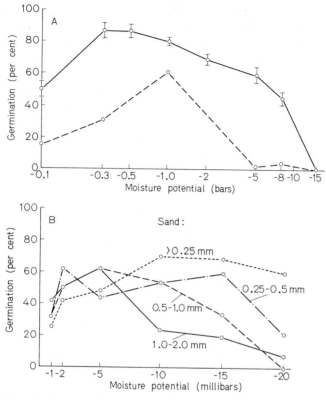

Fig. 4.1. (A) Germination of *Oryzopsis holciformis* as a function of soil moisture potential in loess soil. o---o: in soil for 2 days, o—o: in soil for 8 days. (B) Final germination percentage of *Oryzopsis holciformis* as a function of soil moisture potential in sand of different particle sizes (shown in mm). After Dasberg and Mendel, 1971 [14]

ultimately by the impedance of the soil matrix (due chiefly to surface and colloidal forces) and the degree of contact of the seed with the soil moisture [48], irrespective of how great the initial difference is between the water potential of the seed interior and the soil.

The effects of defined soil moisture potentials in loess soil (53% sand, 30% silt, 17% clay, 14% $CaCO_3$, 0.53% organic matter, pH 7.8) on germination of the grass *Oryzopsis holciformis* are shown in Figure 4.1 A. The onset of germination is delayed at soil matric potentials higher or lower than -1.0 bar. The final germination percentage attained after eight days is highest between -0.3 and -1.0 bar. The effects of high matric potentials on germination of *Oryzopsis* in sand fractions of various particle sizes are shown in Figure 4.1 B. As in loess soil there is an optimum potential for germination, which shifts according to particle size of the medium. With the coarsest sand the optimum is -0.005 bar—compared with -0.5 bar for loess soil. Obviously then, soil moisture potential is not the only factor governing germination and other factors must be involved to explain the differential reaction of the grain to its medium.

One suggestion is that the low germination at low matric potential is due to a decrease in hydraulic conductivity of the soil, severely limiting the amount of water arriving at the surface of the germinating seed. More important however appears to be the area of contact between the seed and the soil.

Harper and Benton [25] have studied the effects of soil moisture tension (i.e. impedance of the soil matrix) on various groups of seeds with common characteristics and considered their results in relation to the importance of seed-soil moisture contact. Some of their observations are presented in Table 4.1. From these we can see that there is no effect of water tensions of 0–200 cm (achieved by using sintered glass plates to approximate soil water tensions) on the final germination percentage of the two species in Group 1, both of which possess copious mucilage that bursts out from the epidermal cells as they imbibe water; the germination rate is slower at higher tensions though. Seeds of Group 2 have less abundant mucilage and those of two species are quite sensitive to increasing tensions. The 3rd group of species have sculptured testas (with tubercules), being particularly pronounced on seeds of *Agrostemma*, but less so on the other two species. They are all sensitive to tensions of 50 cm and above. Group 4 seeds have smooth testas and no mucilage, but vary in size (see mg seed weight) from the small seeds of *Clarkia* to the large *Vicia*

Table 4.1. The percentage germination of the seeds of various species sown on sintered glass plates at controlled water tensions

Species and seed weight (g)	Standard (filter papers in Petri dishes)	Sintered glass plates under tension			
		0 cm	50 cm	100 cm	200 cm
Group 1 Seeds with copious mucilage					
Lepidium sativum (0.0027)	100	98	95	96	95
Camelina sativa (0.0012)	99	98	96	96	94
Group 2 Seeds with less copious mucilage					
Linum usitatissimum (0.0059)	96	92	88	84	73
Plantago major (0.0002)	49	46	30	30	30
Sinapis alba (0.0074)	92	96	73	55	24
Group 3 Tuberculate seeds					
Reseda alba (0.0005)	35	45	8	7	5
Agrostemma githago (0.028)	70	71	4	0	0
Melandrium rubrum (0.0006)	57	39	0	0	0
Group 4 Seeds with smooth testa — no mucilage					
Clarkia elegans (0.0004)	81	73	47	23	10
Brassica napus (0.0038)	97	87	9	0	0
Pisum sativum (0.23)	98	11	0	0	0
Vicia faba (0.66)	100	0	0	0	0

After Harper and Benton, 1966 [25]

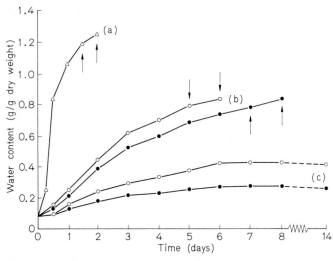

Fig. 4.2. The time course of water uptake by seeds of *Vicia faba* (*a*) immersed in water; (*c*) on sintered glass plates at 0 cm (○) and 200 cm (●) tension exposed to the atmosphere; (*b*) as (*c*) but with the plates protected from evaporation. *Arrows* on (*a*) and (*b*) represent the period over which radicles emerge. After Harper and Benton, 1966 [25]

seeds. Water sensitivity is inversely related to size and *Vicia* seeds do not germinate at all on the sinters, even when there is no tension of water.

These results, when considered as a whole, demonstrate that there is no simple relationship between germination and tension (matric impedance) of the available water. It appears that size, mucilage content and nature of the seed surface also have marked influences on germination. The feature which best explains the behaviour of these seeds on the sintered glass plates is the degree of contact between the seed and the water in the pores of the plate. Mucilage from *Lepidium* and *Camelina* spreads around the seed and greatly extends the pathways for water flow from sinter to seed, even when the applied water tension causes the water menisci to retreat within the pores of the sinter. Such is also the case for *Linum* and *Plantago,* although to a lesser extent because the mucilage layer is less well developed. Even seeds of *Sinapis* in which mucilage rarely escapes from the retaining cells are less sensitive to tensions than non-mucilaginous *Brassica.* Seeds with tubercules have a low area of contact with the menisci of water in the sinters so that at tensions over 50 cm seeds of *Agrostemma* and *Melandrium* cannot take up water sufficiently fast to germinate. The effect of tension becomes greater with increasing seed size in those with smooth coats—small seeds of *Clarkia* germinate more readily on sinters than larger *Brassica* seeds, which in turn do better than the large pea (*Pisum*) and broad bean (*Vicia*) seeds. For these larger seeds contact becomes particularly important because of the large seed surface from which water loss may occur. As shown in Figure 4.2 for *Vicia,* even when the rate of water loss is restricted by atmospheres of high relative humidity the rate of water uptake by seeds from sinters remains less than by seeds immersed in liquid water. Hence contact is still important in determining germination of the seed.

The effect of varying the area of contact between seeds of Meteor pea and wet fine sand is shown in Table 4.2. Here it can be seen that the rate, though not so much the final percentage of germination, increases with increasing area of contact between seed and its moist substrate. In effect, the surface of the seed is partly in contact with the soil particles and partly in contact with the soil pores—at the seed-soil interface the distribution of pores is such that those with a larger effective diameter predominate. These are the pores from which least energy is required to withdraw water and will be the first to drain as soil moisture is depleted. Therefore, relatively small reductions in soil-moisture content are liable to result in disproportionately large reductions in contact area between seed and soil water to which movement of water is restricted.

While it is generally agreed that the solute (osmotic) component (ψ_π) of soil water potential plays a negligible role in the dynamic water relations of all but saline soils, the relative importance of seed-soil contact and soil matric potential has been disputed. We have presented evidence above that the seed-soil contact area appears to be highly important in determining the germinability of many seeds, but this view does not attract complete support. Collis-George and co-workers [10, 11] argue that soil matric potential is an important factor in seed germination over and above the effect of the wetted contact area between seed and its supporting medium. Certainly, the environmental factors which determine the relationship between seed and soil moisture are complex. Physical properties of the soil determine the retention of water and its conductivity, and decreases in soil moisture content result in shrinkage of the soil matrix as colloidal particles are pulled together, resulting in an increase in isotropic mechanical forces which oppose expansion of the imbibing seed (but particularly the expanding seedling). Soluble matter in the soil and climatic factors will also determine the rate of supply and loss of moisture from a soil. Finally, the nature of the contact between the soil liquid and the surface of the seed, which is a function of pore geometry and surface tension, the evaporative

Table 4.2. The effect on germination of the surface area of Meteor pea seeds in contact with wet fine sand

Surface area in contact (mm^2)	Days to radicle emergence of a seed population		Number germinated of 20 seeds
	Beginning	Completion	
0.8	9	21	18
3	8	14	18
9	5	10	20
28	5	9	20
Control—on liquid in Petri-dishes	2	3	20

After Manohar and Heydecker, 1964 [38]

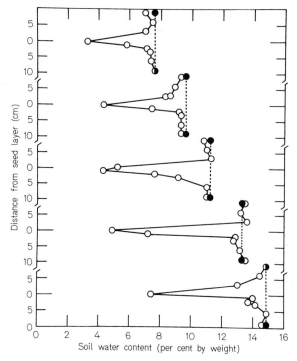

Fig. 4.3. Water uptake by grains of *Oryzopsis holciformis* after three days at different soil water contents. Seeds were arranged as a layer in soils of different initial water contents (●···●). After 72 h, the water content on either side of the seed layer was determined (o—o). It can be seen from some of the curves that seeds removed water from only about 10 mm away from the layer (at 0 cm). After Dasberg, 1971 [13]

surface of the seed, and ambient relative humidity, all play a role in water uptake and germinability.

The distance over which water flows to a seed through soil often does not exceed 10 mm, irrespective of the soil water content [13], as shown in Figure 4.3 for *Oryzopsis*. Water uptake by the seed is determined by water content of the immediately surrounding soil however, and as soil water content increases, the seeds' water uptake and germination also increase in a manner which is species specific (Table 4.3). We can see, for example, that most of the species attain full germination only at a high soil water content, while wheat germinates successfully at much lower water contents. Subsequent shoot growth can be seen to be more affected by the soil water status than is root growth. Over a range of soil water contents the final germination percentage of a population of seeds of some species might not be decreased, but the time taken to achieve full germination might be longer [31].

In order to increase the moisture content of the substrate surrounding a seed, and to improve "seed-soil" contact, the horticultural practice of sowing seeds with accompanying mucilage has arisen. The effect of supplying mucilage to imbibing seeds at controlled moisture tensions on sintered plates is shown

Table 4.3. Water uptake, germination and initial growth of some range species after seven days at different soil water contents

Species	Soil water content (%)	Germina- tion (%)	Water uptake (percent of initial weight)	Length of root (cm)	Length of shoot (cm)
Oryzopsis holciformis	7.6	0	28	—	—
	9.6	0	41	—	—
	11.2	52	99	3.0	—
	13.3	58	169	4.2	—
	14.9	62	250	5.2	6.3
Vicia dasycarpa	11.6	0	76	—	—
	12.5	23	102	3.8	1.7
	13.4	27	103	3.9	2.1
	15.5	90	199	4.2	3.6
Medicago hispida	8.1	4	65	—	—
	9.8	8	89	—	—
	11.6	34	194	2.4	2.8
	13.6	86	637	2.5	4.3
Agropyron elongatum	9.5	16	39	—	—
	10.3	55	90	—	—
	10.7	87	181	5.0	3.2
	11.0	67	131	5.0	3.4
	11.3	90	168	4.8	3.8
	13.0	90	206	4.9	4.0
	13.1	97	224	4.8	4.8
	15.2	90	277	4.6	6.1
Triticum aestivum	8.1	90	54	2.0	0
	9.8	93	91	4.6	0.5
	11.6	93	145	6.0	2.3
	13.6	95	232	5.9	4.0

After Dasberg, 1971 [13]

in Figure 4.4. The reader will recall from Table 4.1 that *Agrostemma* is a tuberculate seed and that *Lepidium* seeds are already producers of copious mucilage on imbibition. Seeds of *Agrostemma* germinate relatively well above 50 cm tension in the presence of mucilage compared with seeds imbibed in its absence. *Lepidium* shows no response to the addition of extra mucilage. It is worth noting that the maximum germination of *Agrostemma* obtained at 0 cm tension in the presence of mucilage is less than in its absence—this could be a consequence of restricted flow of gases (oxygen) from the atmosphere through the mucilage. *Lepidium* being a less dense seed floats on the surface of the mucilage and is more accessible for gaseous exchange. The effects of such restricted gaseous exchange on germination will be discussed in the environmental section of the second volume of this book, as will other important factors which can also restrict germination, e.g. excess moisture, hard or impermeable seed coats, soil temperature, salinity stress, inhibitor content, and inherent dormancy mechanisms.

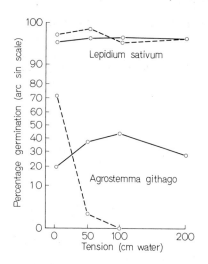

Fig. 4.4. The ultimate percentage germination of seeds of *Lepidium sativum* and *Agrostemma githago* on sintered glass plates at controlled moisture tensions with (—) and without (---) added mucilage. Plant mucilage obtained from seeds of *Plantago psyllium* and called "Coreine" was used. After Harper and Benton, 1966 [25]

4.1.3. Seed Germination in Petri Dishes

Returning to our expression of water flow in terms of Ohm's law (Sect. 4.1.1) let us consider the process of water uptake by seeds imbibing under common laboratory conditions, i.e. sprinkled on one or more layers of moist filter paper in Petri dishes. Here a fair proportion of the seed surface is in contact with the aqueous medium (which is often distilled water) and so the external (contact) impedance (I) and the internal impedance of the matrix (i_1) are more or less eliminated. The improved germination of seeds (particularly the large ones) imbibed on filter papers over that achieved on sintered glass plates even at zero tension can be seen from the data of Table 4.1. Optimum germination on filter papers requires an optimum amount of water — too little creates external and internal impedances and too much can restrict oxygen diffusion into the seed [14, 26, 29]. The effects of imbibing lettuce seeds in different volumes

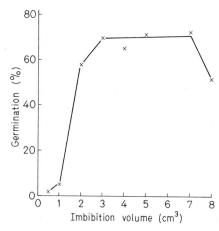

Fig. 4.5. Volume of imbibition liquid and percent germination of lettuce seeds cv. Grand Rapids imbibed on one layer of filter paper in 9 cm Petri dishes in the light at approx. 25°C. Germination percentage counted 24 h after the start of imbibition. By the authors

of distilled water on one layer of filter paper in 9 cm Petri dishes are shown in Figure 4.5 — there is a broad range of imbibition volumes over which germination will occur. In 8 cm³ of liquid the seeds were actually submerged but a considerable percentage still germinated — see Chapter 5 for further details on oxygen supply to immersed seeds.

4.1.4. Kinetics of Water Uptake by Seeds

The water potential of a mature, dry seed is considerably lower than that of the surrounding moist substrate and so water moves in the direction of decreasing water potential, i.e. from substrate to seed. The initial water potential of a seed is extremely low [37] and can exceed − 1000 bars owing to its enormous matric potential [50]. As a seed starts to take up water it exerts a large swelling pressure, and many seeds double in size during this initial period, called imbibition. The swelling forces decrease sharply (the water potential rises), as seeds absorb more and more water, from as high as − 4000 bars during initial imbibition of rape (*Brassica napus*), wheat (*Triticum aestivum*) and maize (*Zea mays*), to somewhat less than − 10 bars at the time of radicle protrusion [49]. Potential exerted by the root after its emergence is probably higher still. Recent studies on imbibing peas have shown that a wetting front is formed as water permeates the seed and that there is an abrupt boundary of water content between wetted cells and those about to be wetted [52]. Furthermore the average water content of the wetted area increases as a function of time. This initial pattern of water uptake, which might be common to many if not all seeds is therefore marked by three characteristics: (1) a sharp front separating wet and dry portions of the seed; (2) continued swelling as water reaches new regions; and (3) an increase in water content of the wetted areas.

There are complicating factors to consider, however. Water uptake might not occur evenly over the whole surface of an intact seed. In a number of seeds there is, at least initially, a greater uptake through the micropyle than through the rest of the testa (e.g. in species of *Vicia* and *Phaseolus*), and in some hard-coated seeds of the Papilionaceae (e.g. *Melilotus alba, Trigonella arabica* and *Crotalaria egyptica*) a plug covering a special opening — the strophiolar cleft — must be loosened or removed before water uptake can occur [24], and then only through that region.

Many seeds placed in distilled water in Petri dishes under optimum conditions for germination show a triphasic pattern of water uptake as shown in Figure 4.6. Initial uptake of water in Phase I (i.e. imbibition) is a consequence of the matric forces (ψ_m) of the cell walls and cell contents of the seed, and this uptake occurs irrespective of whether a seed is dormant or non-dormant, viable or non-viable. Phase II is the lag period of water uptake, when the matric potential is high (less negative), as is the solute or osmotic potential (ψ_π). Dead and dormant seeds maintain this level of hydration typical of Phase II, but unlike germinating seeds they do not enter Phase III, which is associated with radicle protrusion.

Let us consider what is likely to occur in the cells of the radicle during these 3 phases. Phase I (imbibition) is probably very rapid in a tissue as small

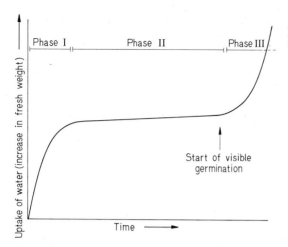

Fig. 4.6. Triphasic pattern of water uptake by germinating seeds

as the radicle and although some metabolic events commence soon after rehydration of enzymes and their substrates (see Chap. 5), the major metabolic events which take place in preparation for germination undoubtedly occur during Phase II. This is shown by recent work on a variety of seeds (e.g. *Allium cepa, Daucus carota, Apium graveolens, Pastinaca sativa, Impatiens, Antirrhinum*) which are allowed to imbibe on solutions of polyethylene glycol of different osmolarities. Certain concentrations allow Phase I to occur but not Phase III, and seeds are therefore held in Phase II. When such seeds are surface dried and later introduced to water, they again rapidly imbibe, and show accelerated radicle emergence, i.e. their Phase II is now markedly shortened. This indicates that certain metabolic processes occurred during the seeds' first experience of Phase II and were conserved. A reiteration of these processes is therefore not necessary and radicle emergence can now occur much more quickly on introduction to water. The horticultural uses of this treatment are being investigated [27, 28].

Cell expansion to elicit radicle elongation marks the final stages of germination (see later), and this presumably occurs because water enters the cells, increases the turgor pressure (ψ_p) and causes expansion of the cell walls. The act of radicle protrusion then is possibly, in part, associated with a reduction in solute potential (ψ_π) of the radicle cells, thus attracting more water from the surrounding medium and increasing ψ_p, and in part a consequence of the loosening of the cell wall to allow expansion to take place. For a further consideration of these events see Section 4.3.

After the initial imbibition phase and hydration of the cell walls and cell contents the matric potential (ψ_m) plays a minor role in attracting water into the cell. Indeed, some of the matrices, e.g. protein and carbohydrate storage products in the storage organs, are later hydrolysed to low molecular weight, osmotically-active substrates which decrease (make more negative) the ψ_π of both the seed as a whole, and, when transported there, of the growing embryonic axis (Chap. 6). The increase in water uptake in Phase III is thus initially associated with decreases in water potential due to some unknown changes related to germination, and then by decreases in ψ_π due to post-germination reserve hydrolysis.

The lengths of each of these phases depends upon certain inherent properties of the seed (content of hydratable substrates, permeability of seed coats, oxygen uptake, seed size, etc.) and upon the conditions during exposure to water (e.g. moisture levels, composition of substrate, temperature, etc.). Different parts of a seed, particularly a larger seed, will pass through these phases at different rates, e.g. an embryo or an axis located near the surface of a seed might commence elongation (i.e. enter Phase III of water uptake) before its associated bulky storage tissue has become fully imbibed (i.e. completed Phase I or II). Some imbibing seeds do not show a distinct lag in water uptake (Phase II), e.g. various species of oak acorns [7], probably for this reason.

Some confusion has arisen in the literature because of attempts to relate the phases of water uptake to the metabolic status of a seed. Let us therefore attempt to clarify the situation and summarize our discussion by restating that Phase I (imbibition) occurs equally well in dead and living tissues and is therefore independent of the metabolic activity of the seed, although metabolism commences rapidly as a consequence of this hydration. Phase II is a period of active metabolism in preparation for germination in non-dormant seeds, of active metabolism in dormant seeds, or inertia in dead seeds. Phase III is associated only with germination and subsequent growth, and during this latter event there is also, obviously, metabolic activity including the commencement of mobilization of the stored reserves. The phases of water uptake observed for a whole seed must be regarded as an "average" for all parts of the seed and care should be taken in relating them solely to the metabolic activity of the germinating embryo or axis. As an example, it has been observed that when the water content of dent corn (*Z. mays*) grains reaches 75%, the water content of the embryo on a dry weight basis is 261% but that of the remainder of the grain is only 50% [6]. Likewise it should be noted than in endosperms and "non-persistent" (hypogeal) cotyledons, when water saturation has been reached the water content remains level (i.e. at Phase II) because these fail to expand further: eventually their water content declines as degeneration occurs. In small seeds or isolated embryos the "average" (at least prior to the completion of germination) more closely reflects the stages of water uptake by all cells.

4.1.5. Soaking Injury and Solute Leakage

Initial water uptake by seeds is accompanied by the release of a large volume of gas (see Chap. 5) and by a rapid leakage of substances, e.g. sugars, organic acids and amino acids. Under field conditions these substances might stimulate the germination and growth of fungal pathogens present in the soil [16, 47], e.g. exudates of bean (*Phaseolus vulgaris*) stimulate the germination of *Fusarium* chlamydospores. Furthermore, there is evidence that varieties of seeds which are more susceptible to damping-off fungi release more exudate than those which are less so. Seeds with cracked [16, 47] or scarified [41] seed coats, and seeds

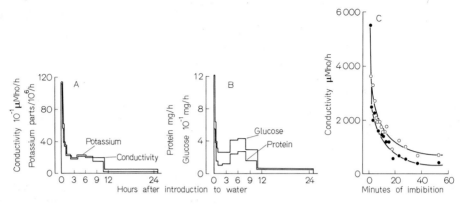

Fig. 4.7A–C. Time course of leakage of (A) electrolytes and potassium and (B) sugar and protein from pea embryos immersed in water. In (C) electrolyte leakage is followed from pea embryos immersed in water (●) and from embryos removed after 60 min in water, dried over calcium chloride and then returned to water (○). The pea embryo is axis and cotyledons with testa removed. After Simon and Raja Harun, 1972 [51]

with seed coats completely removed [36] leak more solutes than those with seed coats intact.

Figure 4.7A shows the rapid leakage into the surrounding water of electrolytes and potassium from immersed pea seeds from which the testa has been removed; Figure 4.7B shows the similar rapid leakage of sugar and protein. That the initial leakage of electrolytes actually lasts less than 30 min is shown in Figure 4.7C—this leakage occurs only from the outermost cell layers at this time. Embryos which have been imbibed for 60 min, dried, then returned to water again show a rapid initial leakage (Fig. 4.7C). On the other hand, seeds which do not undergo drying during their maturation do not leak electrolytes when placed in water [51]—such seeds do not imbibe, however, since they are already hydrated. These observations suggest that the selectively-permeable membranes of the tonoplast and plasmalemma which normally retain solutes within cells lose their integrity during drying and do not act as retentive barriers when seeds are first placed in water. The membranes presumably become re-established within minutes to prevent further leakage. Membranes can be switched from the leaky to the retentive (intact) state rapidly through several cycles of wetting and drying, which suggests that the changes they undergo are physical, rather than those of metabolic breakdown and resynthesis. Inadequate hydration of the membrane components in the dry seed probably results in disruption of the ordered membrane structure, which takes several minutes to re-orientate on further hydration. Changes also occur in the membrane structures of cell organelles during drying and rehydration—see Chapter 5 for details on mitochondrial changes. Here, there appears to be more permanent structural damage due to drying which takes longer (many hours) to repair; this might require the development of an effective biochemical repair mechanism.

Removal of, or damage to seed coats prior to imbibition in water not only results in increased leakage of solutes from the seed, but may also subse-

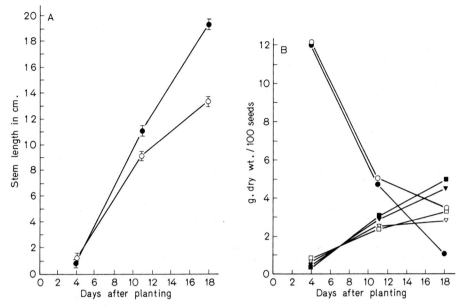

Fig. 4.8. (A) Stem height of pea seedlings grown from seeds imbibed with (●) or without (○) seed coats. (B) Dry weight of roots, stems and cotyledons of seedlings grown from seeds imbibed with and without seed coats. ●—●: cotyledons from seeds with coats, ○—○: cotyledons from seeds without coats, ▼—▼: roots from seeds with coats, ▽—▽: roots from seeds without coats, ■—■: stems from seeds with coats, □—□: stems from seeds without coats. After Larson, 1968 [36]

quently reduce the growth of the seedling (Fig. 4.8A), diminish dry weight increases in roots and stems, and lower degradation of the cotyledons (Fig. 4.8B). Presumably the seed coat normally restricts or regulates water uptake by acting as a barrier to diffusion, and in its absence the increased velocity of water uptake into the seeds (see Fig. 5.1 C) might cause irreversible membrane damage and solute leakage in a significant enough number of cells adversely to affect later germination and growth processes.

The effect of the environment on imbibition, germination and growth is a topic reserved for the second volume of this book. Nevertheless we will briefly mention here the striking effects which the temperature during imbibition of the seed appears to have on subsequent growth of the plant. Initial imbibition of many seeds at moderately low temperatures (3°–15°C) appears to cause some kind of injury which is afterwards expressed as reduced growth and development — see [30, 42, 43, 45] and references therein, and Table 4.4. Low-temperature imbibition does cause an increase in leakage of U.V.-absorbing (264 nm) materials (e.g. amino acids, nucleotides, etc.) from lima bean (*Phaseolus lunatus*) axes (Fig. 4.9); this indicates the occurrence of changes in membranes due to chilling which are not readily reversible. Whether such changes and the accompanying loss of solutes have a direct bearing on the later growth of the seedling, or whether other metabolic processes are critically affected,

Table 4.4. Effect of germination temperature on subsequent growth of pea plants

Germi-nation Temp. °C	Node of 1st flower	Flowers per plant	Pods per plant	Seeds per plant	Final height (mm)	Maximal growth rate (mm/day)	Days to 1st open flower
3°	10.4±0.7	3.7±0.7	1.8±0.6	3.3±1.0	187±22.6	9.4±1.7	24.8±1.6
7°	9.6±0.5	4.2±0.6	1.8±0.7	3.7±1.6	206±16.2	10.1±2.0	24.6±1.2
11°	10.0±0.4	4.7±0.8	2.0±0.7	4.0±1.2	233±31.0	10.8±3.4	23.1±1.5
19°	9.5±0.6	5.1±0.4	2.6±0.7	5.6±1.8	290±38.5	14.9±3.2	24.1±1.0
23°	9.2±0.4	5.2±1.0	1.9±0.6	4.0±1.8	237±22.8	10.7±2.0	23.8±1.5
27°	9.1±0.4	5.0±0.5	2.4±0.7	5.2±1.4	265±29.3	13.0±2.2	24.7±0.9
31°	9.3±0.4	5.0±0.6	2.6±0.7	5.7±1.8	293±48.9	15.3±3.7	24.4±0.8

From Highkin and Lang, 1966 [30]

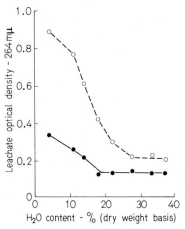

Fig. 4.9. Leaching from embryonic axes of *Phaseolus lunatus* during the period 0.5 to 10 h following the beginning of imbibition. ○---○: 5° imbibition, ●—●: 25° imbibition. After Pollock, 1969 [42]

has still to be resolved. The possibility of an indirect effect due to enhanced activity of microbial contaminants by leached substances cannot be ignored. It is pertinent to note that seeds of some species must be subjected to a period of low-temperature treatment (stratification) to overcome an inherent dormancy mechanism and thus allow germination. Their metabolic responses to such treatment are obviously of interest.

4.2. Radicle Expansion—Cell Elongation or Cell Division?

From their studies on lettuce seeds Evenari and co-workers [15] have concluded that the start of mitoses, the beginning of cell elongation, and protrusion of the radicle are correlated events during germination. Subsequent studies by

Haber and co-workers however [17, 20, 21] have shown that cell division is neither correlated with nor necessary for cell expansion, and that protrusion of the radicle during lettuce seed germination is caused by cell expansion alone, mitotic cell division contributing little or nothing. On the other hand, seeds prevented from germinating, e.g. by high temperature and osmotica, will undergo mitosis and localized expansion—obviously here again is a separation of mitosis from germination. Cereal grains irradiated with large doses of γ-rays to effect chromosome breakage still germinate and develop into seedlings [19]. This apparently occurs without any detectable mitoses or increases in tissue-cell numbers, so germination and early growth *can* occur in the absence of cell divisions. Implicit in these irradiation studies is the assumption that the chromosomes of all cells of the grain are permanently damaged and that no repair occurs later—this might be so, but it is interesting to note that other metabolic activities of the irradiated cells, including RNA synthesis, appear to be functioning.

In a number of other seeds it is clear that cell extension precedes cell division e.g. *Z. mays, Hordeum, Vicia faba* and *Pisum* [4, 8]. Two papers have sometimes been quoted as showing that cell divisions occurs before cell expansion in seeds of two species, namely *Pinus thunbergii* [18] and *Prunus cerasus* (sour cherry) [44]. The studies on *Pinus* do not warrant such interpretation, however, and subsequent work on another pine species, *Pinus lambertiana* apparently shows that cell division and cell elongation occur simultaneously [4]. The studies on *Prunus* do show that the number of cells per axis increases by four weeks of after-ripening at 5°C, before any increase in axis length. The next measurement, taken after eight weeks of after-ripening shows an increase in axis length and no significant further increase in cell numbers. Cell division thus apparently precedes expansion of the axis. We should note, though, that embryos of *Prunus* (and of a number of other species—see Volume 2) grow to maturity during after-ripening at low temperature (i.e. stratification), and this phenomenon should be considered as a special case, not related to radicle expansion during germination itself.

There have been surprisingly few studies to determine the mode of radicle elongation in germinating seeds, but there is some evidence that the elongation occurs in two phases; an initial phase of slowly accelerating elongation, followed by a more rapid one. No dry weight changes occur during the first phase but mobilization of reserves and translocation of catabolites to the axes is characteristic of the second phase. Thus, in *Vicia* seeds the initial phase of radicle elongation, with its slow increase in fresh weight and length, nevertheless shows no increase in dry weight, and no cell division (Fig. 4.10). At this stage elongation of the radicle cell walls is possibly occurring without any net synthesis of new wall material. An abrupt increase in growth rate occurs after 46 h to mark the beginning of the second phase, and soon afterwards mitosis begins (in the apical meristem of the radicle) and an increase in dry weight occurs. Emergence of the radicle through the seed coat is visible in some seeds 48 h after the start of imbibition, so in *Vicia* there is elongation of the radicle prior to its emergence, and penetration through the seed coat is accompanied by mitosis. A similar biphasic mode of radicle elongation has been reported for barley and peas [8], although in barley the radicle emerges from the grain

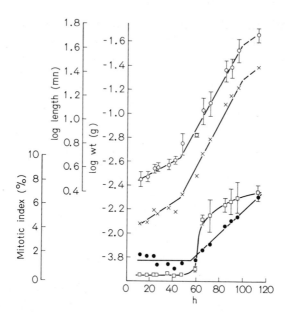

Fig. 4.10. Length (o), fresh weight (×), dry weight (●) and mitotic index (□) of radicles of *Vicia faba* seeds germinating at 20°C. After Rogan and Simon, 1975 [46]

prior to the onset of the second phase of elongation (i.e. before cell division commences).

Thus, on the basis of observations on the germination of very few seeds, it appears that cell elongation precedes cell division during radicle elongation. In some seeds the radicle elongates prior to puncturing the seed coat, and emergence of the radicle from the seed is due to cell elongation or, in some cases, to simultaneous division and elongation. The synthesis of the DNA is discussed in the next chapter.

4.3. The Control of Germination

In the foregoing account we have described some of the physiological events associated with germination—events occurring pre- and post-radicle emergence—and in Chapter 5 we will consider some of the biochemical changes taking place at these times. We have nevertheless not touched upon two important questions: (1) What *specific* events are essential for visible germination to occur? (2) What internal controls, if any, are responsible for setting these events in motion? Unfortunately, definitive answers cannot yet be given to these fundamental questions but instead partial answers and some speculation must suffice.

Germination, as we have defined it, culminates in radicle emergence which in most cases comprises only cell enlargement and not necessarily cell division. Perhaps, then, when searching for specific germination events we should look to phenomena connected with and leading to cell extension. Rapid water uptake is an obvious requirement during which the expanding cells increasingly vacuolate until they acquire the large central vacuole characteristic of elongated cells. Increased water uptake depends upon changes in water potential in the cells possibly due to decreased (i.e. more negative) solute potential (ψ_π). Although we have virtually no direct knowledge about the mechanism of cell expansion in radicle cells (or in the hypocotyl if this starts growth first) our appreciation of the process in other growing organs, especially coleoptiles, would lead us to expect that "loosening" of the cell wall must be a critically important component [9]. We now enter an extensively researched area of plant physiology which we cannot consider here in detail. The mechanism of cell-wall plasticization (to allow of subsequent expansion) is not yet fully understood but one hypothesis which is gaining support is that invoking hydrogen-ion secretion. According to this, protons are secreted into the wall, possibly in exchange for cations, this acidification breaking the hydrogen bonds between chains of wall carbohydrates, or optimizing the conditions for enzyme action upon the wall. The consequent change in wall rigidity, i.e. the increased stretchability, thus allows of cell extension. Control of growth and hydrogen ion extrusion in coleoptiles and stems is achieved by auxin and can also be influenced by other chemicals such as the fungal product, fusicoccin. When treated with this chemical (or with auxin) certain plant tissues show an enhanced growth rate and are stimulated to secrete hydrogen ions to such an extent that not only are the cell walls acidified but the surrounding medium also displays a significant drop in pH (e.g. [40]). Of great interest is the discovery that fusicoccin also initiates or enhances seed germination, again accompanied by the acidification of the medium in which the seeds or embryos are held [33, 34], though these initial findings remain to be fully established. This, then, suggests that radicle emergence is stimulated by acidification (presumably of the cell walls) and raises the possibility that proton extrusion may be an essential prelude to visible germination. Continued synthesis of protein (perhaps critical enzymes or cell wall components) as well as respiration are known to be necessary for cell expansion in stems or coleoptiles. In this context, we may speculate as to the importance of the nucleic acid, protein and respiratory metabolism occurring prior to radicle emergence, which is discussed in Chapter 5.

Cell extension in plant tissues is generally held to be regulated by hormones, especially auxins and gibberellins. There is relatively little evidence that such substances play a key regulatory function in radicle emergence. It is well known, however, that application of certain growth regulators to seeds promotes their germination; this has stimulated much research, mostly in connection with seed dormancy, aimed at demonstrating a role for endogenous hormones. The dormant seed seems to offer a potentially useful situation for studying control processes in germination for here radicle emergence fails to occur unless specific triggering factors such as light, chilling or after-ripening have been experienced (see Vol. 2). Often, however, a requirement for these factors is wholly or partially eliminated by supplying gibberellins, cytokinins or ethylene to the seeds. Many investigations have therefore been performed to see if changing levels of endogenous regulators in the seed result from treatment with the normal dormancy-breaking factors. The findings which will be discussed in detail in Volume 2 of this book indicate that in a few cases these promotive hormonal regulators do increase in amount in treated seeds before radicle emergence, suggesting that germination may result from their appearance.

There is only limited information about the production of growth regulators during germination of non-dormant seeds. Perhaps the best studied case is the barley embryo which has been investigated in connection with the control of storage-food mobilization in the grain (Chaps. 6 and 7). Here, gibberellin levels increase as radicle protrusion and growth of the embryo proceed but there is no evidence of gibberellin production before growth takes place [53]; nevertheless, the germination rate of embryos does respond positively to applied gibberellin. In other seeds, while the role of endogenous gibberellin has not been conclusively demonstrated, it has been found that substances which are thought to inhibit gibberellin biosynthesis (e.g. CCC and AMO 1618) also prevent germination [5, 32]. The production of gibberellin may therefore be a prerequisite for radicle emergence; we must note, however, that the specific biochemical action of the above inhibitors on seeds has not been rigorously proved.

No satisfactory answer can therefore be given to the key questions—what biochemical events are specific to germination and how are they controlled? These are the central questions and at present most of our knowledge is marshalled around them.

4.4. Seedling Development

Water uptake, increasing respiration and the biochemical events described in detail in Chapter 5 culminate in the growth of the embryo and the development of the seedling, provided of course that the seed is not dormant. The first

Table 4.5. Some examples of hypogeal and epigeal germinators

	Hypogeal	Epigeal
Endospermic	*Triticum aestivum*	*Ricinus communis*
	Zea mays	*Fagopyrum esculentum*
	Hordeum vulgare	*Rumex* spp
	Phoenix dactylifera	*Allium cepa*
	Tradescantia spp	
	Hevea spp	
Non-endospermic	*Pisum sativum*	*Phaseolus vulgaris*
	Vicia faba	*Cucumis sativus*
	Phaseolus multiflorus	*Cucurbita pepo*
	Aponogeton spp	*Sinapis alba*
	Tropaeolum spp	*Crambe abyssinica*
		Arachis hypogaea
		Lactuca sativa

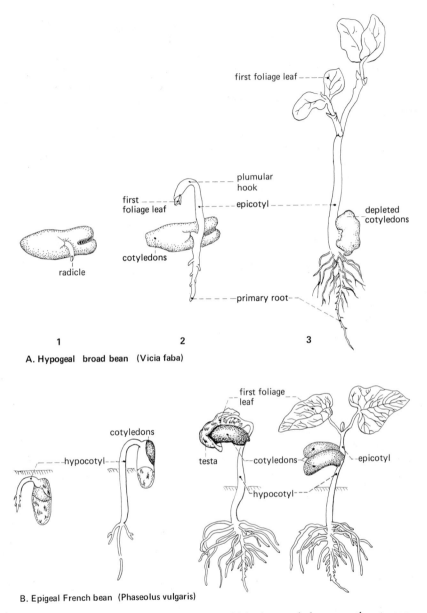

first foliage leaf

plumular hook

first foliage leaf

epicotyl

depleted cotyledons

cotyledons

radicle

primary root

1   2   3

A. Hypogeal broad bean (Vicia faba)

first foliage leaf

cotyledons

hypocotyl

testa   cotyledons   epicotyl

hypocotyl

B. Epigeal French bean (Phaseolus vulgaris)

Fig. 4.11. Seedling development I. Two species in which the cotyledons are the storage organs but one showing hypogeal germination and the other epigeal germination

visible and easily measurable signs of germination are commonly the increase in length and fresh weight of the radicle. In very many cases the radicle bursts through the covering structures virtually as soon as growth commences but in others (e.g. *V. faba*) appreciable growth proceeds before the testa is ruptured. There are, however, some species of seed in which growth of the hypocotyl

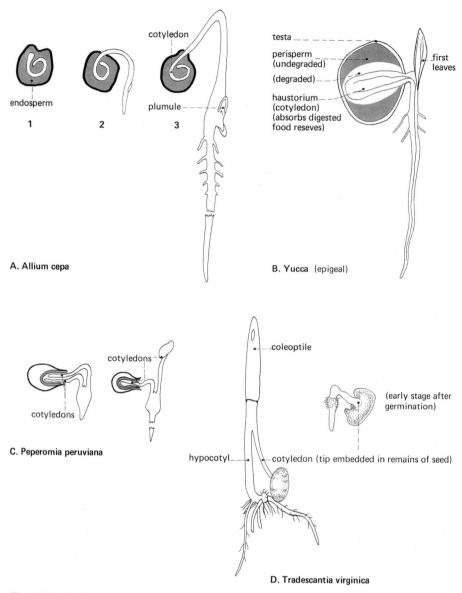

Fig. 4.12A–D. Seedling development II. Note the role of the cotyledon(s) (sometimes modi-fied for a haustorial function)

is the first visible manifestation of germination. This pattern of emergence is found in members of several families including the Bromeliaceae, Palmae, Chenopodiaceae, Onagraceae, Saxifragaceae and Typhaceae [35]. A case has been reported of a Graminaceous grain, *Oropetium thomaeum*, whose germination

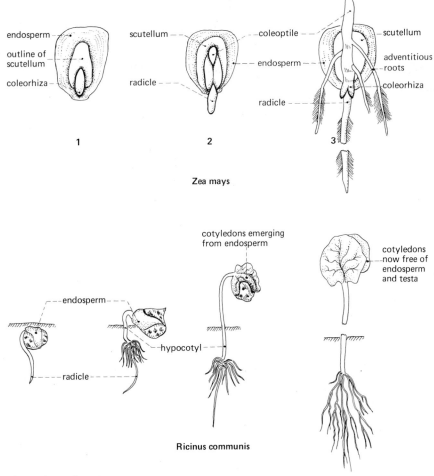

Zea mays

Ricinus communis

Fig. 4.12 continued

is characterized by growth of the coleoptile prior to that of the radicle [12]. This can also happen in certain genetic lines of wheat (cytoplasmic male sterile lines) when they begin to sprout on the mother plant.

An important difference among germinated seeds of different species concerns the fate of the cotyledons. We can divide seedlings into two types: (1) epigeal (or epigeous) in which the cotyledons are raised out of the soil and generally become green and photosynthetic, and (2) hypogeal (or hypogeous) in which the cotyledons remain underground. In dicotyledonous seeds this distinction can be readily understood on the basis of the different degrees of extension of the hypocotyl. The cotyledons of the epigeal germinators are carried up by the great extension growth of the hypocotyl; whereas this organ remains short and compact in the hypogeal seedlings. Here, after the cotyledonary stalks expand somewhat to force the cotyledons apart the epicotyl elongates to raise

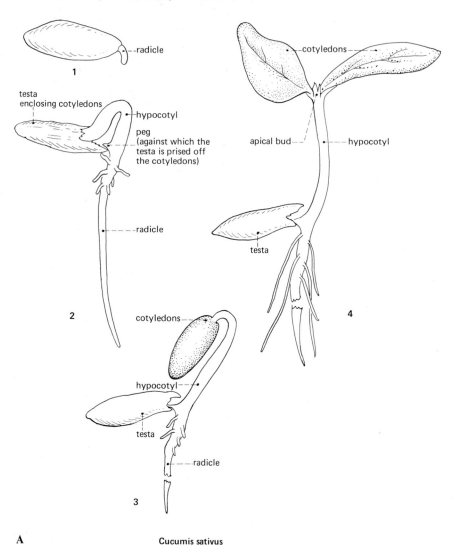

A   Cucumis sativus

Fig. 4.13A and B. Seedling development III. Two epigeal germinators with foliate cotyledons. Note that the testa is automatically removed from the cotyledons in *Cucumis* but that the seed coat (pericarp) in *Helianthus* remains, enclosing the cotyledons until the latter are well developed

the first true leaves out of the soil (Fig. 4.11A). It is generally agreed that the scutellum of the Gramineae represents a modified cotyledon as does the haustorium of other monocots. Neither of these is elevated above the endosperm during seedling emergence. Epigeal and hypogeal germinators are found among endospermic and non-endospermic seeds. A few common examples of these are listed in Table 4.5 and illustrated in Figures 4.11, 4.12 and 4.13. Hypogeal, endospermic seeds are rare among dicots.

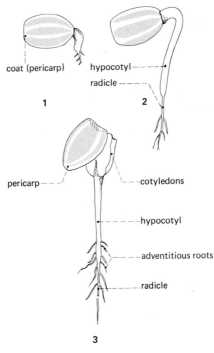

coat (pericarp)

hypocotyl

radicle

1

2

pericarp

cotyledons

hypocotyl

adventitious roots

radicle

3

Helianthus annuus

Fig. 4.13 B

The cotyledons of non-endospermic seeds in most instances store the reserves which are used during the early stages of seedling growth. On the other hand these food reserves are in the endosperm of the endospermic seeds or less commonly in the perisperm of others. In these cases, the cotyledons serve a most important function in mobilization of the reserves—they are responsible for absorbing the digested protein, carbohydrate, fat and other stores. This is clearly seen in the castor bean where the endosperm is carried up by the cotyledons as they consume its food stores. It is illustrated also by another epigeal germinator—*A. cepa,* the onion—where the absorptive tip of the single cotyledon remains embedded in the endosperm while the rest of the cotyledon becomes green (Fig. 4.12). The cotyledon in monocots may in fact become highly specialized for absorption. In *Tradescantia,* the cotyledon tip remains in the endosperm and is joined to the base of the coleoptile (Fig. 4.12). Highly developed haustorial cotyledons are found in the Palmae. The small embryo of the date palm (*Phoenix dactylifera*) is embedded in the hard, "hemicellulosic" endosperm. When embryo growth begins the cotyledon tip enlarges greatly into an umbrella-shaped body still buried within the endosperm and it absorbs the degraded reserves of this tissue. Similarly, the absorptive cotyledon of *Cocos nucifera* (coconut) enlarges to invade and occupy the endosperm. The highly-modified cotyledon of the Gramineae—the scutellum—absorbs the products of endosperm dissolution and transfers them to the embryonic axis.

An interesting germination pattern is exhibited by certain species of the dicotyledonous *Peperomia* in which only one cotyledon is withdrawn from the seed, the other remaining as an absorptive organ (Fig. 4.12). This case has been invoked to explain the possible origin of the monocotyledonous condition which has been held by some to be extremely advanced heterocotyly, the beginnings of which are shown by this species. This is an interesting conjecture but beyond the scope of our present discussion to pursue further.

Some Articles of General Interest

1. Currie, J.A.: The seed-soil system. In: Seed Ecology. Heydecker, W. (ed.). London: Butterworth, 1973, pp. 463–480
2. Koller, D.: Environmental control of seed germination. In: Seed Biology. Kozlowski, T.T. (ed.). New York: Academic Press, 1972, Vol II, 1–101
3. Spurny, M.: The imbibition process. In: Seed Ecology. Heydecker, W. (ed.). London: Butterworth, 1973, pp. 367–389

References

4. Berlyn, G.P.: In: Seed Biology. Kozlowski, T.T. (ed.). New York: Academic Press, 1972, Vol I, pp. 223–312
5. Black, M.: In: Dormancy and Survival. Woolhouse, H.W. (ed.). Cambridge: Univ. Press. Symp. Soc. Exp. Biol. **23**, 193–217 (1969)
6. Blacklow, W.M.: Crop Sci. **12**, 643–646 (1972)
7. Bonner, F.T.: Botan. Gaz. **129**, 83–85 (1968)
8. Brown, R.: In: Plant Physiology. Steward, F.C. (ed.). New York: Academic Press, 1972, Vol VIC, 3–48
9. Cleland, R.E.: Ann. Rev. Pl. Physiol. **22**, 197–222 (1971)
10. Collis-George, N., Hector, J.B.: Australian J. Soil Res. **4**, 145–164 (1966)
11. Collis-George, N., Williams, J.: Australian J. Soil Res. **6**, 179–192 (1968)
12. Dakshini, K.M.M., Tandon, R.K.: Ann. Botany (London) **34**, 423–425 (1970)
13. Dasberg, S.: J. Exp. Botany **22**, 999–1008 (1971)
14. Dasberg, S., Mendel, K.: J. Exp. Botany **22**, 992–998 (1971)
15. Evenari, M., Klein, S., Anchori, H., Feinbrun, N.: Bull. Res. Counc. Israel **6 D**, 33–37 (1957)
16. Flentje, N.T., Saksena, H.K.: Australian J. Biol. Sci. **17**, 665–675 (1964)
17. Foard, D.E., Haber, A.H.: Planta (Berl.) **71**, 160–170 (1966)
18. Goo, M.: J. Japan For. Soc. **34**, 3 (1952)
19. Haber, A.H., Foard, D.E.: Am. J. Botany **51**, 151–159 (1964)
20. Haber, A.H., Luippold, H.J.: Pl. Physiol. **35**, 168–173 (1960)
21. Haber, A.H., Luippold, H.J.: Pl. Physiol. **35**, 486–494 (1960)
22. Hadas, A.: Israel J. Agr. Res. **20**, 3–14 (1970)
23. Hadas, A., Stibbe, E.: In: Ecological Studies. Analysis and Synthesis. Hadas, A. et al. (eds.). Berlin: Springer, 1973, Vol IV, pp. 97–106
24. Hamly, D.H.: Botan. Gaz. **93**, 345–375 (1932)
25. Harper, J.L., Benton, R.A.: J. Ecol. **54**, 151–166 (1966)
26. Heydecker, W., Orphanos, P.I.: Planta (Berl.) **83**, 237–247 (1968)
27. Heydecker, W., Higgins, J., Gulliver, R.L.: Nature (London) **246**, 42–44 (1973)
28. Heydecker, W., Higgins, J., Turner, Y.J.: Seed Sci. Technol. **3**, 881–888 (1975)
29. Heydecker, W., Orphanos, P.I., Chetram, R.S.: Proc. Intern. Seed Test. Assoc. **34**, 297–304 (1969)
30. Highkin, H.R., Lang, A.: Planta (Berl.) **68**, 94–98 (1966)
31. Kamra, S.K.: Svensk. Botan Tidskr. **63**, 265–274 (1969)
32. Knypl, J.S.: Acta Soc. Botan. Pol. **36**, 235–250 (1967)
33. Lado, P., Rasi-Caldogno, F., Colombo, R.: Physiol. Plantarum. **31**, 149–152 (1974)

34. Lado, P., Rasi-Caldogno, F., Colombo, R.: Physiol. Plantarum **34**, 359–364 (1975)
35. Lang, A.: In: Encyclopedia of Plant Physiology. Ruhland, W. (ed.). Berlin: Springer, 1965, Vol XV/2, pp. 848–893
36. Larson, L.A.: Pl. Physiol. **43**, 255–259 (1968)
37. Manohar, M.S.: J. Exp. Botany **17**, 231–235 (1966)
38. Manohar, M.S., Heydecker, W.: Nature (London) **202**, 22–24 (1964)
39. Manohar, M.S., Heydecker, W.: Univ. Nottingham School Agr. Report 1964, pp. 55–62 (1965)
40. Marrè, E., Lado, P., Rasi-Caldogno, F., Colombo, R., De Michelis, M.: Pl. Sci. Lett. **3**, 365–379 (1974)
41. McDonough, W.T., Chadwick, D.L.: Pl. Soil **32**, 327–334 (1970)
42. Pollock, B.M.: Pl. Physiol. **44**, 907–911 (1969)
43. Pollock, B.M.: In: Viability of Seeds. Roberts, E.H. (ed.). London: Chapman and Hall, 1972, pp. 150–171
44. Pollock, B.M., Olney, H.O.: Pl. Physiol. **34**, 131–142 (1959)
45. Pollock, B.M., Toole, V.K.: Pl. Physiol. **41**, 221–229 (1966)
46. Rogan, P.G., Simon, E.W.: New Phytologist **74**, 273–275 (1975)
47. Schroth, M.N., Cook, R.J.: Phytopathology **54**, 670–673 (1964)
48. Sedgley, R.H.: Australian J. Agr. Res. **14**, 646–653 (1963)
49. Shaykewich, C.F.: J. Exp. Botany **24**, 1056–1061 (1973)
50. Shaykewich, C.F., Williams, J.: J. Exp. Botany **22**, 19–24 (1971)
51. Simon, E.W., Raja Harun, R.M.: J. Exp. Botany **23**, 1076–1085 (1972)
52. Waggoner, P.E., Parlange, J.-Y.: Pl. Physiol. **57**, 153–156 (1976)
53. Yomo, H., Iinuma, H.: Planta (Berl.) **71**, 113–118 (1966)

Chapter 5. Biochemistry of Germination and Growth

The cellular processes which have been studied most extensively in the germinating seed are respiration, enzyme and organelle activity and RNA and protein synthesis. This is not surprising for it is reasonable to expect that such fundamental aspects of cellular activity should play an important role in the preparation of seeds for germination; we will concentrate on these in this chapter. Other metabolic events (such as steroid biosynthesis, changes in alkaloid content) have also been studied but since at the present time these do not appear to be important for the initiation and control of germination they will be ignored. For the sake of continuity, biochemical events associated with both germination (radicle emergence) and subsequent growth will be considered together in this chapter although we will indicate which occur prior to visible germination and which are post-germination events. The reader should be careful to distinguish between these. Such a distinction does not always appear in the literature on seed physiology and workers and reviewers often refer loosely to germination when they are in fact considering seedling growth. The exclusively post-germination events in connection with food mobilization are discussed in Chapter 6.

5.1. Respiration — Pathways and Products

Three respiratory pathways, namely glycolysis, the pentose phosphate pathway and the citric acid cycle are widely assumed to be active in the imbibed seed. There is only limited evidence from a small number of species of seeds for the existence of some of the enzymes of these pathways, but at present we shall assume that the pathways are active to varying degrees in most, if not all viable, imbibed seeds. They are important for the production of key intermediates in metabolism, energy in the form of ATP, and reducing power as the reduced pyridine nucleotides NADH and NADPH. Details of their biochemistry may be found in the appropriate textbooks, and will not be considered here.

Glycolysis, catalysed by cytoplasmic enzymes, can operate under aerobic and anaerobic conditions to produce pyruvate, but in the absence of oxygen this is reduced further to ethanol, plus CO_2 (or to lactic acid if no decarboxylation occurs). Anaerobic respiration (or fermentation) produces only two ATP molecules per molecule of glucose respired: contrast this with six ATPs (though some authors claim eight ATPs) produced during pyruvate formation under aerobic conditions. In the presence of oxygen, further utilization of pyruvate occurs within the mitochondria. Here, oxidative decarboxylation of pyruvate to acetyl co-enzyme A (acetyl CoA), followed by complete oxidation of the latter to CO_2 and water via the citric acid cycle, yields a further 30 ATP molecules per glucose molecule respired. The generation of ATP molecules occurs during oxidative phosphorylation when electrons are transferred down an electron transport (redox) chain from the reduced coenzymes NADH and FADH (associated with specific enzymes of the citric acid cycle) via a series of electron carriers (cytochromes) located on the inner membrane of the mitochondrion, to molecular oxygen.

The pentose phosphate pathway is important because it produces NADPH, the currency of readily available reducing power within the cytoplasm which can serve as a hydrogen and electron donor in reductive biosynthesis, especially of fatty acids. Intermediates in this pathway are important starting compounds for various biosynthetic processes, e.g. synthesis of certain aromatics, and perhaps nucleotides and nucleic acids. Reactions of the pentose phosphate pathway in combination with one reverse step of the glycolysis pathway (F-6-P → G-6-P) may, under aerobic conditions, form a cycle whereby glucose can be completely oxidized to CO_2 and water, with NADPH being produced, and also ATP if the reducing power is transferred from NADPH to NAD and then to the electron transport chain. This cycle is the pentose oxidative cycle, or the oxidative pentose phosphate pathway. Doubts have been expressed concerning its existence in plants.

5.1.1. Requirement for Oxygen by Seeds

At the turn of this century it was widely believed that seeds of most land plants would not germinate when submerged under water, i.e. under conditions of reduced oxygen tension [83]. But studies by Morinaga published in 1926 [88] showed that seeds of 43 out of 78 genera of land plants (representing 24 families) could germinate under water. Of those that germinated, 18 genera did so as well as in aerated conditions on moist paper. Since these initial studies other successful "anaerobic" germinators have been found. Seeds of certain aquatic species actually germinate better under conditions of reduced oxygen tension— probably an adaptation to their habitat. *Typha latifolia* (cat-tail or bullrush) seeds, for example, germinate poorly in air [114] unless the coats are removed, but do so when intact in a mixture of 99% hydrogen in air. Rice (*Oryza sativa*) grains germinate 80% in 0.3% oxygen; in comparison, an equivalent amount of germination is shown by wheat grains in a minimum of 5.2% oxygen [129]. In both cereals germination at these reduced oxygen tensions results in an inhibition of root growth more than shoot growth. In stagnant water rice grains produce vertical shoots which through their Schnorkel action supply sufficient oxygen to the seedling for normal root development [69].

Seeds of many terrestrial species which fail to germinate below water quickly lose their viability when maintained under these conditions. Quite the opposite is the case with wild rice (*Zizania aquatica*) however. Here, air drying of the grains for 90 days causes loss of viability, whereas after-ripening at low temperatures in stagnant water actually accelerates their subsequent rate of germination [116]. This represents an adaptation to habitat, for the grains of this particular wild rice normally fall to the mud layering the bottom of a body of water and then pass through a long winter period in submerged, near-freezing, non-aerated conditions before germination in the following spring. Not all seeds appear to show such an apparently elegant adaptation to habitat; those of *Phragmites communis* will not germinate below 5 cm of water, and yet the plant is often established at depths of 1 m [120]!

One seed which carries its own oxygen supply is that of Indian lotus [98]. Full germination can be achieved in 100% nitrogen, hydrogen, or carbon dioxide because a supply of oxygen is available to the embryo through an internal cavity and from intercellular spaces of the seed tissues. An analysis of the trapped internal gas, about 0.2cc volume, reveals 18.3% oxygen, 0.74% CO_2 and 80.93% nitrogen.

5.1.2. Respiration of Germinating Seeds

Low levels of respiration have been reported in "dry" seeds, though we mean dry in a relative sense since such seeds do contain moisture, e.g. cereal grains in dry storage have a 10–15% water content. Respiration of stored seeds will be discussed further in the second volume of this book during our consideration of viability and longevity.

Seed and fruit coats can act as barriers to the uptake of water and thus prevent or retard germination. They can also affect respiration by limiting the permeability to oxygen. Coat effects on germination and the restriction of gaseous exchange is another topic reserved for the second volume.

The introduction of dry seeds to water results in an immediate and rapid evolution of gas, an event which may last for several minutes. This burst upon wetting is not related to respiration; the gas is not predominantly CO_2, and could be that released from colloidal adsorption as water is imbibed. Further, this release occurs from dead seeds and non-living seed parts [54].

A very early metabolic event during imbibition, occurring in less than 15 min, appears to be the reformation of keto acids from amino acids by deamination and transamination reactions [37]. Keto acids important for respiratory pathways (e.g. α-ketoglutarate and pyruvate) may be absent from dry seeds such as wheat [70] and peanut cotyledons [138]. They are known to be chemically unstable and it has been suggested [37] that they are stored in the dry seed as the appropriate amino acid, and then reformed on rehydration. This might occur also in barley embryos, seeds of *Sinapis alba*, and axes and cotyledons of *Phaseolus vulgaris* [36, 37].

The pattern of consumption of oxygen by imbibed pea seeds is shown in Figure 5.1A. Basically, the uptake of oxygen can be divided into two pre- and two post-visible germination phases, although it should be noted that over 90% of the respiration by this seed is contributed by the non-growing cotyledons and not the germinating axis itself. Respiration is considered to involve four phases:

Phase I. This is characterized by a sharp rise in respiration lasting about 10 h and is attributed in part to activation and hydration of mitochondrial enzymes associated with the citric acid cycle and electron transport chain (see

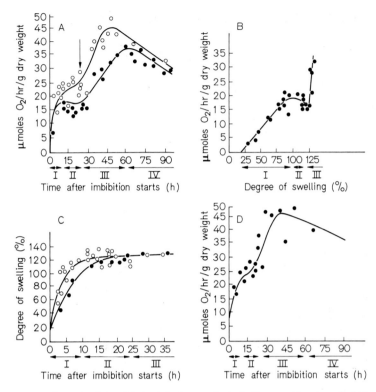

Fig. 5.1. (A) The course of respiration of intact dark-germinated *Pisum sativum* seeds (●) cv. Rondo and of cotyledons with seed coat dissected away (○). Arrow indicates approximate time of radicle protrusion. (B) The relation between the degree of swelling and the respiration rate of cotyledons in intact germinated seeds. (C) The swelling of cotyledons from intact seeds (●) and of cotyledons from seeds imbibed without a seed coat (○). (D) Respiration of excised cotyledons. See text for explanation of Phases I–IV. After Kollöffel, 1967 [67]

the section on mitochondrial changes later in this chapter). Respiratory quotient (R.Q.) values at this time are slightly above 1.0; the major respiratory substrate is probably sucrose [123]. The rise in respiration during this phase progresses linearly with the degree of swelling of the cotyledon tissue (Fig. 5.1 B).

Phase II. Here, there is a lag in respiration between hours 10 and 25 after the start of imbibition. Hydration of the cotyledons is now completed and all pre-existent enzymes activated. At this time R.Q. values have been observed to rise above 3.0, which is indicative of some anaerobic respiration [123]. It is widely believed that the limiting factor for respiration at this stage of germination is the supply of oxygen, the restriction being attributed to the intact seed coat. There appears to be some evidence to support this hypothesis in pea seeds [123], and removal of the seed coat does increase both the rate of water uptake and swelling (Fig. 5.1 C) as well as reduce the lag phase (Fig. 5.1 A and D). It still remains to be explained, however, why there is rapid oxygen uptake into seeds with intact testas during early stages of imbibition (Phase I) when the same testas apparently restrict oxygen uptake in Phase II. Does the

Table 5.1. The respiratory lag phase in seeds

Seeds with a lag phase	Seeds without a lag phase
Pisum sativum	*Avena fatua*
Phaseolus vulgaris	*Hordeum vulgare*
Phaseolus mungo	*Ricinus communis*
Phaseolus aureus	*Sinapis arvensis*
Lactuca sativa	*Phaseolus lunatus* (axes)
Raphanus sativus	*Oryza sativa*
Arachis hypogaea (axes)	
Pinus densiflora	
Pinus thunbergii	
Pseudotsuga menziesii	
Trifolium incarnatum	
Zea mays	
Glycine max	
Lathyrus odoratus	

pea seed testa become less permeable to oxygen as imbibition proceeds? Metabolism during the lag phase in seeds of pea and other species will be discussed in the next section of this Chapter.

Between Phases II and III the radicle penetrates the testa.

Phase III. A second respiratory burst characterizes this phase, attributed in part to an increased oxygen supply through the now punctured testa. Another contributing factor to the continued rise in respiration must be the activity of newly synthesized mitochondria and respiratory enzymes in the dividing cells of the growing axis. During this phase the R.Q. falls to about 1.0 indicating that aerobic respiration of carbohydrate is predominant.

Phase IV. The marked fall in respiration noted at this time coincides with the disintegration of the cotyledons following depletion of the stored reserves. Since seeds used in the experiment outlined in Figure 5.1 were dark-grown, no replenishment of substrates by photosynthesis was possible.

A number of other seeds exhibit a similar pattern of respiration to that of peas (see Table 5.1). The lengths of Phases I–IV are variable from species to species, e.g. in *Phaseolus mungo* they are much abbreviated, and radicle emergence and Phase III commence after only 6 h from the start of imbibition [89]. On the other hand, in *Pinus densiflora* Phase I alone takes two days for completion [57]. Even in pea, the lengths of these phases vary as a function of temperature, moisture availability and ambient oxygen concentration.

5.1.3. The Lag Phase in Respiration—Possible Causes

Both ethanol and lactic acid, products of anaerobic respiration, accumulate in the pea seed prior to the penetration of the radicle through the enclosing testa [38, 39]. Following radicle emergence their levels diminish as they are metabolized under the conditions of increasing aerobiosis. This event is accompanied by increased oxygen uptake, CO_2 output and a fall in R.Q. Alcohol

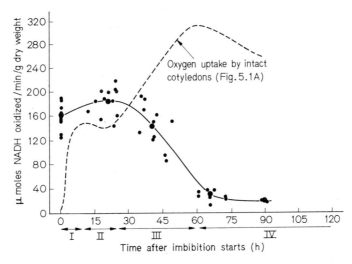

Fig. 5.2. Alcohol dehydrogenase activity (●) in, and oxygen uptake by cotyledons of dark-germinated intact pea seeds cv. Rondo. Germination time as Figure 5.1 A. After Kollöffel, 1968 [68]

dehydrogenase (ADH), the enzyme responsible for ethanol formation from acet-aldehyde (which itself is derived from pyruvate), is present in dry pea seeds and through Phase II of respiration (Fig. 5.2). It only begins to disappear during Phase III, when the ethanol is being utilized.

Shortening or removal of the lag phase (i.e. Phase II) can be achieved in several species by removal of the testa, presumably allowing of oxygen entry. Anaerobiosis is unlikely, however, to account for the appearance of the respiratory lag phase in all germinating seeds. For example, isolated peanut axes still exhibit a marked lag in oxygen uptake from about the 2nd to the 16th hour after imbibition starts (Fig. 5.3A), at which time they start to grow, and respiration increases again [141]. Accompanying this growth is the development of mitochondria which have greater respiratory efficiency and control than those present in the axes during early imbibition (also see later, Sect. 5.1.6). There is indirect evidence from studies using respiratory inhibitors that in some seeds respiration prior to visible germination is not totally dependent upon oxygen and/or its utilization by the terminal oxidation step of the normal cyto-chrome-mediated electron transport chain. That many seeds can germinate under water, i.e. at reduced oxygen tensions, certainly suggests that efficient oxidative phosphorylation may not be a pre-requisite for germination. In fact, it is claimed that the germination of some seeds is even enhanced when some respiratory steps—particularly the terminal oxidation process (cytochrome oxidase)—are inhibited [105]. Although chemicals commonly considered to uncouple ATP production from the electron transport chain appear to stimulate the germination of some seeds (e.g. *Trifolium subterraneum* [13]) it has never critically been shown that these substances in fact act as uncouplers in the seed to reduce ATP synthesis, nor has it been shown that there are alternative pathways for

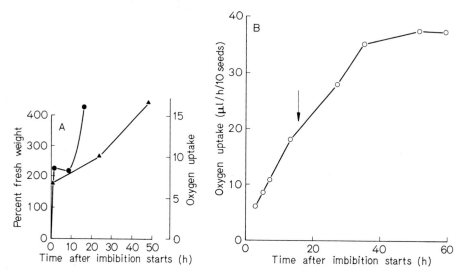

Fig. 5.3. (A) Changes in water content and oxygen uptake (arbitrary units) during the first 48 h after imbibition by peanut axes. ▲——▲: fresh weight, ●——●: oxygen uptake. After Wilson and Bonner, 1971 [141]. (B) Time course of respiration of non-dormant grains of *Avena fatua*. Arrow indicates approximate time of radicle emergence. After Chen and Varner, 1970 [29]

ATP synthesis. In those instances where ATP production is known to be inhibited by other means (e.g. nitrogen, Fig. 5.5), no germination occurs. Prevention by inhibitors of terminal oxidation may make more oxygen available within a seed and somehow promote its germination, but inhibition of ATP synthesis probably always inhibits germination. A fuller account of oxidative processes and the control of dormancy and germination will appear in the second volume.

The precise alternative pathways through which oxygen is used, and their importance for the production of energy and reducing power during the early stages of imbibition when the metabolism of the embryo is becoming established have never been fully elucidated. However, (1) oxidative (mitochondrial) pathways which are insensitive to cyanide (a potent inhibitor of cytochrome oxidase, the normal terminal oxidase of the electron transport chain); and (2) the pentose oxidation cycle, have both been implicated [104, 105, 144]—see later, in the section on Reducing Power (Sect. 5.1.7).

Thus, in some seeds the lag phase could perhaps represent a time of temporary anaerobiosis due to restricted oxygen supply by the surrounding structures, although the available evidence is not always particularly convincing. In others, the lag phase before radicle emergence could reflect, to a greater or lesser extent, the time taken for the development of a second respiratory system (i.e. mitochondria with efficiently coupled oxidative phosphorylation) to replace the less efficient initial system which operates at earlier times. Mitochondria present in dried seeds often appear deficient in cristae, indicating internal damage. Such damage might be repaired during or after imbibition, or new mitochondria synthesized, but until this is effected we might expect inefficient (i.e. partially uncoupled) respiration. To what extent this respiration is initially enhanced through the operation of alternative pathways remains to be determined.

5.1.4. Seeds Without a Lag Phase of Respiration

An analysis of the respiratory pattern published for *Avena fatua* during imbibition, germination and subsequent growth reveals the absence of a lag phase in oxygen uptake (Fig. 5.3 B). Several other seeds exhibit a similar feature (Table 5.1). There is no simple explanation for this lack of a lag phase, and we are not aware of any attempts in the literature to distinguish critically between seeds with and without a lag phase. It could be that the latter do not pass through a period of temporary anaerobiosis and there are no restrictions imposed by the seed coat; or perhaps their efficient mitochondrial respiratory systems become established early to effect continuity of oxygen uptake. The matter is obviously worthy of further research, even if only to make sure in those instances where the lag phase seems to be absent that this is not due to a lack of measurements at the critical time period.

5.1.5. Sources of Substrate for Respiration Prior to Reserve Mobilization

Respiration during initial water uptake, germination and early growth requires a supply of readily-available substrate other than that derived from hydrolysis of the major stored reserves, since the latter only become available after embryo growth has commenced. Duperon [44] analysed the dry seeds of 81 species of monocots and dicots and found that all of them contained sucrose, 70 contained the trisaccharide raffinose (galactose-glucose-fructose) and 35 contained the tetrasaccharide stachyose (galactose-galactose-glucose-fructose). The distribution and levels of these sugars within seeds is variable. For example, in *Pinus thunbergii* the content of stachyose, raffinose and sucrose is 0.3, 0.22 and 0.12 g/100 g dry seed, respectively [58], whereas in soybean the proportions are 5.2, 1.9 and 8.0 g/100 g dry seed [102]. Differences in distribution of these sugars are even shown between different varieties of the same species, e.g. in coffee seeds the ratio of sucrose:stachyose may vary as much as from 90:1 to 40:1, although the amount of stachyose present is consistently 1.5 times greater than the amount of raffinose [113].

Changes in respirable substrates in germinating barley grains are outlined in Chapter 6—see Table 6.1. Sucrose and raffinose in the embryo are depleted during the first 24 h of imbibition, sucrose levels declining rapidly within the first 6–9 h, followed by utilization of raffinose after 14 h [73]. This sequence of di- and oligo-saccharide hydrolysis and oxidation has also been found in a number of dicot and gymnosperm seeds, often with the sucrose utilized before radicle emergence and then raffinose or stachyose during subsequent early growth, prior to mobilization of the major stored reserves. There are exceptions though. For example, in soybean both raffinose and stachyose levels decline by about 50% over the first 48 h after imbibition starts, during which time germination occurs and rootlets are formed. Sucrose, which is more abundant than the other two oligosaccharides, only declines between the second and fourth day after imbibition starts [45]. It has been shown that in several species of seeds the activity of α-galactosidase increases coincidentally with the decline in raffinose and stachyose levels [71]. There is little direct evidence that any of these sugars is utilized for respiration, although the circumstantial evidence is highly suggestive.

Free fructose and glucose accumulate within seeds during sucrose, raffinose and stachyose breakdown (e.g. in soybean [102]), but no build-up of galactose has ever been observed. This is indicative of its rapid utilization, perhaps by incorporation into cell walls or into galactolipids of the newly-forming organelles in the cells of the developing seedling.

5.1.6. Mitochondrial Activity and ATP Synthesis

Many of the studies on the development of mitochondrial activity in seeds have been carried out using dicot storage tissues. Fewer studies have been made on the imbibed embryo or axis. It is unfortunate that many of the changes in mitochondrial activity observed over several days in the cotyledon and endosperm cells of dicot seeds have been fallaciously quoted as evidence for changes during "germination." It is important to discriminate between those events taking place in the axis or embryo prior to radicle elongation and those taking place in non-persistent, non-growing storage tissues to accompany mobilization of reserves. Thus, we will deal with germinating tissues and storage tissue quite separately.

(i) Germinating Tissues

Mitochondria from dry and early-imbibed peanut embryonic axes (which exhibit a lag phase of respiration during germination) are deficient in cytochrome c, the penultimate electron acceptor in the electron transport chain. Furthermore, malate dehydrogenase levels are low (preventing conversion of malate to oxaloacetate in the citric acid cycle) and phosphorylation is only loosely coupled to oxygen consumption. Respiration is therefore inefficient and less than maximal ATP production occurs [141]. Such deficiencies are probably a legacy of the drying of the seed terminating its development (Chap. 3), for it is known that some respiratory enzymes decline in activity at this time (Fig. 3.20). These deficiencies persist until about the 16th h after the start of imbibition by the embryo, when respiration again increases and radicle elongation begins (Fig. 5.3A). The mitochondria now start to assume the normal characteristics of those found in the adult plant, with a normal cytochrome c and dehydrogenase content and more efficient, coupled respiration. Although only a low level of cytochrome c is present in the mitochondria during early imbibition, and through the lag phase, it might still be important in preventing completely uncontrolled loss of respiratory substrate at this stage and some oxidative phosphorylation may occur. Also, or alternatively, another terminal oxidase [109] may operate at these early times. This might be possible if the electron transport chain is branched, one branch being through the deficient cytochrome c/cytochrome oxidase pathway which takes time to develop, and the other branch using an alternative (cyanide insensitive) oxidase which is immediately effective.

It has been suggested that increased respiration and respiratory efficiency during growth of the peanut embryo is associated with the synthesis of new mitochondria, rather than with the modification of those functioning during early imbibition [141]. Electron-microscopic studies on the germinating rye embryo also suggest biosynthesis of mitochondria after water uptake. Here an increase in respiration over the first 6 h is apparently accompanied by a substantial increase in the number of mitochondria in the cells of the root primordium region [55]. Moreover, the number of cristae per mitochondrion increases with time after the start of imbibition, which indicates more ordered mitochondrial activity. Cristae also become better defined in the mitochondria of pea embryos prior to visible germination [145]. In fact, increasing orderliness of mitochondria before visible germination appears to be a common feature of many embryos, although electron-microscopic studies on lima bean axes [66] have been interpreted as showing that there are no changes in either structure or numbers of mitochondria prior to radicle protrusion. As yet, none of the observations made from electron-microscopic studies have any substantial biochemical evidence to support them, and these observations alone do not provide sufficient evidence for either mitochondrial biogenesis or mitochondrial repair. It seems likely that some development of mitochondrial activity takes place after imbibition has occurred, but the mode of this development is still unknown.

Synthesis of ATP has only been studied in a few germinating seeds, but, as far as we can tell, never in conjunction with studies on the changing structure and/or activity of mitochondria. Water uptake by lettuce seeds is completed within 2 h of wetting (Fig. 5.4A). Oxygen consumption rises over this time period, and then shows a lag phase which is followed by a further rise in respiration to accompany radicle emergence and growth. A rapid increase in ATP production occurs over the first 4 h after imbibition starts and then, like oxygen consumption, this reaches a plateau. No further increase in ATP levels occurs until after radicle protrusion has commenced (Fig. 5.4B). The increase in ATP in the seed is accompanied by an increase in the total pool of adenine nucleotides, which suggests an early requirement for their de novo synthesis. Total ATP levels in the seed remain constant from the 4th to 16th h after the start of imbibition because the rate of ATP synthesis is balanced exactly by its rate of utilization. If ATP synthesis is inhibited during this lag phase, e.g. by placing seeds in an atmosphere of nitrogen to prevent oxidative phosphorylation, then the ATP pool is rapidly used up, and the levels diminish (Fig. 5.5). The total nucleotide pool is maintained at a high level and AMP, rather than ADP, increases. Obviously then, ATP synthesis which is coupled to oxygen consumption must be taking place in the imbibed lettuce seed prior to radicle protrusion, although we do not know whether the mitochondria which we presume to be involved have any of the deficiencies apparent in those of the peanut embryos. The extent of the participation of an alternative oxidase system in

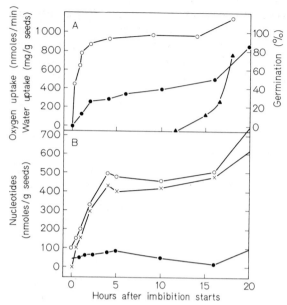

Fig. 5.4A and B. Time course of increases in (A) water uptake, oxygen consumption and germination and in (B) total adenine nucleotides, ADP and ATP in lettuce seeds. Symbols in (A): o——o: water uptake; ●——●: oxygen uptake, ▲——▲: germination. Symbols in (B): ●——●: ADP level, ×——×: ATP level, o——o: total adenine nucleotide level. Based on Pradet et al., 1968 [103]

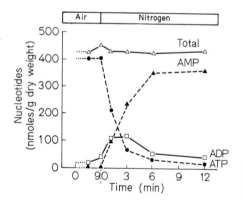

Fig. 5.5. Changes in adenine nucleotides in response to anaerobiosis in lettuce seeds. After Pradet et al., 1968 [103]

lettuce seeds is not known, although the initial level of cytochrome oxidase present in the dry seed seems to be more than adequate to account for the observed oxygen uptake taking place prior to radicle protrusion [82]. Mitochondria in the dry lettuce embryo contain few cristae [124] but whether they lack any particular components or enzymes, or undergo any changes prior to radicle emergence is not known; they do become more organized after many hours of embryo growth.

The ATP content of isolated wheat embryos has been found to increase substantially during the first hour of imbibition, followed by only a small increase over the next 15 h (Table 5.2). Likewise, a large increase in ATP occurs within an hour of wetting radish seeds; ATP levels then remain constant until radicle emergence at 16 h [87]. In contrast, it is claimed that in intact germinating peas there is no rise in ATP levels until after 4 h of imbibition [22], an observation which is not consistent with the fact that respiratory activity and oxygen uptake develop rapidly in these seeds at the same time.

In recent years a relationship has been demonstrated between ATP levels in an imbibed seed and vigour of that seed. For example, it is known that the ATP content of 4-h-imbibed lipid-, starch- and protein-containing seeds can be positively correlated with seed size, with the size, weight and length of the developing seedling [33, 35] and also with the viability of different seed lots (Table 5.3). This is an important observation, for it suggests one way by which seedling vigour might be predicted without the requirement for a long period of germination and growth.

Table 5.2. Adenine nucleotide content of wheat embryos during germination

Experiment	Time after imbibition starts (h)	Adenine nucleotide content (nmol/125 mg embryos)				Energy charge
		ATP	ADP	AMP	Total	
1	6.5	106	32	9	147	0.83
	13.5	112	34	10	156	0.83
	16.5	97	24	7	128	0.85
2	0.0	8				
	1.0	97	77	42	216	0.63
	6.5	108	39	10	157	0.82
3	0.0	4				
	0.5	37				
	1.0	85	79	39	203	0.61
	6.5	98	28	7	133	0.84

After Obendorf and Marcus, 1974 [97]

Table 5.3. Effects of ageing seeds at different temperatures on ATP content of imbibed seed, germination percentage and seedling size of *Trifolium incarnatum* cv. Dixie (Dixie crimson clover)

	Seed age			
	6 months		15 years	
	22° C	3° C	22° C	38° C
Seed weight, mg	2.9	3.8	3.8	3.2
ATP, nmol/seed	2.28	2.16	0.92	0.02
Germination, %	98	96	69	0
4-day seedling length, cm	4.4	3.0	0.9	0

Seeds were placed in water at 20° C for 4 h before ATP content was measured. After Ching, 1973 [33]

Metabolism of the germinating seeds themselves may not depend upon or be controlled by ATP content alone. Levels of ADP and AMP may also play a role. Protein synthesis is an early event during seed imbibition, and while it is a process which requires ATP, it can also be inhibited by high levels of ADP and AMP. Both of these latter adenine nucleotides decrease in wheat embryos after the first hour of imbibition (Table 5.2), at a time when protein synthesis is increasing markedly [76]. Similarly in both pea and lettuce seeds there is a depletion of AMP coincidental with ATP synthesis, with ADP levels remaining constant [22, 103].

The amount of metabolically-available energy which is momentarily stored in the adenylate system of a living cell is linearly related to the mole fraction of ATP plus half of the mole fraction of ADP. Expressed in terms of concentration of the nucleotides, this available energy is called the energy charge [26], which can be calculated from the following formula:

$$\frac{[ATP] + \frac{1}{2}[ADP]}{[ATP] + [ADP] + [AMP]} = \text{Energy Charge (E.C.)}$$

When the concentrations of ATP, ADP and AMP within a cell are such that the value for energy charge is above 0.5, then ATP-utilizing systems increase their activities, and above 0.8 cells may actively metabolize and multiply. Energy charges of less than 0.5 are indicative of senescent, or quiescent cells. The concept of energy charge was originally developed to express the metabolic status of bacterial cells, but some researchers have also applied it to plant cells. The energy charge of germinating wheat embryos increases during early imbibition and is adjusted from 0.6 to 0.8 between the first and sixth hour after imbibition begins (Table 5.2). This must mean that the increase in metabolic activity of the embryo due to decreased ADP and AMP levels, and hence increased energy charge, occurs at a time when there is no (or little) net production of ATP. Thus cells of the imbibing wheat embryo must first rapidly establish a high ATP level, and then adjust their adenine nucleotide ratio, increasing the energy charge to favour optimum cellular activity.

(ii) Storage Tissues

Interest in the development of mitochondria in storage tissues has arisen because of their potential role in the modification and utilization of the stored reserves during the mobilization phase. Two distinct patterns of mitochondrial development appear to be emerging, one involving the repair and activation of the organelles already existing in the dry seed (e.g. in pea cotyledons), and one involving biogenesis of new mitochondria (e.g. in peanut cotyledons). In many dicot storage tissues it is recognized that an increase in the activity of mitochondria coincides with the mobilization of food reserves, but in very few cases has it been conclusively shown whether biogenesis or activation (or both) is involved.

Mitochondria in unimbibed cotyledons or endosperms of many seeds are poorly differentiated and lack cristae. At the time of reserve mobilization they have a better defined structure, and a greater number of cristae. Some internal

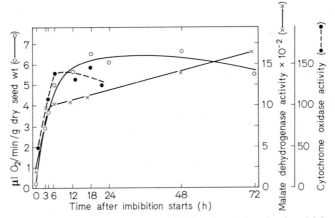

Fig. 5.6. Increase in the respiratory rate of the mitochondrial fraction (○), mitochondrial malate dehydrogenase (×) and cytochrome oxidase activity (●) in intact cotyledons of dark-germinated Alaska peas. Based on Nawa and Asahi, 1971 [94]

development of mitochondria occurs in pea seeds within the first day after imbibition begins, and by the end of the second day cristae appear more or less to be fully developed [11]. The structurally deficient mitochondria of dry cotyledons of Alaska pea are also enzymically deficient, having only low levels of malate dehydrogenase and cytochrome oxidase. Both of these rise rapidly in activity during the first few hours of imbibition (Fig. 5.6), with malate dehydrogenase activity rising slowly thereafter over the next three days; cytochrome oxidase activity reaches a plateau sooner. In another cultivar of pea (Homesteader) it has been observed that both cytochrome oxidase and succinate dehydrogenase activities increase during the first four days after dark-imbibition, the latter enzyme increasing in activity over 20-fold between the 3rd and 12th h [118]. Thus, in both cultivars it is apparent that there is a rapid rise in activity of storage-organ mitochondria following imbibition, which is reflected in an increase in oxygen uptake at this time (Fig. 5.1A). Over the first day after imbibition starts there is an increase in respiratory control by the mitochondria, with a tighter coupling of ATP production to oxygen utilization; this increased efficiency is maintained for several days [94, 118, 134]. In watermelon cotyledons there is an increase in mitochondrial malate dehydrogenase activity during the first four days after imbibition and this is due to de novo synthesis of the enzyme [134].

Incorporation of applied, radioactive leucine into the mitochondrial proteins of pea cotyledons has been observed within 6 h of the start of imbibition, and it may continue for at least another 18 h [95]. Inhibition of protein synthesis on cytoplasmic ribosomes prevents this synthesis of mitochondrial proteins, whereas inhibition of mitochondrial ribosomes does not. Therefore these new mitochondrial proteins are synthesized in the cytoplasm and not in the mitochondria themselves. But even in the absence of synthesis of new mitochondrial proteins, cytochrome oxidase and malate dehydrogenase levels rise and the mitochondria increase in their respiratory efficiency [74, 95]. Thus protein synthesis is not essential for the development of mitochondrial activity: in fact, in the absence of de novo synthesis of proteins there appears to be an enhancement of membrane development within the mitochondria. It thus appears that preformed proteins, both structural and enzymic, are transferred into preformed, immature mitochondria following imbibition, resulting in active, efficient mitochondria with complete membrane and respiratory systems. That reassembly rather than synthesis of mitochondria occurs is supported by the observation that ^{3}H-thymidine is not incorporated into pea mitochondrial DNA over a three-day period after imbibition [74].

Cytochromes b, c and cytochrome oxidase, succinic dehydrogenase, succinoxidase (the system whereby succinate dehydrogenase passes its electron via cytochromes to oxygen) and mitochondrial proteins all increase in imbibed peanut cotyledons [19, 20]. The number of particles with a density characteristic of mitochondria (but could include glyoxysomes?—see Chap. 6) also increases, as does the level of mitochondrial DNA (Table 5.4). These latter observations appear to be more consistent with mitochondrial biogenesis than with activation, although protein-inhibition studies of the type conducted on pea cotyledons have not been repeated on peanuts. So at the present time we cannot rule out the possibility that here there might be some activation of pre-existing mitochondria as well as (or even instead of) de novo synthesis.

As the cotyledonary reserves are consumed the mitochondria become disorganized and gradually lose their respiratory efficiency, enzyme complement and activity. In cotyledons of dark-grown Alaska peas this is marked by a decline in respiratory control (Fig. 5.7A), a loss of efficiency of oxidative phosphorylation (shown by the fall in ADP/O ratio) (Fig. 5.7B) and in cytochrome oxidase activity (Fig. 5.7C). Another mitochondrial enzyme, malate dehydrogenase does not decline however. The gradual loss of mitochondrial activity is accompanied by the disruption of cell structure (see Chap. 6).

This fall in respiratory activity which accompanies the overall decline in stored reserves might not occur uniformly in all parts of dicot cotyledons. There are indications, for example in *Cucurbita maxima* (squash) cotyledons which persist and become photosynthetic, that the mitochondria deteriorate

Table 5.4. Comparison of various constitutive and functional changes in peanut cotyledon mitochondria occurring after imbibition has taken place

	Ratio of amount or rate at 9.5 days to that at indicated time	
	0.5 day	2.5 day
Cytochrome a-a$_3$	—	5.6
Succinoxidase	13	5.5
Succinic dehydrogenase	6.0	2.4
Mitochondrial DNA	4.1	3.1
No. of particles	—	2.9

After Breidenbach et al., 1967 [19]

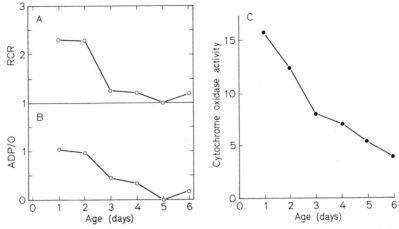

Fig. 5.7A–C. Changes in (A) respiratory control ratio (*RCR*); (B) ADP/O ratio; and (C) mitochondrial cytochrome oxidase activity in intact cotyledons of dark-grown Alaska peas. Based on Nawa et al., 1973 [96]

in storage cells as the reserves are expended and chloroplasts are formed, but not in the cells close to the veins [72]. Thus, there is a redistribution of respiratory activity in the cotyledons with age — fairly uniform during reserve mobilization but becoming associated with the vascular tissue as the organs become photosynthetic. The role of the vein-cell mitochondria might be to provide energy for the vein-loading of catabolites and the subsequent products of photosynthesis. In dark-grown seedlings vein localization is less distinct.

In intact cotyledons of Meteor peas ATP levels increase over the first two days after imbition and then gradually decline during the following six days [21], an observation which is more or less in line with those outlined in Figure 5.7. There is, though, one report for whole pea seeds which shows an initial 250% increase in ATP levels up to 16 h and then a 50% decline prior to visible germination at 40 h [22]. Changes in levels of the three adenine nucleotides in the megagametophyte of stratified and germinated ponderosa pine seedlings are shown in Figure 5.8A. The increase in ATP coincides with the initiation of radicle emergence and marks the time when reserve mobilization commences. Biogenesis of mitochondria is claimed to occur from about the third day. The decline after 12–14 days coincides with the emergence of the cotyledons, the dependence of the embryo on photosynthesis for the provision of respirable metabolites, and the depletion of reserves from the gametophyte. Comparative changes in the developing seedling are also shown (Fig. 5.8 B). The energy charge fluctuates between 0.65 at the stage of cell expansion and cell division of the developing seedling to 0.85 following cotyledon establishment. It remains constantly around 0.75 in the gametophyte throughout the pre-mobilization and mobilization stages.

Control of aerobic glycolysis and the reverse process, gluconeogenesis (hexose synthesis), in storage tissue can be mediated through the availability of ATP and the energy charge. An early step in glycolysis is the production of fructose-

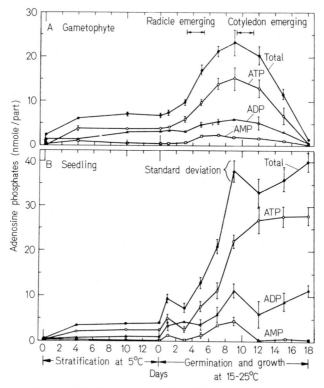

Fig. 5.8A and B. Changes in AMP, ADP, ATP and total adenosine phosphate in the (A) gametophyte and (B) embryo or seedling of *Pinus ponderosa* during stratification and subsequent germination and growth. After Ching and Ching, 1972 [34]

1,6-diphosphate (FDP) by phosphofructokinase (PFK). ATP is an allosteric inhibitor of PFK and at elevated concentrations will limit the action of this enzyme, depressing the rate of glycolysis. It is possible that in imbibed storage tissues the development of mitochondria leads to enhanced respiration, ATP accumulation and a high energy charge, which in turn causes a decline in glycolysis through reduced PFK activity. Such changes would favour hexose, and hence sucrose synthesis, thus promoting the production of the substrate for transport to the growing axis. Control of PFK by ATP (in conjunction with orthophosphate) seems possible in cell-free extracts from imbibed pea seeds [50], although the level of extractable PFK itself in pea cotyledons does in fact decrease during the period of seedling growth [21].

5.1.7. Reducing Power—the Synthesis and Utilization of Pyridine Nucleotides

Reduced pyridine nucleotides (NADH, NADPH) are essential coenzymes for a number of key metabolic pathways and therefore they are likely to play an important role in the germinating embryo, developing seedling and storage organs. They are involved, for example, in glucose respiration, amination reactions, deoxynucleotide synthesis and fatty acid metabolism. Control of metabolic pathways might also be effected through these coenzymes, through regulation of their concentration and availability within the cell.

The oxidized pyridine nucleotide NADP is the coenzyme of glucose-6-phosphate dehydrogenase (G6Pdh) and is a key link between the glycolysis pathway and the pentose phosphate pathway. The former requires a supply of NAD for one of its intermediate steps, as outlined below:

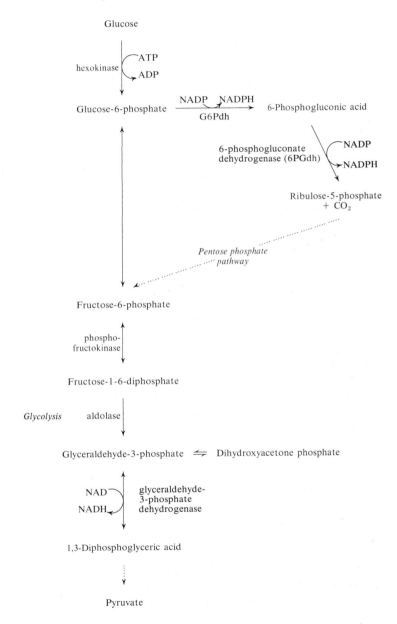

When NADP is in limited supply glycolysis proceeds at the expense of the pentose phosphate pathway, whereas in conditions of limiting NAD the reverse will be the case. Consequently, the

$$\frac{NADP + NADPH}{NAD + NADH}$$

ratio of a tissue gives an indication of the relative availability of these coenzymes and also the predominant respiratory pathway in operation.

On the basis of fragmentary evidence from work on cotyledons of pea and *Vigna sesquipedalis* [21, 143] the following sequence of events occurs during and after imbibition:

1. Activity of hexokinase, the enzyme which makes glucose-6-phosphate available for either the glycolytic or pentose phosphate pathway, rises as the total metabolic activity of the cotyledons rises, then falls as the reserves are depleted.

2. NAD levels rise rapidly — presumably due to synthesis from nicotinic acid — and then they decline.

3. Aldolase and phosphofructokinase (glycolysis enzymes) have high activities at the early stages after imbibition, and then they decline.

4. NADP and/or NADPH levels rise, but never to more than 10–20% of the level of NAD. In *Vigna* NADP levels rise as those of NAD decline, but the rise and fall of NADPH and NAD coincide in pea cotyledons.

5. Activities of the pentose phosphate pathway enzymes glucose-6-phosphate dehydrogenase and 6-phosphogluconate dehydrogenase increase as activities of the glycolysis enzymes decline.

Interpretation of these observations in terms of storage-tissue metabolism and development is difficult because there is no good correlation between the rise and fall of enzymes of glycolysis and the pentose phosphate pathway and the extractable levels of NAD/NADH and NADP/NADPH. On the basis of the enzyme data alone it appears that glycolysis predominates early after imbibition (e.g. 1–3 days in pea), favouring pyruvate formation. This is converted to ethanol under anaerobic conditions, but in the presence of oxygen the citric acid cycle operates and oxidative phosphorylation occurs. Later, the pentose phosphate pathway predominates. Thus in these seeds, initial anaerobic respiration might be followed by a period of energy production for increased metabolic activity associated with reserve mobilization, and then NADPH production occurs later to provide reducing power.

In storage organs of other seeds a similar pattern of events can occur. In cotyledons (and embryos) of *Cicer arietinum* (chick-pea), enzymes of the glycolysis and fermentative pathways rise over the first 24 h after imbibition starts and then decline [41]. Enzymes of other pathways were not measured. In radish cotyledons, however, enzymes of the oxidative pentose phosphate pathway are present in the dry seed to a high level and rise even further over several days after imbibition. In contrast, the activity of the glycolytic enzyme phosphofructokinase never rises above the low level present in dry seed [108].

In germinating and growing tissues the situation is not much clearer. Changes in the levels of pyridine nucleotides in imbibed wheat and rice grains have been followed, but without any correlative enzyme studies. NAD levels rise in both grains (presumably mainly in the axis) during the first 24 h after the start of imbibition, which suggests increased glycolysis. Then NADPH levels (which are high from the start of imbibition) increase after 24–48 h, probably reflecting increased pentose phosphate pathway activity in the growing seedling. Indeed the pentose phosphate pathway appears to play an increasingly important role in a number of seedling tissues during their establishment [143]. For example, in juvenile roots of *Vigna* the

$$\frac{NADP + NADPH}{NAD + NADH}$$

ratio is less than 1.0, favouring energy production via glycolysis and the citric acid cycle, but as the roots become established the ratio increases to 15–18 in the mature regions. This presumably is reflected in the metabolism of the growing tissues, with increased reducing power being available for reduction of nitrate, glutathione and sulphate, the synthesis of deoxynucleotides and fatty acids, and the amination of organic acids to amino acids.

An approximate estimation of the relative importance of the glycolytic and pentose phosphate pathways of glucose metabolism in a tissue can be achieved experimentally. The technique involves the application of glucose-6-^{14}C and glucose-1-^{14}C to separate samples of the same tissue followed by collection and estimation of evolved $^{14}CO_2$ from each: the results are expressed as a C_6/C_1 ratio. The success of the technique depends upon the fact that during glycolysis a glucose molecule is split into two 3C units and the carbons in positions 1 and 6 of the glucose molecule both end up in the same position in pyruvate. Consequently both units are then decarboxylated in an identical fashion in the citric acid cycle. Thus if all glucose-6-^{14}C and glucose-1-^{14}C applied to a tissue is respired solely via glycolysis, then the C_6/C_1 ratio will be unity. On the other hand, if the pentose phosphate pathway is operative the carbon in position 1 of glucose becomes removed by the activity of 6-phosphogluconate dehydrogenase during the conversion of 6-phosphogluconate to ribulose-5-phosphate (see page 149). Carbon 6 is not removed. Hence the operation of the pentose phosphate pathway in a tissue reduces the C_6/C_1 ratio due to the greater release of $^{14}CO_2$ from glucose-1-^{14}C.

The C_6/C_1 ratios of dormant and non-dormant barley grains following imbibition are compared in Table 5.5. A lower ratio, hence a more active pentose phosphate pathway, is apparent in the non-dormant grains prior to radicle emergence. At the time of emergence there is an increase in glycolysis, but the pentose phosphate pathway may assume ascendency once again as growth of the seedling tissues proceeds [143]. It has been suggested [104] that one cause for the failure of the dormant barley grains to germinate is the limited activity of the pentose phosphate pathway. Similarly in excised embryos of wild oat it has been found that the C_6/C_1 ratio of dormant embryos is 0.90 after 10 h from imbibition, but of non-dormant embryos it is 0.55 [115]. Thus, in these

Table 5.5. The evolution of $^{14}CO_2$ from glucose-6-^{14}C and from glucose-1-^{14}C (C-6/C-1 ratios) when applied for 3-h periods to Pallas barley grains at different times from the beginning of imbibition

Period from start of imbibition (h)	C-6/C-1 ratio	
	Dormant	Non-dormant
0.5–3.5	0.18	0.13
3.5–6.5	0.29	0.15
6.5–9.5	0.28	0.13
12.5–15.5	0.42	0.14
21.5–24.5	0.33	0.12
33.5–36.5[a]	0.35	0.28

[a] Radicle appeared during this period in non-dormant grains. After Roberts, 1969 [104]

two cereals, and possibly also in rice [93], the pentose phosphate pathway (probably the pentose oxidation cycle) appears to predominate early after imbibition, and its activity might be important for the maintenance of some as yet undefined event(s) important for germination.

Further work is now needed to determine the levels of pyridine nucleotides, respiratory enzymes of the alternate pathways, and the C_6/C_1 ratios of the embryos/axes of a variety of monocot and dicot seeds during the time prior to visible germination and during early post-germination to clarify the situation.

5.1.8. Special Oxidation Systems

A number of oxidation systems other than the major oxidative phosphorylation system of the mitochondria have been reported in seeds and seedlings. Their role in germination is obscure; they may interfere with normal respiration or they may play some role in the recycling of reducing equivalents. The most commonly studied systems are the phenol oxidase (phenolase) and ascorbic acid (AA) oxidase systems. An in vitro system can be constructed in peas to link reduced substrate through NADPH and glutathione (GSSG) to dehydro-ascorbic acid (DHA) [75] in the following manner:

$$\text{Substrate-H}_2 + \text{NADP} \xrightarrow{\underset{\text{dehydrogenase}}{\text{Substrate}}} \text{NADPH} + \text{Substrate}$$

$$\text{NADPH} + \text{GSSG} \xrightarrow{\underset{\text{reductase}}{\text{glutathione}}} \text{NADP} + \text{GSH}$$

$$2\,\text{GSH} + \text{DHA} \xrightarrow{\underset{\text{reductase}}{\text{DHA}}} \text{AA} + \text{GSSG}$$

$$\text{AA} + \tfrac{1}{2}\text{O}_2 \xrightarrow{\text{AA oxidase}} \text{DHA} + \text{H}_2\text{O}$$

No energy is produced by this system, but NADP is recycled, suggesting a tentative link with the pentose phosphate pathway. The significance of these linked pathways for respiration in vivo is unknown.

5.2. Protein and Nucleic Acid Synthesis

It is widely accepted that protein synthesis is a pre-requisite for radicle emergence. There is some contention as to whether or not this protein synthesis is dependent upon prior RNA synthesis, but there can be no doubt that DNA synthesis only occurs after germination as an integral part of growth of the axes.

As in previous sections of this chapter we will distinguish carefully between protein and nucleic acid synthesis occurring within growing tissues and that in the storage organs, although all evidence so far suggests that the biochemistry of the synthetic mechanisms at both sites is identical. Techniques for the study of protein and nucleic acid synthesis have undergone marked improvements over the past decade, and new ones are still being developed and applied to plant systems. The new techniques have sometimes confirmed earlier work but equally often they cast doubt upon it. We do not intend to present an exhaustive or critical review of protein and nucleic acid synthesis in seeds, nor shall we extensively discuss the strong and weak points of the techniques used, even though critical reviews on different aspects of these topics are overdue. The account presented here describes a sequence of events which, on the basis of published results, has the best prima facie evidence in its favour. Definitive proof is often missing though, and when equivocation seems necessary this is explained.

5.2.1. The Mechanism of Protein Synthesis

Our understanding of the mechanism of protein synthesis in plants comes mainly from studies on cell-free (in vitro) systems derived from seeds—by far the most popular material has been the isolated wheat embryo and commercially-produced wheat germ.

Protein synthesis does not occur in the dry seed of course, but commences when cells are sufficiently hydrated to allow cytoplasmic (eukaryotic 80S) ribosomes to associate with messenger RNA (mRNA). Initiation of protein synthesis in the wheat embryo involves the attachment of the small (40S) ribosomal subunit and the initiating transfer RNA (tRNA) molecule (methionyl tRNA) to the starting point—the initiation site—on the mRNA. After formation of this initiation complex the large (60S) ribosomal subunit becomes attached and protein synthesis commences [80, 128, 140]. Formation of the initiation complex requires soluble protein factors, known to be present in the cytoplasm of the dry wheat embryo [111], GTP, and perhaps ATP. Other soluble protein factors are required for transfer of aminoacylated tRNAs to their appropriate codon on the ribosome-mRNA complex, and to move the message through the ribosome: these are the transfer, or elongation factors, and they require GTP for their activity [51, 130]. Although less is known about the nature of protein synthesis in dicots, what is known is consistent with the findings obtained from work on the wheat embryo.

5.2.2. Protein Synthesis in Imbibing Embryos and Axes

Protein synthesis has been detected within minutes of the start of imbibition in seeds of some species, but only after hours in others. Since techniques used for detection of protein synthesis depend upon the incorporation of radioactive amino acids (usually ^{14}C- or ^{3}H-leucine) into growing polypeptide chains, it is obvious that availability of the amino acids at the sites of synthesis within a seed can be a limiting factor. Thus, in large intact seeds the time taken for imbibition of water and the accompanying distribution of exogenous radioactive precursors throughout all tissues might be considerably longer than for smaller seeds, excised embryos, or axes. Permeability of seed coats and other surrounding structures will also be a factor affecting uptake.

Dissected axes of lima bean (*Phaseolus lunatus*) [66], *Phaseolus vulgaris* [49], and isolated embryos of rye [112], rice [15] and wheat [79] commence protein synthesis within 30–60 min of being introduced to water. Polysomes (or polyribosomes: the active protein synthesizing complex comprised of several ribosomes attached to mRNA) are absent from dry seed tissues, and their formation on imbibition is accompanied by a decrease in the number of single (free or unattached) ribosomes. Polysome activity in wheat embryos has been assessed by monitoring the ability of an extracted ribosomal fraction (containing both single ribosomes and polysomes) to catalyse in vitro protein synthesis. For example, it has been found that the ribosomal pellet extracted from dry embryos is incapable of carrying out cell-free protein synthesis (Fig. 5.9). Uptake of water into the wheat embryo occurs rapidly, and after only a 10–15 min lag period the protein-synthesizing capacity of the extracted pellet is evident. Free ribosomes, not polysomes, are present in the dry embryo, but polysomes form within 15 min of imbibition.

That ribosomes present in dry embryos, dry whole seeds and dry dicot storage organs still retain their potential for protein synthesis is now widely

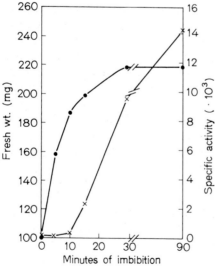

Fig. 5.9. Comparison of water uptake and development of in vitro ribosomal activity during early imbibition of wheat embryos. ●——●: fresh weight, ×——×: ribosomal activity. After Marcus et al., 1966 [79]

Table 5.6. Conditions for the in vitro incorporation of radioactive amino acids into protein by ribosomes isolated from dry peanut seeds

Condition of system	Incorporation (counts/min/mg protein)
1. Complete system	2188
2. Complete system + cycloheximide	312
3. ATP omitted	90
4. Supernatant (cytoplasmic factors) omitted	411
5. mRNA omitted	413
6. mRNA omitted from complete system from 48-h imbibed seeds	2480

The complete system contains: 0.3 mg ribosomes from dry seeds (lines 1–5) or 48-h imbibed seeds (line 6), 0.8 mg supernatant (cytoplasm) protein, 125 µg "mRNA," 0.5 µCi ^{14}C-leucine, 10–15 µmol/ml Mg^{2+}. Incubation at 37° for 30 min. Cycloheximide (15 µg), an inhibitor of protein synthesis on 80S ribosomes, is added to one tube. The effect of omitting ribosomes was not tested. Based on Jachymczyk and Cherry, 1968 [62]

known. Provided that a messenger RNA is added, as well as certain other components, ribosomes extracted from dry seeds can support in vitro protein synthesis. Necessary additives include ATP (because mitochondria are lost or destroyed during ribosome preparation) and essential cytoplasmic components (see later) within the supernatant fraction. (To illustrate these points see Table 5.6.) In the absence of added mRNA the ribosomal pellet extracted from dry peanut seeds, for example, is inactive (cf. lines 1 and 5): contrast this with the incorporation by the ribosomal fraction from 48-h imbibed seeds (line 6). These results are easily explained by the fact that dry seeds contain no mRNA associated with the ribosomes and added mRNA becomes essential for their activity. Forty eight hours after imbibition, however, the ribosomes are already associated with mRNA in a polysome complex, and further addition of exogenous message in vitro is not needed for protein synthesis. These results were obtained using whole peanut seeds where the major source of ribosomes and polysomes is the cotyledons, but similar results have been obtained with isolated axes of peanut, and embryos of wheat [78].

In vitro protein synthesis in the presence of added ribosomes and mRNA (or polysomes) and ATP/GTP can be supported by supernatant (cytoplasmic, or post-ribosomal) fractions obtained from dry peanut seeds (Table 5.6) and dry wheat embryos [76]. Consequently, components of the cytoplasm essential for protein synthesis (e.g. initiation and elongation factors, tRNA, amino acids, and aminoacyl tRNA-synthesizing enzymes—synthetases) must be present in the dry seed, presumably in sufficient quantities to permit resumption of protein synthesis in the seed upon imbibition.

5.2.3. Messenger RNA — Conserved, Synthesized, or Both?

One of the most disputed issues concerning protein synthesis in the newly-imbibed seed is whether or not prior RNA synthesis, particularly messenger

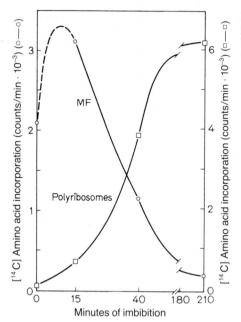

Fig. 5.10. Levels of activity of messenger fraction and extractable polysomes during the early phases of wheat embryo imbibition. After Weeks and Marcus, 1971 [139]

RNA synthesis, is necessary. A cellular fraction has been obtained from *dry* wheat embryos which can stimulate in vitro protein synthesis [139]. This fraction, having characteristics of mRNA, has been called the messenger fraction (MF). During the first few minutes of imbibition there is a transient increase in MF activity in the embryo, which then begins to decline rapidly (Fig. 5.10)—preparations extracted from three-hour imbibed embryos lack any active MF. As the activity of the extractable MF falls there is a concomitant rise in polysomes (Fig. 5.10). The implication from these observations is that conserved mRNA is present in the dry embryo in the MF (probably associated with proteins), and this quickly becomes attached to ribosomes during early imbibition, forming the active protein-synthesizing polysome complex and depleting the MF pool. Other possibilities are not precluded however, e.g. the MF might simply be degraded at the same time as polysomes are being formed using newly-synthesized mRNA. Indeed, recent studies have shown that isolated wheat embryos can synthesize mRNA very soon after they start to imbibe water [121], and the rate of synthesis is maintained at a constant level throughout the first 18 h after imbibition (radicle elongation commencing at the 6th h). Furthermore, analysis of the extracted ribosomal/polysomal pellet from wheat embryos imbibed for 0.5–1.5 h in radioactive uridine (an RNA precursor) shows newly synthesized mRNA associated with polysomes at this time [122]. Despite these observations, however, it appears that the new mRNA synthesis is not essential for resumption of protein synthesis during the first hour of imbibition [121]. For example, polysomes are still formed when over 90% of the initial RNA synthesis (including mRNA) is prevented by chemical inhibitors.

The addition of a sequence of polyadenylic acid residues (poly A) to newly synthesized mRNAs appears to be a necessary step for their activation prior to involvement in cellular protein synthesis. Cordycepin (3′-deoxyadenosine) is a drug which inhibits RNA synthesis and suppresses the polyadenylation reaction, thus preventing activation of mRNA and its association with ribosomes to form polysomes. Imbibition of wheat embryos in effective concentrations of cordycepin does not, however, prevent rapid polysome formation [121], which suggests that the mRNA conserved in the dry embryo and utilized for early protein synthesis is already activated and does not require further adenylation. Thus, to summarize the situation in imbibing wheat embryos, it appears that new mRNA is synthesized (and activated) as an early event but protein synthesis can resume without its involvement. The time at which newly-synthesized mRNA does become essential for continued protein synthesis is not known.

A similar situation appears to occur within intact cotton seeds. Here, protein synthesis during the first six hours after the start of imbibition (in the axis and cotyledons) is not inhibited by cordycepin, again suggesting that neither new RNA synthesis nor processing of conserved mRNA is a pre-requisite [56]. RNA with poly A sequences has been extracted from mature cotton seeds, located mainly in what seems to be a nuclear fraction—this raises the possibility that mRNA is stored in the nucleus in dry seeds. For continued protein synthesis in the whole cotton seed from the sixth hour after the beginning of imbibition, however, both new mRNA synthesis and activation appear to be essential. In contrast, as was pointed out in Chapter 3, the production of an active mRNA for proteinase in cotton cotyledons following germination may require the processing (polyadenylation) of a purportedly stored, preformed message—thus in this case the message is apparently stored in a non-adenylated state.

While, for the sake of convenience, we have so far used the term mRNA in the singular, there is good evidence from recent work by Gordon and Payne [52] that a number of polyadenylated messages are stored intact within dry seeds. They have extracted mRNA fractions from dry seeds of three species (rye embryos and intact seeds of pea and rape), placed them in an in vitro protein-synthesizing system and analysed the protein products. The variety of stored mRNA-stimulated proteins can be seen in Figure 5.11.

It is pertinent to speculate here on how certain mRNA molecules might become stored in the dry seed during maturation. It is generally presumed (though not proven) that many messages required for protein synthesis during seed development, particularly those in the storage organs (e.g. mRNA for storage proteins and for storage product-synthesizing enzymes) are destroyed prior to seed maturation. Yet, as we have seen, other mRNAs are conserved in mature dry seeds, and in the embryo or axis these messages may be important for the establishment of protein synthesis for germination. Why are these not destroyed also? One possible reason is that messages to be conserved are protected by being stored in the nucleus during maturation (either in their activated state, or ready for activation) and those remaining in the cytoplasm are degraded. Reactivation or release of conserved messages could then occur following drying.

At the present time we do not know why mRNAs should be stored in seeds for later utilization during imbibition, germination and growth. It is possible that in the embryo storage makes them readily available for protein synthesis

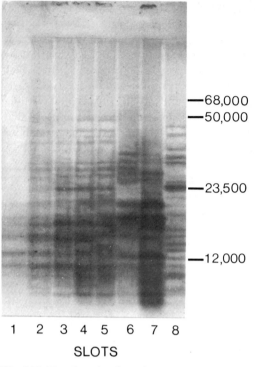

—68,000
—50,000

—23,500

—12,000

1 2 3 4 5 6 7 8

SLOTS

Fig. 5.11. Fractionation by polyacrylamide-gel electrophoresis of the products of translation of template-active RNA from dry seeds. Total RNA from embryos of rye and seeds of rape and pea was fractionated by oligo (dT)-cellulose chromatography and two fractions (Fractions II and III), were obtained consisting of poly A-rich RNA species (i.e. polyadenylated messenger RNA). After translation of these message-rich fractions in a cell-free protein-synthesizing system containing ^{35}S-methionine the protein products were fractionated by polyacrylamide gel electrophoresis and incorporated radioactivity detected by autoradiography. The bands in this figure are therefore those of proteins synthesized in vitro by extracted messages from dry seeds. Numbers on the right are the positions to which proteins with the indicated molecular weights migrated. *1:* Endogenous activity of in vitro system minus mRNA, *2:* High M.W. RNA from rye embryos (contaminated with some polyadenylated RNA and/or non-adenylated messages), *3:* Fraction II from rye embryos, *4:* Fraction III from rye embryos, *5:* Fraction I (95% of total RNA) from rye embryos reintroduced on to an oligo (dT) column to obtain a Fraction III, *6:* Fraction II from rape seeds, *7:* Fraction III from rape seeds, *8:* Fraction II from pea. There are 23 bands of proteins synthesized in the presence of polyadenylated messages from rye embryos, 15 from those of rape seeds and 20 from pea seeds. From Gordon and Payne, 1976 [52]

immediately on rehydration, though it is not apparent why this should be important when mRNA synthesis can also rapidly resume. Obviously we need to know if the mRNA transcribed during early imbibition is for the same, or different proteins than are the stored messages, and if the latter are in any way unique. Rapidity of utilization cannot explain why mRNAs for food reserve hydrolytic enzymes are stored in the cotyledons (as in cotton, see Sect. 3.5.1), for these are not required until many hours after germination has occurred.

5.2.4. Ribosomal and Transfer RNA Synthesis

Synthesis of ribosomal RNA (rRNA) in the imbibing wheat embryo is detectable as early as that of mRNA. Its rate of synthesis increases during and after imbibition, particularly after the sixth hour (the time of radicle elongation), and at 16–18 h it reaches a rate some 12-fold greater than at the earlier stages [122]. Although rRNA synthesis is linear between 0.5 and 3.5 h after imbibition commences, the incorporation of newly-synthesized RNA into ribosomes is very limited over this time period. The rate-limiting step effecting this delay in utilization of new rRNA could be the processing which the nuclear rRNA precursor must undergo to make it available for assembly into ribosomes. Indeed, there is some evidence to suggest that while rRNA synthesis begins very early in imbibition of the wheat embryo it might take up to 12 h before it is assembled into mature ribosomes [28].

In rye embryos maturation of rRNA appears to occur much faster, the pattern of synthesis being [112]:

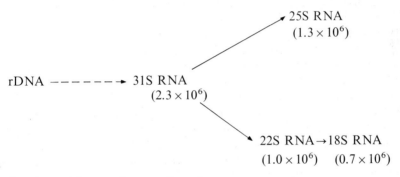

(molecular weights are in parentheses)

Twenty minutes after imbibition starts a heterogeneous nuclear RNA (hnRNA) is evident in rye embryos, which could be a de novo precursor of mRNA. Also at this time 4S RNA (probably transfer RNA) and 5S RNA (the low molecular weight RNA of ribosomes) is synthesized. After 40 min of imbibition the 31S ribosomal RNA precursor is evident and by 60 min it has begun to mature into the 25S rRNA of the large ribosomal subunit and 18S rRNA of the small subunit, the latter via an immature 22S form. By the third to sixth hour after imbibition begins, processing of the rRNA is essentially complete and large amounts of mature 25S and 18S rRNAs are present in the embryos. However, it is yet to be established when the newly-synthesized rRNA becomes associated with ribosomal subunits actively participating in cellular protein synthesis. Studies on rRNA synthesis in pea embryos show that a large RNA precursor (approx. 2.3×10^6, i.e. 31S) accumulates in the nucleolus prior to migration of the mature rRNA species into the cytoplasm [127]: the precise timing of this event is unfortunately not recorded.

It should be noted here that there are various claims in the literature that ribosomal, or even total, RNA synthesis commences only several or many hours after the beginning of imbibition of dry seeds. These reports should be treated with caution, however, for in many instances the techniques used were inadequate and in some instances precluded detection of anything less than the massive RNA synthesis which ensues only after germination has occurred. We might also note that the suggestion now seems implausible that transcriptional events are activated in a cascade manner [43], with sequential initiation of mRNA, rRNA, and tRNA synthesis occurring at intervals of many hours.

Amplification of rRNA genes occurs during certain developmental processes in animals, e.g. during amphibian oocyte development. Since seeds also undergo dramatic metabolic changes during development, maturation and subsequent germination, studies have been carried out to determine if such changes are accompanied by an increase in the number of copies of genes from which rRNA can be transcribed. Table 5.7 shows at different stages of wheat embryo development and germination the percentage of DNA capable of hybridizing with rRNA from the large ribosomal subunit (MW 1.3×10^6) and with that from the small subunit (0.7×10^6). Note that DNAs from developing, mature and germinated embryos all hybridize to a similar extent with both types of rRNA, showing that, contrary to a previous report [27], at no time is there amplification or deletion of rRNA genes.

Synthesis of RNA during early imbibition necessitates the presence of an active RNA polymerase. Both dry wheat [64] and rye [46] embryos are rich sources of DNA-dependent RNA polymerases, and these enzymes appear to be available in sufficient quantities to catalyse RNA synthesis immediately upon imbibition. In fact, RNA polymerase activity in imbibing, isolated wheat embryos even declines after the second hour from the start of imbibition [84]. This might not be a characteristic of the intact embryo, however, for in intact wheat grains RNA polymerase activity increases after germination [85]. Whether this increase is in the growing embryo (seedling) or in the aleurone layer has not been determined.

There is a sufficient supply of tRNAs and their aminoacylating enzymes present in the supernatant of dry seeds to support in vitro protein synthesis

Table 5.7. rRNA genes during development and germination of wheat embryo

Source of DNA	% DNA hybridized	
	1.3×10^6 rRNA	0.70×10^6 rRNA
Ovule, 5 days after pollination	0.101	0.066
Embryo, 10 days after pollination	0.086	0.067
Embryo, 15 days after pollination	0.096	0.069
Embryo, from dry grain	0.087	0.068
Embryo, 3 days after imbibition	0.089	0.068

DNA was prepared from complete ovules 5 days after pollination, from the embryos 10 and 15 days after pollination, from the embryo of the dry grain and 72 h after imbibition. RNA was prepared from pea roots, which had grown in the presence of 100 µCi/ml ^{32}P-ortho-phosphate for 24 h. The specific activity of the RNA used was 58,000 cpm/µg, and the values are calculated from means of triplicate samples. 1.3×10^6 rRNA is 25S RNA from the large ribosomal subunit and 0.7×10^6 rRNA is 18S RNA from the small subunit. After Ingle and Sinclair, 1972 [61]

(Sect. 5.2.2). Transfer RNA synthesis does, however, begin within 20 min in imbibing rye embryos [112] and within an hour in dissected axes of *P. vulgaris* [133]. When and if this tRNA becomes involved in early protein synthesis is not known. The total amount of low molecular weight RNA (combined tRNA and 5S rRNA in molar proportions 15:1) present in the isolated axes of *P. vulgaris* during the first 15 h after imbibition remains more or less constant. An increase then occurs between the 18th and 25th h to coincide with the initiation of cell division in the axes and expansion of the hypocotyl. A markedly different pattern of events is reported for imbibing wheat embryos [132]. Here it is claimed that tRNA levels decrease significantly during the first 15 h after imbibition and only regain the levels present in the dry embryo by the 20th h, increasing thereafter. It is difficult to reconcile the decrease in tRNA with the fact that synthesis of proteins and of other components of the protein-synthesizing system (viz. mRNA and rRNA) is increasing in the embryos at the same time. At the moment we can only presume that tRNA is present in the dry embryo in excess of its requirement for protein synthesis on imbibition, and its new synthesis is only necessary during later stages—perhaps as a pre-requisite for growth. Since tRNA synthesis does commence within another cereal embryo (that of rye) during early imbibition it would appear that additional studies are required to correlate more closely tRNA synthesis with tRNA breakdown. The ability should also be tested of stored and newly-synthesized tRNA to be aminoacylated throughout imbibition and effectively to translate mRNA.

Aminoacylation of tRNA is catalysed by aminoacyl-tRNA synthetases. In an exhaustive study of the activity of the synthetases for 20 tRNA species from the radicle, plumule and cotyledons of *P. vulgaris* it was found that specific enzyme activity does not change prior to axial growth, but activity in the plumule and radicle increases thereafter over several days (Fig. 5.12). Hence it is probable that enzymes present in the dry seed are sufficient for aminoacylation at all stages before radicle emergence and that increased enzyme activity is only necessary when seedling growth commences. The eventual decline in total activity in the cotyledons must be related to the senescence of this organ following leaf expansion. A similar decrease in synthetase activity occurs in cotyledons of two germinated pea cultivars [7].

In summary we can say this: the pattern of RNA and protein synthesis which is emerging is that embryos and axes have the capacity to synthesize all types of RNA soon after tissue hydration and long before radicle expansion. Even so, definitive studies have been conducted on only a few seeds and in any one seed on only limited aspects of these events. It appears that mRNA synthesis is not an essential prelude to the synthesis of certain proteins although synthesis and translation of mRNA as a pre-requisite for germination itself cannot be discounted. As yet, there is no unequivocal evidence to suggest that mRNA conserved in the dry embryos is sufficient to code for all the proteins synthesized de novo prior to germination. Claims that mRNA synthesis is not required because seeds will germinate in the presence of a variety of RNA synthesis inhibitors (e.g. actinomycin D) can generally be disregarded, for they mostly lack accompanying biochemical proof that mRNA synthesis is indeed inhibited. The fact that new mRNA is formed during the very early stages

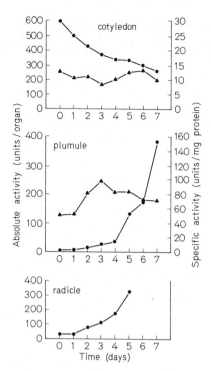

Fig. 5.12. Activity of the aminoacyl tRNA synthetases in various organs of *Phaseolus vulgaris* during imbibition. ●——●: activity per organ, ▲——▲: specific activity. After Anderson and Fowden, 1969 [7]

of imbibition is probably significant but, as pointed out earlier, it is important to know if such messages differ from the conserved ones and from ones synthesized later in imbibition. In the intact seed, control of protein synthesis during germination through lack of availability of ribosomes, tRNA, amino acids, synthetases and cytoplasmic factors is not yet suggested by existing studies and neither is regulation of protein synthesis by modulation of specific tRNAs. Nevertheless, we cannot ignore the possibility that certain of the components present in the dry seed might not be in sufficient quantity to support protein synthesis throughout the germination period, and that their formation might be required. We might also note our ignorance on a very fundamental question — what specific proteins (enzymes) must be synthesized in order for germination to take place?

5.2.5. RNA and Protein Synthesis in Storage Tissues

Whereas RNA and protein synthesis in embryonic axes precedes and accompanies cell enlargement, division and differentiation, the development of enzymatic activity in storage organs occurs in the absence of such changes. Protein and RNA metabolism in storage organs is nevertheless a complex process, for synthesis and/or activation of hydrolytic enzymes to mobilize and modify reserves is accompanied by catabolism of other (stored) protein and RNA. For an account of those events associated with reserve degradation and hydrolytic enzyme activity the reader is referred to Chapter 6; in this section we will concern ourselves mainly with proliferation of the protein-synthesizing system, i.e. the establishment of active protein synthesis in storage tissues. Since the cells of the mature cereal starchy endosperm are functionally incapable of de novo syntheses they will receive no attention: changes in the aleurone layer are considered in Chapter 7.

At least one messenger RNA, that for carboxypeptidase, is claimed to be conserved in the cotyledons of dry cotton seeds and is then activated and translated following imbibition (see Chap. 3). Unfortunately, this observation has often been cited erroneously as evidence for the conservation of an mRNA necessary for, or for the use during germination. This is not so — the putative message is for a hydrolytic enzyme in the storage organ of the cotton seed, an enzyme which is only synthesized after radicle elongation has taken place. Several seeds (including cotton), imbibed in the RNA synthesis inhibitor actinomycin D, germinate and/or produce enzymes associated with reserve mobilization. As indicated previously, however, this cannot be used as evidence for the utilization only of long-lived mRNAs for protein synthesis during germination, growth or reserve hydrolysis because it has never been shown in these studies that this drug inhibits new mRNA synthesis completely or even partially. Experiments using cordycepin are subject to similar criticisms.

A number of enzymes in storage organs of other species are activated from a pre-formed or zymogen state following imbibition, or synthesized de novo from messages of unknown "vintage" (see Chap. 6). Some studies have been published which suggest that new "mRNA-like" or "D-RNA with mRNA function" is synthesized in cotyledons during the first few hours after imbibition begins, e.g. in red bean seeds (*Phaseolus angularis*), but what the putative mRNA fractions code for is not known [136]. On the whole though, there have been surprisingly few studies conducted in recent years, using modern techniques, to follow RNA synthesis in storage organs from the time of *initial* imbibition. In contrast, there is ample evidence to show that RNA is synthesized in cotyledons many hours after imbibition, at a time when the axis is growing, e.g. in cotton cotyledons after 40 h [137] and in those of peas within two days [59].

The relationship between degradation, synthesis and turnover of RNA in storage organs is poorly understood. Total RNA levels in imbibing cotyledons may (e.g. peanut, Fig. 5.15) or may not (e.g. *Pisum sativum* [14, 53], *Vicia faba* [101]) rise before eventually declining. The ribosomal and soluble RNA content of the megagametophyte of red pine (*Pinus resinosa*) seeds does not change over 14 days after initial imbibition, even though radicle emergence occurs on the 5th day, cotyledons expand on the 10th and by the 14th the megagametophyte and seed coat are attached to the end of the expanding cotyledons and are about to be shed. In the embryos, however, both RNA types increase by the second day after the start of imbibition [107]. Likewise, in the Austrian pine (*Pinus nigricans*) embryo RNA content increases steadily over the first 96 h after initial imbibition, but no changes occur in the megagametophyte [12].

Not all regions of a storage organ necessarily exhibit the same pattern of metabolism during and after imbibition. For example, nuclear RNA appears

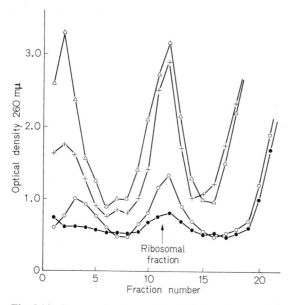

Fig. 5.13. Sucrose density gradient of the ribosomal fraction from endosperms of castor bean seeds imbibed from 0 to 72 h. ●——●: zero time, ○——○: 24 h after the start of imbibition, +——+: 48 h after the start of imbibition, △——△: 72 h after the start of imbibition. After Marrè et al., 1965 [81]

to increase in the outer storage tissues of *P. arvense* cotyledons during the first 5 days after imbibition and then decline to a low level before any changes occur in the nuclei of the inner storage tissues (Fig. 5.17 B). Similarly, the initial levels of cytoplasmic RNA appear to be high in the abaxial epidermis and hypodermis and underlying storage tissue, but low throughout the rest of the cotyledon [117]. Thus, measurements made on whole cotyledons must be regarded as an aggregate of changes throughout the organ, with potentially important localized changes being masked (see also Sect. 6.10).

During maturation and drying of the castor bean endosperm there is a marked loss of RNA, including ribosomes [126] (Chap. 3). Massive synthesis of ribosomal and soluble (transfer) RNA occurs on rehydration and ribosomes are re-formed. Some increase occurs during the first 24 h after imbibition has begun (total extractable RNA from the endosperm doubles [81]) but the major increase occurs during the subsequent 24-h period (Fig. 5.13). Polysomes also increase in numbers up to 72 h indicating increased protein synthesis at this time. Lipid hydrolysis begins at the second day from imbibition, although some enzymes do increase in activity prior to this time (Chap. 6). Unpublished but quoted [81] data suggest that RNA synthesis in castor bean might begin within the first hour of imbibition. Autoradiographic studies on the endosperm of onion (*Allium cepa*) have shown that RNA precursors become associated with the nucleolus in some cells within 15 min of imbibition, and with that of most cells by the second hour [100]. This suggests that rRNA synthesis occurs at

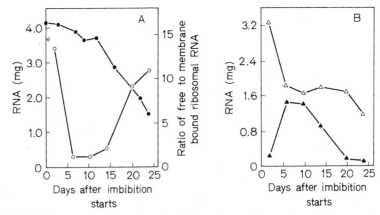

Fig. 5.14. (A) Total RNA content (●) and ratio of free to membrane-bound ribosomal RNA (○) in cotyledons of *Vicia faba*. (B) The contents of free ribosomal RNA (△) and membrane-bound ribosomal RNA (▲) in cotyledons of *V. faba*. From Payne and Boulter, 1969 [101]

these early times, although it is not known if ribosomes are conserved in the endosperm of onion and resynthesized on imbibition.

On the other hand, the cotyledons of dry seeds do retain their ribosomes, and rehydration not only elicits the formation of polysomes but also the redistribution of ribosomes within the cells. Electron microscopic studies on cotyledons of peas, soybean, *V. faba* and *P. vulgaris* have shown that ribosomes aggregate into polysomes over a period of several days after imbibition, and that these polysomes are often associated with the endoplasmic reticulum. Membrane-bound ribosomes in *V. faba* appear to be synthesized de novo, at least during the first four days after imbibition and do not originate by the attachment of pre-existing free ribosomes. This synthesis is not accompanied by any net increase in total RNA levels in the cotyledons, however, which suggests degradation of free ribosomes at the same time as synthesis of membrane-bound ones (Fig. 5.14A and B). As cotyledons senesce the ratio of free to membrane-bound ribosomes rises, due mainly to a steep decline in the latter—total RNA also decreases during senescence (Fig. 5.14A and B).

Dry cotyledons also appear to contain all the tRNA species and appropriate aminoacyl tRNA synthetases necessary for protein synthesis to commence on hydration. Several synthetases have been isolated from dry cotyledons of *Aesculus* [8], and there is substantial evidence that dry cotton cotyledons contain a full complement of tRNA species (including isoacceptors) capable of being aminoacylated [86]. Six leucyl and three tyrosyl tRNAs have been extracted from two-day-imbibed soybean seeds. Between the 2nd and 15th day, however, two of the leucyl tRNAs increase about three-fold while the levels of the others fluctuate only slightly. Of the tyrosyl tRNAs, two decline and one rises over the same time period [17]. The significance of these changes can still only be guessed but one possibility is that protein synthesis might be regulated at the translational level in cotyledons of different ages by the availability of certain aminoacyl tRNAs. In differentiated seedlings of peas and soybean differences occur in the synthetases and tRNA species between different organs [9, 65, 142]. For example, in four-day-old soybean cotyledons there are three leucyl tRNA synthetases, one of which charges two of the six leucyl tRNA species while the other two synthetases charge the remaining four leucyl tRNA species. In contrast, the hypocotyls contain only the latter two synthetases, perhaps because only four leucyl tRNA species occur in appreciable quantities in this organ. During senescence of the soybean cotyledons there are also changes in the activities of leucyl tRNA synthetases [16]. Whether cell differentiation is regulated in any way through these changes in specific tRNAs and/or synthetases is still unknown: there is no evidence available yet to suggest that a cell's capacity for protein synthesis is directly linked with these types of changes.

Cotton cotyledons persist and become green as the stored reserves are consumed. The cells do not divide over the first five days of germination and growth of the seedling, while the levels of cytosol ribosomes and tRNA remain constant. On the other hand, the levels of plastid ribosomes and tRNA increase eight-fold, these increases occurring equally well in light- and dark-grown seedlings. Thus, the RNA changes in the plastids of cotton cotyledons appear to be pre-programmed as part of seedling development and do not require environmental control by light. Similarly, five species of chloroplast tRNA synthetases increase during the first few days after germination even in dark-grown non-greening seedlings [18]. It is beyond the intended scope of this book to discuss plastid development in persistent cotyledons—suffice it to mention that in darkness the proplastids present in young cotyledons develop into etioplasts (which contain some of the enzymes associated with chloroplasts and photosynthesis), which in turn become transformed into chloroplasts by light.

Finally, a brief word is necessary about some of the early studies on RNA synthesis in peanut cotyledons. That unimbibed peanut cotyledons appear to contain all the cytoplasmic components necessary for in vitro protein synthesis in the presence of ribosomes, mRNA and an energy source has already been demonstrated (Table 5.6). It is also known that polysomes form in cotyledons of this, and other species, on imbibition. Total RNA content of peanut cotyledons increases by up to 40% over the first few days after imbibition begins, stabilizes and then declines after the 8th day to almost zero (Fig. 5.15). Synthesis of RNA was followed over this time period by taking slices of cotyledons at two-day intervals and incubating for 30 min in radioactive phosphorus (^{32}P). The results (Fig. 5.15) do tend to suggest increased capacity for RNA synthesis by the cotyledons with increasing time after first imbibition—and have often been quoted as evidence for such. This might be so, but unfortunately the work was conducted in conditions where contaminating bacteria could have contributed wholly or partially to the apparent incorporation of ^{32}P into cotyledon RNA, and so these experiments need confirming. Additional ones would be useful to determine just how soon after imbibition RNA synthesis commences. It is not acceptable to quote these experiments, as some workers have done, as evidence for increased RNA synthesis during "early imbibition" and certainly not for evidence of events during "germination." Indeed, it is worth noting that many early studies on the incorporation of radioactive precursors into

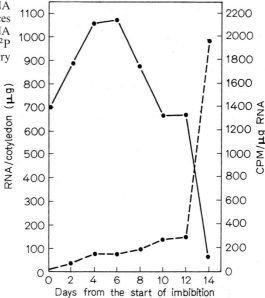

Fig. 5.15. RNA content and RNA synthesis in peanut cotyledon slices after imbibition. ●——●: total RNA content, ●---●: specific activity of ^{32}P incorporation into RNA. After Cherry et al., 1965 [31]

protein and RNA in storage organs did not take into account the possibility that a major contribution could have been made by contaminating bacteria.

5.3. DNA Synthesis, Germination, and Growth

Initial expansion of the radicle within the seed generally occurs by cell elongation, and emergence through the seed coat might or might not be accompanied by cell division (Chap. 4). Thus, the following discussion on DNA synthesis and cell division is largely concerned with the post-germination period of seedling development.

There exists some confusion in the published literature because some workers have used the terms "DNA synthesis" and "mitotic cell division" synonymously. This is incorrect for as we shall see some cells which never divide synthesize DNA; others synthesize DNA many hours before cell division commences. The latter case is easily explained if we consider the cell cycle, which is simplified in the following diagram.

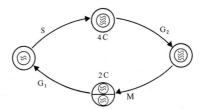

Following mitosis (M), when chromosomes are in their 2-stranded (2C) configuration there is a period of normal cell growth (Gap 1 or G_1) during which synthetic events, including those for subsequent DNA synthesis, take place. DNA synthesis (S phase) results in a doubling of the chromosomes to a 4-stranded (4C) configuration without cell division, and a second preparative growth period (G_2) occurs before mitosis. Clearly, if the G_2 phase is long, then so will be the period between DNA synthesis and cell division. If all cell nuclei of a dry embryo contained chromosomes in the 4C state then DNA synthesis would not be expected before cell division; in all seeds studied so far, however, there has been no recorded instance of all nuclei being exclusively 4C. The embryos or radicles of some species do contain both 2C and 4C nuclei and others only 2C (Table 5.8).

The proportion of nuclei in the radicle tips in the 2C and 4C stage varies from species to species, and also within species. For example, some workers [23] claim as many as 11–27% 4C nuclei in *V. faba* major, 53% 4C in *V. faba* minor and even a few per cent 8C in each. This is disputed by others [63] who claim 96% of the radicle tip cells of *V. faba* major are in the G_1 state (i.e. 2C). Such differences could be due to differences in techniques used by the workers, but in any population of seeds there could be variation among seeds depending upon how the metabolism of any seed was arrested during development and drying, or upon the conditions under which seeds matured in different seasons on the mother plant, or there might simply be cultivar differences. In those species where the nuclei of the dry embryo/axes cells are exclusively in the G_1 phase there must presumably be some rigid control of nuclear events so that all nuclei enter final development and drying stages of maturation at the same stages of their cell cycle.

Events associated with mitosis following germination have been clearly demonstrated in *A. cepa* seeds [24]. The first visible change in the nuclei is their enlargement—nuclei in the quiescent state appear uniformly circular and have a diameter of about 7 μ, but during their activation this increases to 15 μ. DNA synthesis (S phase) has been followed using the technique of autoradiography. Basically this involves imbibing the seed in a radioactive precursor of DNA (usually ^{3}H-thymidine) for a given length of time, collecting, fixing and sectioning the tissue under study, washing out unincorporated precursor, and then localizing the remaining radioactivity by laying photographic stripping

Table 5.8. Some species of seeds with embryo or radicle nuclei in the 2C, or 2C and 4C stage of the cell cycle

2C only	2C and 4C
Pinus pinea	*Triticum durum*
Lactuca sativa	*Zea mays*
Allium cepa	*Hordeum vulgare*
Tradescantia paludosa	*Pisum sativum*
	Crepis capillaris
	Vicia faba

film over the sections and leaving the two in contact long enough for the radioactivity to change the photographic emulsion so that it blackens on subsequent development. Thus, the regions of blackening locate the radioactivity and hence its site of incorporation into macromolecules. As seen in Figure 5.16A, radioactive thymidine becomes incorporated into the DNA of the nuclei of *A. cepa* after the roots have elongated to 1.4 mm, at which time the nuclei have enlarged to 15 μ diameter. Mitosis is clearly separated from DNA synthesis, because this does not commence until the roots are at least 2.8 mm long (Fig. 5.16B). From DNA synthesis to mitosis (S and G_2 phases) there is a lapse of approximately 9.2 h. Stable indices are attained at about 7 mm root length for DNA synthesis, and about 10 mm for mitosis (Fig. 5.16A and B). DNA synthesis precedes mitosis in the roots of *V. faba* also [40]. Here DNA synthesis commences in some cells 28 h after the start of imbibition and in most cells by the 40th h. Cell division barely commences after 36 h (frequency only 0.02%) and only 5% of the cells are dividing by hour 56.

The onset of DNA synthesis in different tissues of a germinated seed can be quite variable. For example, in *Zea mays* embryos thymidine incorporation into DNA occurs first in some cells of the root cap, coleorhiza and scutellar node about 30 h after imbibition begins [125]: these tissues do not undergo mitosis, however. Some root cells commence mitosis by the 40th h, with the DNA becoming labelled slightly later. This could mean that DNA synthesis is not necessary for the first mitotic division (i.e. cells are already in the G_2 stage), but it is also possible that the thymidine precursor is not penetrating the roots cells and hence cannot be incorporated into newly synthesized DNA, or else it is penetrating but being used other than as a DNA precursor. Mitosis and DNA synthesis in the mesocotyl, shoot apex and leaves occur even later.

An example of rapid synthesis of DNA following imbibition is found in radicles of barley embryos [10]. Here ³H-thymidine fed to embryos becomes associated with some nuclei in the roots (shown by autoradiography) within six hours of the start of imbibition. Division of some cells occurs after ten hours (this is after visible germination), but the earliest labelled mitotic figures cannot be observed until the 26th h. This could indicate that the earliest mitotic divisions are by 4C nuclei present in the dry embryo, which are unlabelled because DNA synthesis is already completed and that 2C nuclei carry out DNA synthesis early after imbibition starts but do not divide for up to 20 h thereafter. This interpretation assumes that during imbibition thymidine penetrates all cells equally well, for failure by a cell to take up thymidine would allow it to divide without mitotic figures becoming labelled. Such are the problems of interpretation of results from this type of experiment. In another variety of barley germinated under different conditions, other workers have observed germination after 16–19 h and DNA synthesis after 22 h [47]. Total DNA increases in the plumule and radicle of rye grains after germination as cell division proceeds, falls in the scutellum, and rises and then falls in the endosperm [60]. The changes in the latter tissue might be related to metabolism of the aleurone layer, although in mature seeds this no longer undergoes cell division.

A number of reports have been published which suggest that DNA synthesis in the cytoplasm precedes that in the nucleus. For example, in castor bean

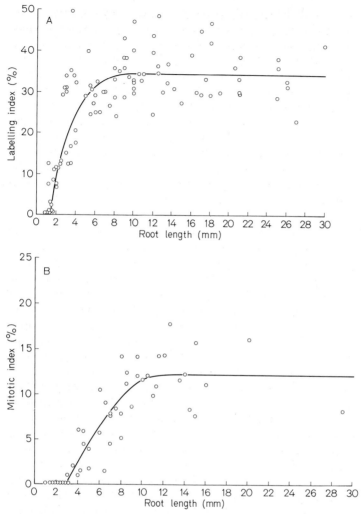

Fig. 5.16. (A) Labelling index of thymidine incorporation into DNA as a function of root length following emergence from the quiescent *Allium cepa* seed. Each point represents one root, approximately 1000 cells counted per root. (B) Mitotic index as a function of root length. Each point represents a random count of approximately 1000 cells from one root. After Bryant, 1969 [24]

the cytoplasm in radicle, cotyledons and endosperm cells becomes labelled with radioactive thymidine within 12 h after imbibition [119] whereas nuclear DNA becomes labelled in the radicle only after 24 h, in the cotyledons after 72 h and after 36 h in the endosperm. Similarly, in corn roots [131] there is exclusive incorporation of thymidine into cytoplasmic DNA 9 h after imbibition and into nuclear DNA after 36 h. It has been suggested that in wheat embryos the cytoplasm is the initial site of nuclear DNA synthesis, using RNA primers instead of nuclear DNA [25], but this has been questioned [42], and at the present time there is no convincing evidence that nuclear DNA synthesis is initiated in the cytoplasm. Some of the observed cytoplasmic DNA synthesis might reflect radioactivity being incorporated into the DNA of cytoplasmic organelles, as occurs in *Haplopappus gracilis* root tips (Table 5.9), but in a number of instances the "cytoplasmic DNA" is doubtless an artifact [92].

Thymidine must first be converted to its triphosphate (TTP) before incorporation as the deoxyribonucleotide (dTTP) into the DNA molecule. Thus two enzymes must be present for thymidine to be utilized: (1) thymidine kinase (ATP:thymidine 5' phosphotransferase), for phosphorylation of thymidine; and (2) DNA polymerase, for incorporation of deoxyribonucleotides on the DNA template. Limiting levels of either of these enzymes in seeds could obviously restrict DNA synthesis and, as a corollary, DNA synthesis could be regulated by either of these enzymes.

Thymidine kinase activity is very low in dry wheat embryos, but is markedly active by the 36th h after imbibition has started [135] — it is not known when enzyme activity starts to rise. The enzyme consists of two components (P and T), both of which are required for its activity. In corn embryos the T (but not the P) component is absent during the first 36 h after imbibition. Then it starts to be synthesized and the activity of the enzyme rises. Thus, thymidine kinase activity during the early hours after imbibition starts could be limiting DNA synthesis by restricting the utilization of the endogenous pool of thymidine precursors. In germinated peanut axes the activities of thymidine, deoxyuridine and deoxycytidine kinases increase 36 h after imbibition, which also appears to coincide with an increase in DNA synthesis [110].

The activity of a soluble DNA polymerase in wheat embryos starts to increase during the first six hours after imbibition commences and continues to do so for at least the 1st day [91]. DNA replication occurs after radicle expansion, some 15 h after the initial hydration of the embryo [90]. The polymerase appears to be synthesized de novo during germination, for if protein synthesis is inhibited during the first nine hours then subsequent DNA synthesis is inhibited. Inhibition of protein synthesis at times after nine hours is progressively less effective in preventing DNA synthesis, presumably indicating that critical levels of the polymerase are synthesized prior to the ninth hour.

Table 5.9. Cytoplasmic labelling distribution over the various cytoplasmic components of the tips of primary roots of germinated *Haplopappus gracilis*, from electron microscope autoradiographs

Cytoplasmic component	Number of grains	% of total	
Mitochondria	146	43.40	66.31
Plastids	77	22.91	
Remaining cytoplasm	113	33.69	

Results from the analysis of 23 cells. After Galli and Sparvoli, 1973 [48]

A soluble and a chromatin-bound DNA polymerase has been extracted from germinated pea axes [106]. Both increase in activity prior to DNA synthesis, although the rise in soluble polymerase shows better temporal correlation with net DNA synthesis which occurs several hours after germination. Even so, the increased polymerase activity might not be a pre-requisite for DNA replication because even the basal activity of this enzyme appears already to be far in excess of that required to sustain DNA replication.

It is worth re-emphasizing here that while levels of these two enzymes, the kinase and the polymerase, may or may not control DNA synthesis and hence subsequent mitosis, they should not, and cannot be regarded as potential control mechanisms for germination—this latter event takes place without the necessity for either DNA synthesis or cell division.

5.3.1. DNA in Storage Tissues

There is a drop in DNA in the starchy endosperm of maturing cereal grains (e.g. Fig. 3.6) and a further decline occurs following germination. In contrast, DNA levels in the cotyledons of some dicots increase dramatically during maturation due to endoreduplication, and in *V. faba* the average DNA value per cell in the dry seed reaches 16C. Following subsequent imbibition the DNA content decreases after six days, although no measurements were made before that time [99].

In peanut cotyledons DNA levels may double up to the 7th–10th day after imbibition even though no cell division seems to occur; they decrease thereafter over several days [30, 77]. Part of this increase in DNA could be due to synthesis of mitochondrial DNA (Table 5.4), although it appears that this cannot account for more than 1–2% of the new synthesis [19]. Thus, most DNA synthesis is probably nuclear in origin. The reasons for such synthesis in the absence of cell division are obscure—amplification of genes for enzymes involved in reserve degradation is an attractive, but wholly unsubstantiated suggestion.

There are claims for DNA synthesis in cotyledons of *P. sativum*, followed by degradation [14], but in *P. arvense* cotyledons DNA does not increase above the level present in the dry seed (Fig. 5.17A). During seedling development DNA levels decline in different tissues of the cotyledon to different extents, e.g. the DNA content of the epidermal tissue remains constant over 12 days after the start of imbibition whereas that of the hypodermis declines rapidly and that of the inner and outer storage tissue declines more slowly (Fig. 5.17A). Similarly there are localized changes in nuclear RNA levels (Fig. 5.17B, and see earlier). Studies on the biochemistry of metabolism in cotyledons of germinated seeds have in general ignored the possibility of localized changes in different regions of these organs. On the basis of these and other observations by Smith (e.g. Fig. 6.24) and by other workers (see Chap. 6) it is becoming increasingly obvious that more emphasis will have to be placed on changes in specific regions of large differentiated storage organs if we are to obtain a clear, coordinated picture of their metabolism.

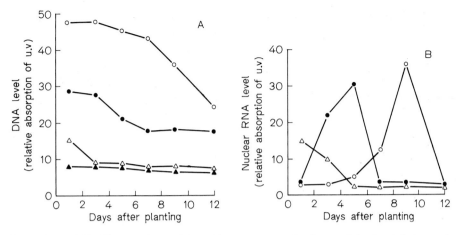

Fig. 5.17A and B. Changes in nucleic acid components in the nuclei of cotyledons of dark-grown *Pisum arvense*. (A) DNA, (B) RNA (nuclear), ▲——▲: epidermis, △——△: hypodermis, ●——●: outer storage tissue, ○——○: inner storage tissue. After Smith, 1971 [117]

In the megagametophyte of gymnosperms DNA content varies but slightly during embryo germination, e.g. *Pseudotsuga menziesii* [32], *Pinus resinosa* [107] and *P. nigricans* [12].

In this chapter we have considered those events essential for the establishment of the metabolism of germinating and growing tissue. In the following chapter we will turn our attention to the post-germination mobilization of stored reserves and the provision of catabolites for continued embryo (seedling) growth.

Some Articles of General Interest

1. Boulter, D., Ellis, R.S., Yarwood, A.: Biochemistry of protein synthesis in plants. Biol. Rev. **47**, 113–175 (1972)
2. Brown, R.: Germination. In: Plant Physiology. Steward, F.C. (ed.). New York: Academic Press, 1972, Vol VIC, pp. 3–48
3. Ching, T.M.: Metabolism of germinating seeds. In: Seed Biology. Kozlowski, T.T. (ed.). New York: Academic Press, 1972, Vol II, pp. 103–218
4. Ching, T.M.: Biochemical aspects of seed vigor. Seed Sci. Technol. **1**, 73–88 (1973)
4a. Payne, P.I.: The long-lived messenger ribonucleic acid of flowering plant seeds. Biol. Rev. **51**, 329–363 (1976)
5. Stiles, W.: Respiration in seed germination and seedling development. In: Encyclopedia of Plant Physiology. Ruh-land, W. (ed.). Berlin: Springer, 1960, Vol XV/2, pp. 465–492
6. Zalik, S., Jones, B.L.: Protein biosynthesis. Ann. Rev. Pl. Physiol. **24**, 47–68 (1973)

References

7. Anderson, J.W., Fowden, L.: Pl. Physiol. **44**, 60–68 (1969)
8. Anderson, J.W., Fowden, L.: Biochem. J. **119**, 691–697 (1970)
9. Anderson, M.B., Cherry, J.H.: Proc. Natl. Acad. Sci. **62**, 202–209 (1969)
10. Arnason, T.J., El-Sadek, L.M., Minocha, J.L.: Can. J. Genet. Cytol. **8**, 746–755 (1966)
11. Bain, J.M., Mercer, F.V.: Australian J. Biol. Sci. **19**, 69–84 (1966)
12. Balevska, L.P., Hristova, R.K.: C.R. Acad. Bulgare Sci. **26**, 109–112 (1973)

13. Ballard, L.A.T., Lipp, A.E.G.: Science **156**, 398–399 (1967)
14. Beevers, L., Guernsey, F.S.: Pl. Physiol. **41**, 1455–1458 (1966)
15. Bhat, S.P., Padayatty, J.D.: Indian J. Biochem. Biophys. **11**, 47–50 (1974)
16. Bick, M.D., Strehler, B.L.: Proc. Natl. Acad. Sci. **68**, 224–228 (1971)
17. Bick, M.D., Liebke, H., Cherry, J.H., Strehler, B.L.: Biochim. Biophys. Acta **204**, 175–182 (1970)
18. Brantner, J.H., Dure, L.S. III: Biochim. Biophys. Acta **414**, 99–114 (1975)
19. Breidenbach, R.W., Castelfranco, P., Criddle, R.S.: Pl. Physiol. **42**, 1035–1041 (1967)
20. Breidenbach, R.W., Castelfranco, P., Peterson, C.: Pl. Physiol. **41**, 803–809 (1966)
21. Brown, A.P., Wray, J.L.: Biochem. J. **108**, 437–444 (1968)
22. Brown, E.G.: Biochem. J. **95**, 504–514 (1965)
23. Brunori, A., Avanzi, S., D'Amato, F.: Mut. Res. **3**, 305–313 (1966)
24. Bryant, T.R.: Caryologia, **22**, 127–137 (1969)
25. Buchowicz, J.: Nature (London) **249**, 350 (1974)
26. Chapman, A.G., Fall, L., Atkinson, D.E.: J. Bacteriol. **108**, 1072–1086 (1971)
27. Chen, D., Osborne, D.J.: Nature (London) **225**, 336–340 (1970)
28. Chen, D., Schultz, G., Katchalski, E.: Nature New Biol. **231**, 69–72 (1971)
29. Chen, S.C.C., Varner, J.E.: Pl. Physiol. **46**, 108–112 (1970)
30. Cherry, J.H.: Biochim. Biophys. Acta **68**, 193–198 (1963)
31. Cherry, J.H., Chroboczek, H., Carpenter, W.J.G., Richmond, A.: Pl. Physiol. **40**, 582–587 (1965)
32. Ching, T.M.: Pl. Physiol. **41**, 1313–1319 (1966)
33. Ching, T.M.: Pl. Physiol. **51**, 400–402 (1973)
34. Ching, T.M., Ching, K.K.: Pl. Physiol. **50**, 536–540 (1972)
35. Ching, T.M., Danielson, R.: Proc. Assoc. Off. Seed Analysts **62**, 116–124 (1972)
36. Collins, D.M., Wilson, A.T.: Phytochemistry **11**, 1931–1935 (1972)
37. Collins, D.M., Wilson, A.T.: J. Exp. Botany **26**, 737–740 (1975)
38. Cossins, E.A.: Nature (London) **203**, 989–990 (1964)
39. Cossins, E.A., Turner, E.R.: Nature (London) **183**, 1599–1600 (1959)
40. Davidson, D.: Am. J. Botany **53**, 491–495 (1966)
41. De La Fuente Burguillo, P., Nicolas, G.: Pl. Sci. Lett. **3**, 143–148 (1974)
42. Delseny, M.: Nature (London) **250**, 792 (1974)
43. Dobrzanska, M., Tomaszewski, M., Grzelczak, Z., Rejman, E., Buchowicz, J.: Nature (London) **244**, 507–508 (1973)
44. Duperon, R.: C.R. Acad. Sci. Paris **241**, 1817–1819 (1955)
45. East, J.W., Nakayama, T.O.M., Parkman, S.B.: Crop Sci. **12**, 7–9 (1972)
46. Fabisz-Kijowska, A., Dullin, P., Walerych, W.: Biochim. Biophys. Acta **390**, 105–116 (1975)
47. Fousova, S., Veleminsky, J., Gichner, T., Pokorny, V.: Biol. Plantarum **16**, 168–173 (1974)
48. Galli, M.G.M., Sparvoli, E.: Caryologia **26**, 495–506 (1973)
49. Gillard, D.F., Walton, D.C.: Pl. Physiol. **51**, 1147–1149 (1973)
50. Givan, C.V.: Planta (Berl.) **108**, 29–38 (1972)
51. Golińska, B., Legocki, A.B.: Biochim. Biophys. Acta **324**, 156–170 (1973)
52. Gordon, M.E., Payne, P.I.: Planta (Berl.) **130**, 269–273 (1976)
53. Guardiola, J.L., Sutcliffe, J.F.: Ann. Botany (London) **35**, 809–823 (1971)
54. Haber, A.H., Brassington, N.: Nature (London) **183**, 619–620 (1959)
55. Hallam, N.D., Roberts, B.E., Osborne, D.J.: Planta (Berl.) **105**, 293–309 (1972)
56. Hammett, J.R., Katterman, F.R.: Biochemistry **14**, 4375–4379 (1975)
57. Hatano, K.-Y.: Pl. Cell Physiol. (Tokyo) **4**, 129–134 (1963)
58. Hattori, S., Shiroya, T.: Arch. Biochem. **34**, 121–134 (1951)
59. Hewish, D.R., Wheldrake, J.R., Wells, J.R.E.: Biochim. Biophys. Acta **228**, 509–516 (1971)
60. Holdgate, D.P., Goodwin, T.W.: Phytochemistry **4**, 845–850 (1965)
61. Ingle, J., Sinclair, J.: Nature (London) **235**, 30–32 (1972)
62. Jachymczyk, W.J., Cherry, J.H.: Biochim. Biophys. Acta **157**, 368–377 (1968)
63. Jakob, K.M., Bovey, F.: Exptl. Cell Res. **54**, 118–126 (1969)
64. Jendrisak, J., Becker, W.M.: Biochim. Biophys. Acta **319**, 48–54 (1973)

65. Kanabus, J., Cherry, J.H.: Proc. Natl. Acad. Sci. **68**, 873–876 (1971)
66. Klein, S., Barenholz, H., Budnik, A.: Pl. Cell Physiol. (Tokyo) **12**, 41–60 (1971)
67. Kollöffel, C.: Acta Botan. Neerl. **16**, 111–122 (1967)
68. Kollöffel, C.: Acta Botan. Neerl. **17**, 70–77 (1968)
69. Kordan, H.A.: New Phytologist **73**, 695–697 (1974)
70. Krupka, R.M., Towers, G.H.N.: Can. J. Botany **36**, 165–177 (1958)
71. Lechevallier, D.: C.R. Acad. Sci. Paris **255**, 3211–3213 (1962)
72. Lott, J.N.A., Castelfranco, P.: Can. J. Botany **48**, 2233–2240 (1970)
73. Macleod, A.M.: New Phytologist **56**, 210–220 (1957)
74. Malhotra, S.S., Solomos, T., Spencer, M.: Planta (Berl.) **114**, 169–184 (1973)
75. Mapson, L.W., Moustafa, E.M.: Biochem. J. **62**, 248–259 (1957)
76. Marcus, A.: In: Dormancy and Survival. Woolhouse, H.W. (ed.). Cambridge: Univ. Press: Symp. Soc. Exp. Biol. 1969, **23**, pp. 143–160
77. Marcus, A., Feeley, J.: Biochim. Biophys. Acta **61**, 830–831 (1962)
78. Marcus, A., Feeley, J.: Proc. Natl. Acad. Sci. **51**, 1075–1079 (1964)
79. Marcus, A., Feeley, J., Volcani, T.: Pl. Physiol. **41**, 1167–1172 (1966)
80. Marcus, A., Weeks, D.P., Leis, J.P., Keller, E.B.: Proc. Natl. Acad. Sci. **67**, 1681–1687 (1970)
81. Marrè, E., Cocucci, S., Sturani, E.: Pl. Physiol. **40**, 1162–1170 (1965)
82. Mayer, A.M.: Seed Sci. Technol. **1**, 51–72 (1973)
83. Mazé, M.P.: Ann. Inst. Pasteur **14**, 350–368 (1900)
84. Mazuś, B.: Phytochemistry **12**, 2809–2813 (1973)
85. Mazuś, B., Buchowicz, J.: Phytochemistry **11**, 2443–2446 (1972)
86. Merrick, W.C., Dure, L.S. III: J. Biol. Chem. 7988–7999 (1972)
87. Moreland, D.E., Hussey, G.G., Shriner, C.R., Farmer, F.S.: Pl. Physiol. **54**, 560–563 (1974)
88. Morinaga, T.: Am. J. Botany **13**, 126–140 (1926)
89. Morohashi, Y., Shimokoriyama, M.: J. Exp. Botany **23**, 45–53 (1972)
90. Mory, Y.Y., Chen, D., Sarid, S.: Pl. Physiol. **49**, 20–23 (1972)
91. Mory, Y.Y., Chen, D., Sarid, S.: Pl. Physiol. **55**, 437–442 (1975)
92. Moutschen, J., Lacks, S.: Exptl. Cell Res. **51**, 462–472 (1968)
93. Mukherji, S., Dey, S., Sircar, S.M.: Physiol. Plantarum **21**, 360–368 (1968)
94. Nawa, Y., Asahi, J.: Pl. Physiol. **48**, 671–674 (1971)
95. Nawa, Y., Asahi, J.: Pl. Physiol. **51**, 833–838 (1973)
96. Nawa, Y., Izawa, Y., Asahi, T.: Pl. Cell Physiol. (Tokyo) **14**, 1073–1080 (1973)
97. Obendorf, R.L., Marcus, A.: Pl. Physiol. **53**, 779–801 (1974)
98. Ohga, I.: Am. J. Botany **13**, 754–759 (1926)
99. Olsson, R., Boulter, D.: Physiol. Plantarum **21**, 428–434 (1968)
100. Payne, J.F., Bal, A.K.: Phytochemistry **11**, 3105–3110 (1972)
101. Payne, P.I., Boulter, D.: Planta (Berl.) **87**, 63–68 (1969)
102. Pazur, J.H., Shadaksharaswamy, M., Meidell, G.E.: Arch. Biochem. Biophys. **99**, 78–85 (1962)
103. Pradet, A., Narayanan, A., Vermeersch, J.: Bull. Soc. Franç. Physiol. Végét. **14**, 107–114 (1968)
104. Roberts, E.H.: In: Dormancy and Survival. Woolhouse, H.W. (ed.). Cambridge: Univ. Press: Symp. Soc. Exp. Biol. 1969, **23**, pp. 143–160
105. Roberts, E.H.: In: Seed Ecology. Heydecker, W. (ed.). London: Butterworth 1973, pp. 189–218
106. Robinson, N.E., Bryant, J.A.: Planta (Berl.) **127**, 69–75 (1975)
107. Sasaki, S., Brown, G.N.: Pl. Physiol. **44**, 1729–1733 (1969)
108. Schnarrenberger, C., Tetor, M., Herbert, M.: Pl. Physiol. **56**, 836–840 (1975)
109. Schonbaum, G.R., Bonner, W.D. Jr., Storey, B.T., Bahr, J.T.: Pl. Physiol. **47**, 124–128 (1971)
110. Schwarz, O.J., Fites, R.C.: Phytochemistry **9**, 1899–1905 (1970)
111. Seal, S.N., Bewley, J.D., Marcus, A.: J. Biol. Chem. **247**, 2592–2597 (1972)
112. Sen, S., Payne, P.I., Osborne, D.J.: Biochem. J. **148**, 381–387 (1975)
113. Shadaksharaswamy, M., Ramachandra, G.: Phytochemistry **7**, 715–719 (1968)
114. Sifton, H.B.: Can. J. Botany **37**, 719–739 (1959)

115. Simmonds, J.A., Simpson, G.M.: Can. J. Botany **49**, 1833–1840 (1971)
116. Simpson, G.M.: Can. J. Botany **44**, 1–9 (1966)
117. Smith, D.L.: Ann. Botany (London) **35**, 511–521 (1971)
118. Solomos, T., Malhotra, S.S., Prased, S., Malhotra, S.K., Spencer, M.: Can. J. Biochem. **50**, 725–737 (1972)
119. Sparvoli, E.: Caryologia **26**, 483–494 (1973)
120. Spence, D.H.N.: In: Vegetation of Scotland. Burnett, J.H. (ed.). Edinburgh: Oliver and Boyd, 1964, pp. 306–425
121. Spiegel, S., Marcus, A.: Nature (London) **256**, 228–230 (1975)
122. Spiegel, S., Obendorf, R.L., Marcus, A.: Pl. Physiol. **56**, 502–507 (1975)
123. Spragg, S.P., Yemm, E.W.: J. Exp. Botany **10**, 409–425 (1959)
124. Srivastava, L.M., Paulson, R.E.: Can. J. Botany **46**, 1447–1453 (1968)
125. Stein, O.L., Quastler, H.: Am. J. Botany **50**, 1006–1011 (1963)
126. Sturani, E.: Life Sci. **7**, 527–537 (1968)
127. Tanifuji, S., Higo, M., Shimada, T., Higo, S.: Biochim. Biophys. Acta **217**, 418–425 (1970)
128. Tarragó, A., Monasterio, O., Allende, J.E.: Biochem. Biophys. Res. Commun. **41**, 765–773 (1970)
129. Taylor, D.L.: Am. J. Botany **29**, 721–738 (1942)
130. Twardowski, T., Legocki, A.B.: Biochim. Biophys. Acta **324**, 171–183 (1973)
131. Van De Walle, C., Bernier, G.: Exptl. Cell Res. **55**, 378–384 (1969)
132. Vold, B.S., Sypherd, P.S.: Pl. Physiol. **43**, 1221–1226 (1968)
133. Walbot, V.: Planta (Berl.) **108**, 161–171 (1972)
134. Walk, R.-A., Hock, B.: Planta (Berl.) **129**, 27–32 (1976)
135. Wanka, F., Vasil, I.K., Stern, H.: Biochim. Biophys. Acta. **85**, 50–59 (1964)
136. Watanabe, A., Nitta, T., Shiroya, T.: Pl. Cell Physiol. (Tokyo) **14**, 29–38 (1973)
137. Waters, L.C., Dure, L.S. III: J. Mol. Biol. **19**, 1–27 (1966)
138. Webb, J.A., Fowden, L.: Biochem. J. **61**, 1–4 (1955)
139. Weeks, D.P., Marcus, A.: Biochim. Biophys. Acta **232**, 671–684 (1971)
140. Weeks, D.P., Verma, D.P.S., Seal, S.N., Marcus, A.: Nature (London) **236**, 167–168 (1972)
141. Wilson, S.B., Bonner, W.D. Jr.: Pl. Physiol. **48**, 340–344 (1971)
142. Wright, R.D., Kanabus, J., Cherry, J.H.: Pl. Sci. Lett. **2**, 347–355 (1974)
143. Yamamoto, Y.: Pl. Physiol. **38**, 45–54 (1963)
144. Yentur, S., Leopold, A.C.: Pl. Physiol. **57**, 274–276 (1976)
145. Yoo, B.Y.: J. Cell Biol. **45**, 158–171 (1970)

Chapter 6. Mobilization of Reserves

In this chapter we shall concentrate on those events taking place after radicle emergence, which are necessary for the maintenance of growth of the germinated seed, i.e. the supply of the breakdown products of stored reserves to the growing axis. For convenience we have divided the chapter such that each section covers the mobilization of one type of stored reserve (carbohydrate, lipid, protein or phosphorus-containing compounds) although it must be remembered that the storage organs of seeds usually contain substantial quantities of at least two major reserves and that degradation of these can occur concurrently. It is unfortunate that individual seed physiologists and biochemists have often concentrated their studies on the mobilization of one of the stored products within a seed and ignored the other(s). When different workers have studied the mobilization of the other product(s), they have often used different varieties or cultivars of the same species, germinated and grown under different conditions. This has made it difficult for us to present an integrated picture of the relationship of the mobilization of one reserve to another, and unless there is more effort made to use one variety of a particular species (preferably from one source and one known harvest), germinated and grown under defined conditions with an event marked by each group of workers for reference purposes (e.g. time of radicle elongation, initiation of synthesis of a particular enzyme, etc.), comparisons will always be difficult and the value of many individual studies will consequently be reduced.

6.1. Stored Carbohydrate Metabolism

Since starch is the most important and widespread reserve carbohydrate in seeds we will begin by surveying the metabolic reactions through which this polysaccharide is mobilized into a form available for the embryo. Details of the breakdown of other stored carbohydrates (hemicellulose, galactomannan, oligosaccharides) are discussed in the appropriate sections.

6.1.1. General Metabolism of Starch

Two catabolic pathways of starch are known. One is hydrolytic, and involves two amylases:

$$\text{Amylose} \xrightarrow{\alpha\text{-amylase}} \text{Glucose} + \alpha\text{-Maltose} + \alpha\text{-Maltotriose}$$

$$\text{Amylopectin} \xrightarrow{\alpha\text{-amylase}} \begin{array}{l} \text{Glucose} + \alpha\text{-Maltose} + \alpha\text{-Maltotriose} \\ + \alpha\text{-Limit Dextrin} \end{array}$$

$$\alpha\text{-Maltotriose} \xrightarrow{\alpha\text{-amylase}} \alpha\text{-Maltose} \xrightarrow[\text{(maltase)}]{\alpha\text{-glucosidase}} \text{Glucose}$$

$$\text{Amylose} \xrightarrow{\beta\text{-amylase}} \beta\text{-Maltose}$$

$$\text{Amylopectin} \xrightarrow{\beta\text{-amylase}} \beta\text{-Maltose} + \beta\text{-Limit Dextrin}$$

$$\text{Limit dextrins} \xrightarrow[\text{limit dextrinase}]{\text{debranching enzymes, } \alpha\text{-glucosidase}} \text{Glucose}$$

Starch degradation in dicot seeds yields more glucose and maltotriose than that in cereals, where more maltose is produced. This is related, in part, to the relative activity of β-amylase in the two classes of seeds.

The other catabolic pathway is phosphorolytic:

$$\text{Amylose} + \text{Amylopectin} + \text{Pi} \xrightarrow{\text{starch phosphorylase}} \begin{array}{l} \text{Glucose-1-P} \\ + \text{Limit Dextrin} \end{array}$$

Let us now consider some details of these breakdown processes.

Amylose and amylopectin in the native starch grain can be attacked by α-amylase, which hydrolyses the α-1,4 glycosidic links between the glucose residues at random points throughout the chains. Oligosaccharides, degraded in turn to maltose and glucose, are the products of amylose degradation, but because of the inability of α-amylase to hydrolyse the α-1,6 branch points of amylopectin, highly branched "cores" of glucose units, namely limit dextrins are produced. These limit dextrins could serve as primers for amylopectin synthesis, but during catabolism a debranching enzyme hydrolyses the branch points to yield small linear oligosaccharides, which are then subject to amylolysis. The combined actions of α-amylase and a debranching enzyme are therefore, in theory, sufficient for complete degradation of starch. Another amylase exists, however. β-Amylase cleaves away successive maltose units from amylose and amylopectin, starting at the non-reducing end of the polymer. Apparently β-amylase cannot attack native starch grains; only the large dextrins released into solution by prior α-amylolytic attack on the grain itself are degraded. β-Amylolysis of amylopectin yields β-limit dextrin. The disaccharide maltose, produced from starch by α- and β-amylase action, is further hydrolysed by α-glucosidase (maltase) to two glucose molecules.

Starch phosphorylase incorporates phosphate, rather than water, across the α-1,4 linkage between the penultimate and last glucose at the non-reducing end of the polysaccharide chain. One molecule of glucose-1-phosphate is released. This enzyme attacks amylose repetitively and amylopectin can be degraded to within two or three glucose residues of a α-1,6 branch linkage.

6.1.2. Sucrose Synthesis

Sucrose is the major form in which the products of carbohydrate (and triglyceride) catabolism are transported into the developing seedling. Glucose-1-phosphate (G-1-P) can be used directly as a substrate for sucrose synthesis, but any free glucose must first be phosphorylated via glucose-6-phosphate (G-6-P) and then isomerized. G-1-P is then combined with a uridine nucleotide (UTP) to yield the nucleotide sugar, uridine diphosphoglucose, UDPG. This transfers glucose to free fructose or to fructose-6-phosphate.

$$\text{G-1-P} + \text{UTP} \xrightarrow{\text{UDPG pyrophosphorylase}} \text{UDPG} + \text{PPi}$$

$$\text{UDPG} + \text{Fructose} \xrightleftharpoons{\text{sucrose synthetase}} \text{Sucrose} + \text{UDP}$$

or

$$\text{UDPG} + \text{Fructose-6-phosphate} \xrightleftharpoons{\text{sucrose-6-phosphate synthetase}} \text{Sucrose-6-phosphate} + \text{UDP}$$

The phosphate group is split off from sucrose-6-phosphate by sucrose phosphatase.

When required, sucrose can be hydrolysed to free glucose and fructose by a β-fructofuranosidase (sucrase and invertase are enzymes of this type).

6.2. Mobilization of Stored Carbohydrate Reserves in Cereals

Of the 300 or so families of flowering plants, none has assumed a greater importance to man than the Gramineae. Wheat and barley formed the agricultural base for the development of early civilization in the Near East; the civilization of the Far East arose with a dependence upon rice; maize in the Americas was grown in large quantities. Currently, a combined total of over 900 million metric tons of these three cereals are produced each year. In addition, substantial quantities of the more recently developed cereals, rye, oats, millet and sorghum, are harvested annually. The stored food within the endosperm of these cereal grains, deposited as a potential energy source for the growing embryo, has been elaborated over the millenia as a source of nutrition for man and his domesticated animals. So it is not surprising that man has advanced in his abilities as a cereal technologist; breeding, processing and using the harvested grain. The development of the agrarian and industrial societies has also brought its pressures: man has sought (at least temporary) relief from these by utilizing the same plants which made civilization possible for the production of fermented beverages (e.g. beer from barley, saké from rice, bourbon from maize, etc.). The malting and brewing industries have provided, in many instances, the incentive for research into how stored starch reserves are mobilized, for it is on the successful manipulation of events, subsequent to germination that these industries depend.

In 1890 the large and important industry of malting was still in its empirical stage when Horace T. Brown and G. Harris Morris published their 70-page treatise *Researches on the Germination of some of the Gramineae* [36] in which they explained many of the events which we now understand to be involved in the mobilization of the insoluble reserves of the cereal endosperm. Although some of the arguments these workers put forward in their discussion are no longer valid, it is an exciting exercise to apply the modern interpretation (slowly being constructed using sophisticated biochemical techniques) to their astute, yet crude, experimentation.

To illustrate how carbohydrate reserves are mobilized during cereal germination and growth we will concentrate principally on those observations made on barley (*Hordeum* spp.), the most widely studied grain. Where events are more clearly understood in other cereals they will be outlined. In some instances it is not easy to relate the results obtained by one worker to those of another, even when both have used the same species of grain. For example, cultivars of barley exhibit large differences in the degree and timing of response to germination stimuli. Furthermore, workers interested in the malting process have used different imbibition conditions (steeping in excess water, partially anaerobic) from those adopted by workers interested in germination (fully aerobic conditions, less water and higher temperatures).

The structure of the cereal grain was outlined earlier (Chap. 2). To recapitulate, the embryo is situated towards the base of the kernel (which is mostly endosperm) on the side opposite the longitudinal furrow. The barley embryo consists of the acrospire, nodal and rootlet region and attached scutellum (this constitutes the definition of the cereal embryo as used in this chapter). The aleurone layer, which surrounds the endosperm, encroaches around the periphery of the scutellum but does not extend between it and the endosperm. Instead, abutting the endosperm is the absorptive epithelial layer of the scutellum, although this layer is separated from the starch-containing cells per se by an intermediate layer containing depleted and compressed cells of the endosperm, their contents having been used by the embryo during its development. This intermediate layer also extends between individual epithelial cells.

We can divide the metabolic changes associated with food mobilization into two groups: (1) mobilization of the reserves of the embryo—a quantitatively minor component of total mobilization; and (2) utilization of the reserves of the endosperm.

6.2.1. The Embryo Reserves

Grains of barley cv. Proctor imbibed at 25°C commence germination within 18 h of being introduced to water. Within 24 h starch begins to accumulate in the scutellum, not at the expense of the starchy reserves of the endosperm but at the expense of fat droplets stored in the scutellum itself. (For details of fat mobilization see Sect. 6.4.) Detached embryos exhibit the same change [88]. These embryos, which are able to germinate, also metabolize their stored carbohydrate. Sucrose and raffinose (a trisaccharide: galactose-glucose-fructose) are depleted over the first 24 h after imbibition starts and are either respired or utilized for the formation of insoluble cell wall carbohydrate (see also Chap. 5). When the soluble reserves of the isolated embryo are exhausted (Table 6.1 A) the insoluble cellulose and hemicellulose of the cell wall are mobilized as an energy source but the embryo consequently eventually dies. Changes in carbohydrate status during germination and growth of dark-grown attached embryos are shown in Table 6.1 B. In the attached embryo a replenishment of carbohydrate, as sucrose, begins on the third day after the start of imbibition due to mobilization of the starch reserves of the endosperm. Maltose levels also increase and this is possibly a catabolite of the starch which is known to increase in the root at this time. Growth of the seedling is reflected by increases in the insoluble cell wall fractions. In the absence of light (for photosynthesis) the food reserves are almost exhausted by about the seventh day and growth, as indicated by synthesis of the cell wall constituents, soon ceases (Table 6.1 B).

Table 6.1. Depletion of carbohydrate reserves from isolated embryos (A) and embryos of intact grains (B) of barley (cv. Plumage Archer) during germination and growth in darkness

Hours (A) or days (B) after the start of imbibition	Sugars				Insoluble carbohydrates	
	Sucrose	Maltose	Raffinose	Maximum hexose	"Cellulose" fraction	"Hemi-cellulose" fraction
A						
2	15.26	0.58	5.86	1.04	2.03	19.63
24	0.07	1.86	1.02	1.05	4.44	23.32
48	0.26	0.66	1.04	1.61	3.52	22.70
96	0.00	0.73	0.72	2.32	2.68	19.75
144	0.13	0.50	0.50	2.72	2.66	13.92
B						
0	15.3	0.58	5.86	1.04	2.30	19.6
1	5.4	—	0.00	—	3.48	22.8
2	0.0	9.33	0.00	2.40	4.92	35.8
3	31.0	4.60	0.00	2.50	5.80	46.8
5	60.0	11.97	0.00	1.04	16.08	91.2
6	58.5	—	0.00	—	22.44	104.4
7	29.0	20.85	0.00	6.52	23.52	177.6
8	24.9	—	—	—	—	243.0
9	39.0	—	—	—	—	282.0
10	33.6	22.00	—	13.15	22.92	266.0
12	10.5	24.03	—	14.02	—	—
14	2.8	2.33	—	10.99	—	—

For the experiments in (A) the embryos were dissected from 2-h imbibed grains and maintained in the isolated state throughout the experiment. For (B) the embryos were only dissected from the intact grain at the times indicated and then analyzed immediately. The results are expressed as milligrams of equivalent hexose per 100 embryos (A) or 100 seedlings (B). After James, 1940 [77]

We will now go on to examine the important mobilization of endosperm reserves which commences about 1–3 days after the grains are set to germinate, depending upon the barley cultivar and germination/growth conditions. For a summary of these events in relation to changes in the embryo see Table 6.2.

6.2.2. The Endosperm Reserves

Utilization of the endosperm reserves is controlled by the embryo which secretes the controlling factor — gibberellin. Details of this control, and the events affected by gibberellin are discussed in Chapter 7, but a brief introduction is now appropriate. Gibberellin is synthesized by the embryo prior to the mobilization of endosperm reserves. Its precise site of synthesis has yet to be resolved, for it has been proposed on the one hand that initially it is the axis [88, 89] (particularly the nodal region) with the scutellum possibly contributing later, and on the other hand that the first site of gibberellin synthesis is the scutellum [118], with the axis contributing later. Auxin is synthesized early after imbibition in the coleoptile tip, and from there it possibly diffuses down to the scutellar node. The role of auxin here might be to promote the extensive irregular lignification of the walls of the cells in the centre of the node and the subsequent deposition of lignin in the elongated elements of the provascular strand of the scutellum [90]. These events begin as soon as 5 h after wetting the grain (Table 6.2),

Table 6.2. Changes in the barley grain during the first two days of germination and growth

This table is an attempt to place in a chronological sequence the events leading up to carbohydrate breakdown. It should be noted that this scheme only applies to work done on Proctor barley, imbibed and germinated under non-malting conditions. The times should be taken as being very approximate since the literature drawn upon to devise this scheme is often equivocal. Related events for which the published chronology is very inconsistent are marked with asterisks

Time (hour)	Event
0	Imbibition of water commences — "dry" grain moisture content: 12%. Fat droplets, but no starch, observed in scutellum. β-Amylase present
1	Moisture content of embryo: 40%
5	Lignification of scutellar node and vascular tract commences
12	Starch granules observed in scutellum. *Shrinkage of fat deposits
16	Moisture content of endosperm: 34%
17	Translocation of synthesized gibberellin to aleurone layer (although see text). Coleorhiza ruptures coat
22–24	Dissolution of intermediate layer begins; scutellar epithelial cells elongate. Endo-β-glucanase and α-amylase released from the aleurone layer
22–29	Dissolution of starch grains in endosperm observed
24	*Isocitrate lyase (see Sect. 6.4.5) present in scutellum
48	Axis accumulates starch

so the completed vascular system is ready at an early stage for translocation of metabolites from the scutellum to the embryo, and/or (if it occurs) for the transport of gibberellins synthesized at the node to the scutellum, whence they can be released. There is evidence, coming from the use of radioactive gibberellic acid, that this hormone can indeed be transported through the vascular strand of the scutellum [112].

6.2.3. Dissolution of the Endosperm and the Role of the Aleurone Layer

Events associated with the initiation of breakdown of the endosperm reserves are incompletely understood. One important event which is certain, however, is that gibberellin released from the scutellum diffuses to the aleurone layer where it stimulates the production and release into the endosperm of a number of hydrolytic enzymes. The first change that the barley endosperm appears to undergo is the dissolution of the intermediate layer adjacent to, and in between the cells of the absorptive epithelium of the scutellum. The cells of this epithelial structure then elongate, thus presenting a larger surface area for absorption of the degradation products of the reserves [4]. The intermediate layer is devoid of starch but an enzyme capable of degrading its cell wall carbohydrates is produced. This task has been assigned to endo-β-glucanase [88], a class of enzymes capable of degrading hemicelluloses (glucan components containing β-1,3 and β-1,4 links [113]) and which includes endo-β-1,3-glucanase, one of the enzymes released from the aleurone layer under the influence of gibberellin [136] (but see later).

Brown and Morris [36] reported that dissolution of the endosperm cell wall material and hydrolysis of the starch reserves commences close to the scutellum on the side away from the longitudinal crease (i.e. near the acrospire) and later spreads laterally and upwards through the grain (see Fig. 6.1). Macleod and Palmer [88] suggest that this asymmetric pattern exists because gibberellins are transported from their site of synthesis in the nodal region along the vascular tract of the scutellum to the acrospire region and then released, thus inducing the aleurone cells here to be first to produce and secrete hydrolytic enzymes. However, studies by Briggs [35] point to a more or less symmetrical release of gibberellins from the scutellum and an induction of hydrolytic enzymes equally on the dorsal and ventral (creased) side. Briggs' contention is that initial liquefaction of the endosperm takes place close to the scutellum, influenced by a complement of enzymes released directly from this organ (or perhaps the small amount of associated aleurone tissue). The degradative process then advances rapidly up the sides of the grain as the aleurone layer starts to release its hydrolytic enzymes, leaving the undegraded endosperm separated from the scutellum and aleurone layer by the liquefied catabolites, but still attached in the ventral crease region where the endosperm cells appear to be resistant to hydrolytic attack. The line of advance of endosperm degradation, observed as the advancing edge of the fluid-filled space, is claimed by Briggs almost exactly to parallel the scutellum, implying a continued, even secretion of gibberellin by this tissue. To resolve which of these proposed patterns

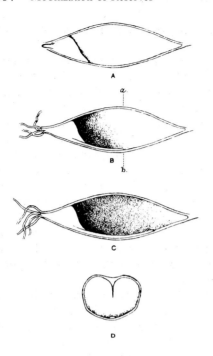

Fig. 6.1. Dissolution of the endosperm of barley as observed by Brown and Morris, 1890 [36]

Plate 2

Diagram shewing the progress of the dissolution of the cell-wall of the endosperm of barley at different periods of growth. The advancing disintegration of the contents of the grain is indicated by shading.

$$A \quad at \quad 3 \ days$$
$$B \quad „ \quad 6 \quad „$$
$$C \quad „ \quad 10 \ „$$

D Transverse section of B at a.b.

of hydrolysis occurs within the germinated grain, further studies must be carried out to determine more precisely the localization of hydrolytic enzymes, both those activated in, and those released into the endosperm. An approach using the cytochemical and immunological techniques now being successfully adopted for studies on plant proteins should be quite profitable.

In their original classic work, Brown and Morris observed that hydrolysis of starch grains never takes place if the enclosing endosperm cells walls are intact. About the same time as α-amylase is released from the aleurone layer, there is also the release of an endo-β-glucanase, the mode of action of which is purported to include the dissolution of the endosperm cell walls as well as the cell walls of the intermediate layer. The action of this enzyme, however, has not rigorously been determined in relation to the structure of the endosperm or intermediate layer cell walls themselves, though it has been reported that a crude mixture of β-glucanases (together with pentosanase) attacks endosperm wall preparations [113a].

Endosperm cell walls contain protein, β-glucan and arabinoxylan. In wheat, wall protein measures about 15%, and polysaccharide about 75%, of which 85% is arabinoxylan [93]. (Compared with primary cell walls there is much less pectin and cellulose). This simplicity in composition may reflect an adaptation to their transient structural role during endosperm development, and for their ready dissolution during reserve mobilization. Degradation of the endosperm walls occurs initially in the sub-aleurone layer and progresses towards the centre of the endosperm (see Fig. 6.2A and B), the disappearance of the cell walls accompanying an increase in the formation of water-soluble arabinoxylans. In barley, and probably in wheat, arabinoxylanases (pentosanases) are synthesized in and released from the aleurone layer. Erosion of the starch granules (Fig. 6.2C) is a late event and begins only 3–4 days after imbibition, compared with the modification of the cell walls which begins after only 1–2 days (Fig. 6.2A and B). In fact, it seems that dissolution of the cell walls renders the enclosed starch accessible to enzymic attack.

Events associated with gibberellin-induced formation and release of hydrolytic enzymes from the aleurone layer have received much attention. These will be considered separately, and in depth, in Chapter 7. We will simply reiterate here that gibberellin (GA) released from the embryo induces the aleurone layer to undergo a series of metabolic changes which result in the release of α-amylase (along with a number of other hydrolases) into the endosperm, where they degrade the reserves stored therein.

Evidence has been accumulated over a number of years to show that there is more than one α-amylase involved in starch digestion. Seven isozymes of α-amylase appear in GA-treated isolated aleurone layers and all appear to be of approximately the same molecular weight (MW 45,000) [28].

The participation of β-amylase in starch digestion has been recognized also: nine isozymes have been separated from the starchy endosperm of barley, four of MW 43,000, and the remainder as multimeric associations of these. β-Amylases are present in the ungerminated grain: in wheat kernels, 80% of this enzyme is present in an inactive, latent form bound to glutenin bodies in the starchy endosperm by disulphide linkages. The extracted enzyme can be activated through release from its bound form, either by digestion of the glutenin with proteolytic enzymes, or by disruption of the binding linkages with sulphydryl-group-containing compounds [140]. In vivo, gibberellins stimulate the activation of β-amylase, probably by increasing proteinase production and release from the aleurone layer, which in turn releases the bound enzyme [124]. Indeed, activation of β-amylase possibly coincides with the appearance of α-amylase in the barley grain because α-amylase and proteinase are simultaneously produced and released from GA-stimulated isolated aleurone layers (see Fig. 7.5A). There is no de novo synthesis of β-amylase during mobilization of the starchy reserves [60]. β-Amylases alone are unable to attack native starch grains and only digest starch previously solubilized by α-amylase.

Complete amylolysis of the amylose component of starch can be achieved by α- and β-amylases, but digestion of amylopectin halts with the production

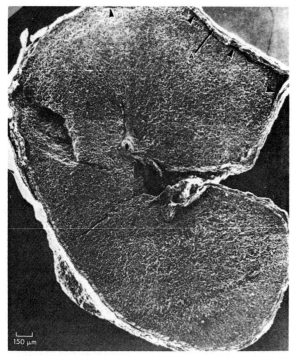

Fig. 6.2A

Fig. 6.2A–C. Scanning electron micrographs of a median cross section of a wheat grain to show modification of the endosperm after (A): 1 day; and (B): 2 days. (C) shows the partially hydrolysed starch grain in the sub-aleurone endosperm of grain modified for 4 days. In (A) a region of weakness has been *arrowed* where the endosperm is becoming modified close to the aleurone (A) layer. In (B) the aleurone layer remains adhered to the grain coat. From Fincher and Stone, 1974 [50]

of limit dextrins. Two debranching enzymes (i.e. attacking α-1,6 linkages) have been reported in cereals — R-enzyme and limit dextrinase. R-enzyme, found in barley malt, debranches amylopectin and β-limit dextrin. Limit dextrinase from Proctor barley (the activity of which is enhanced 10–15 fold between the first and fourth day after imbibition) and ungerminated oats attacks α- and β-limit dextrins, but does not debranch the large amylopectin molecule [91, 92]. It would appear that at least the limit dextrinase is synthesized de novo in the aleurone layer at the time of starch breakdown, rather than activated from a latent form in the endosperm [60]. Phosphorylases appear to play a negligible role in the hydrolysis of cereal endosperm reserves, with the possible contentious exception in the case of rice.

The major products of α- and β-amylolysis are α- and β-maltose respectively, both of which are hydrolysed by α-glucosidase, an enzyme present in the embryo and aleurone tissue of ungerminated barley [78]. The enzyme increases in activity at both sites over the first 5 days after imbibition, but it seems that the α-glucosidase in the embryo is associated with starch utilization in the axis and, unlike the aleurone layer enzyme, is not involved with mobilization of the endosperm

Fig. 6.2B

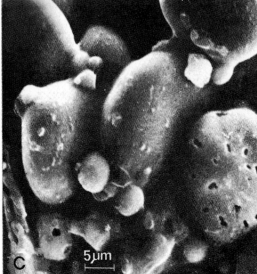

Fig. 6.2C

reserves. Increased activity of α-glucosidase in the region close to the scutellum and in the scutellar (absorptive) epithelium has been reported. Barley half-grains release a de novo-synthesized α-glucosidase from the aleurone layer within 12 h of the addition of GA [60, 116].

In some cereals, many of the carbohydrases are present even before germination is completed. α-Amylase, β-amylase, α-glucosidase, and a number of other hydrolases (proteinase, several phosphatases and ribonuclease) have been isolated in lytic bodies from dry, ungerminated grains of sorghum [14]. The lytic bodies (resembling protein bodies and spherosomes) obtained from the whole grain were found to be rich in α-glucosidase, but those from the embryo and aleurone cells were not. The implication from this observation is that it is the lytic bodies in the endosperm which contain the α-glucosidase. Further experiments suggest that spherosome-type lytic bodies in sorghum and maize (but not wheat or barley) endosperm contain α-amylase and α-glucosidase, and the protein body-type contains β-amylase and α-glucosidase [15]. Thus, in the endosperm of these two species two sets of pre-formed enzymes may be sequestered in the two types of organelle (with both types also containing proteinase, phosphatase and nuclease activity). It is interesting to note that the development of α-amylase activity in incubated, de-embryonated *Z. mays* kernels occurs independently of the presence of the embryo and does not require an exogenous source of GA [54]. In maize therefore (and perhaps also in sorghum) the initiation of the hydrolysis of reserves following germination might be due to the release of hydrolases pre-formed in the endosperm during maturation, rather than to the de novo synthesis of hydrolases, such as occurs in barley, wheat, rice and oats. Further work on the dry and germinated grains of maize and sorghum is required to locate (histochemically and immunologically) the hydrolases and to determine enzyme activities in the isolated, aleurone layer-free, endosperm.

It would obviously be of great interest to know whether two distinct patterns of reserve mobilization really do occur within cereals. It is still possible that there is fundamentally only one (GA-dependent) and that the endosperms of maize and sorghum contain sufficient endogenous GA (produced there during development) to stimulate maximum hydrolase production without a requirement for a further supply from the embryo, or without any additional effect of exogenously added GA. The effect of added GA on maize endosperm (aleurone layer) probably varies in different hybrid or inbred lines because some already have a high endogenous GA level while others, e.g. dwarf maize, are naturally deficient in GA. The latter show a three- to five-fold increase in hydrolase production in response to added GA [64] (see Chap. 7).

6.2.4. The Fate of the Products of Starch Hydrolysis

The embryonic axis and the endosperm in cereals are separated by the shield-shaped scutellum (a modified cotyledon). Its role is to absorb the products of starch digestion from the endosperm, convert them to sucrose and transport this to the growing embryo. Glucose is absorbed by the scutellum both passively and by active transport: the scutellum can absorb maltose also, but if this occurs it is probably hydrolysed soon after uptake. Isolated scutella of maize have an active α-glucosidase situated at the cell surface [71]. Some of the absorbed glucose is converted to fructose-6-phosphate but neither glucose nor fructose-6-phosphate accumulate in the scutellum; they are converted to sucrose [49]. The formation from fructose-6-phosphate of phosphorylated sucrose from which the phosphate is then released by sucrose phosphatase is the favoured reaction in the rice scutellum [107] and it has been shown that sucrose-6-phosphate synthetase is synthesized de novo within this tissue. Enzymes capable of forming sucrose from both free and phosphorylated fructose have been isolated from maize scutella [65] and the levels of extractable sucrose-6-phosphate synthetase, sucrose phosphatase and sucrose synthetase all increase over at least the first three days after the start of imbibition. It must be remembered, however, that changes in activity of extracted enzymes may not necessarily reflect the changes which occur in vivo; there may be other factors which control their activity and hence their relative importance within the tissue itself. Once formed, sucrose is not attacked since enzymes for its hydrolysis are at a low level within the scutellum. The synthesized sucrose is transported to the axis (perhaps through the phloem of the vascular system connecting this with the scutellum) and is utilized there by the developing root and shoot tissues.

A second, probably minor site of sucrose synthesis in the germinated barley grain is the aleurone layer [42]. The mechanism of synthesis is unknown, but presumably as starch hydrolysis occurs glucose is absorbed by the aleurone cells from the endosperm, converted therein to sucrose, and then released back into the endosperm (the release, but not the synthesis, being enhanced two-fold by GA). Sucrose may then be absorbed by the scutellum. The importance of this mode of sucrose formation is unknown because the relative contribution of the aleurone layer and the scutellum to the total production of sucrose for the growing embryo has yet to be fully assessed.

6.3. Mobilization of Stored Carbohydrate Reserves in Legumes

Although carbohydrate depletion in the endosperm and cotyledons of legumes has received considerable attention, the mechanism by which it is achieved is less clearly understood than that in cereals. We will first review the mobilization of reserves in the non-endospermic legumes, i.e. those in which the endosperm is absorbed during seed maturation, the cotyledons then becoming the major storage organs (e.g. pea and dwarf and broad bean). We will afterwards consider members of the tribe Trifolieae, which have endospermic seeds, having retained the endosperm into maturity e.g. fenugreek (*Trigonella foenum-graecum*), crimson clover (*Trifolium incarnatum*), lucerne (*Medicago sativa*), honey locust (*Gleditsia triacanthos*), carob (*Ceratonia siliqua*), soybean (*Glycine max*) and guar (*Cyamopsis tetragonolobus*). For details on the formation and composition of the storage reserves in legumes see Chapters 2 and 3.

6.3.1. Non-Endospermic Legumes

Isolated axes of peas are not dependent on the reserve material of the cotyledons during the initial stages of radicle elongation [17]. Reserves of carbohydrate, protein and fat in the radicle itself must be sufficient for these early events, and sucrose, raffinose and stachyose probably serve as sources of respirable substrate. But after these early events have passed, the further development of the root and shoot systems depends upon the contributions from the cotyledons. Their stored carbohydrate (and the other reserves) are hydrolysed and transported into the axis. It is not yet clear if the initiation of mobilization is actually controlled by the embryonic axis; this is discussed fully in Chapter 7. We can say, however, that the subcellular changes in the cotyledons that precede and accompany reserve mobilization do need the presence of axis, at least over the first 48 h after the start of imbibition.

Mobilization of starch (and protein, Sect. 6.7) reserves in intact legume cotyledons commences after the radicle has started to elongate. The course of depletion of the starchy reserves in the cotyledons of the smooth pea cultivar, Early Alaska, is shown in Figure 6.3A. Breakdown of the starch (and protein) reserves is biphasic, the rate increasing some 5–6 days after planting. Neither free sugars nor dextrins accumulate in the cotyledons at any time during the hydrolysis of starch, indicating a rapid movement of the products into the growing axis. The first enzyme to increase in activity in the cotyledons is a starch phosphorylase, followed several days later by an increase in α- and β-amylase activity

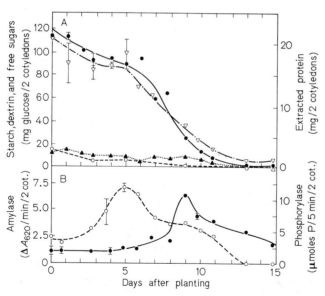

Fig. 6.3. (A) Changes in the levels of starch and dextrin (●), dextrin (○), free sugars (▲) and extracted protein (▽) in the cotyledons of the smooth pea (cv. Alaska). Time of radicle elongation was not reported in these studies. (B) Changes in starch phosphorylase (○) and amylase (●) enzymes in cv. Early Alaska. After Juliano and Varner, 1969 [79]

(Fig. 6.3 B). The initial, slow degradation of starch therefore appears to be a consequence of phosphorolysis but the later, and more rapid degradation also requires amylolysis.

The pattern of starch degradation in the cotyledons of other cultivars may show variations from the one outlined above for the Early Alaska smooth pea. For example, in the wrinkled pea cultivar Victory Freezer, the loss of starch is not biphasic: there is a linear depletion from the 2nd to the 22nd day after planting, with sugars accumulating in the cotyledons until the 7th day, and declining thereafter (Fig. 6.25). In another wrinkled pea cultivar, Fullpod, it is again biphasic with the starch levels declining only slowly over the first 10–11 days after planting but rapidly and completely over the next two days. As in Early Alaska pea, reducing sugars (glucose and maltose) do not accumulate in the cotyledons at any time during starch hydrolysis, but unlike Early Alaska, water-soluble carbohydrates (possibly large oligosaccharides or dextrins) do [13]. Phosphorylase activity in Fullpod cotyledons, like that in Early Alaska, increases before that of α-amylase but declines at the time of maximum starch hydrolysis; in contrast, β-amylase activity is negligible throughout. Whether these differences in the mode of starch hydrolysis are unique or significant features of the individual cultivars or even of wrinkled and smooth peas remains to be determined. Very few cultivars have been tested, and since different workers have used different germination and growth conditions, and even different extraction and assay conditions for the enzymes and carbohydrates, the possibility still exists that some of the observed differences are artifacts of technique rather than a feature of the seeds themselves.

What proportion of the observed increase in phosphorolytic and amylolytic activity shown in Figure 6.3 B is due to de novo synthesis of enzymes and what proportion is due to activation from a latent form remain to be investigated thoroughly. It has been suggested (but not proven) that α-amylase increases due to de novo synthesis [135]; whether β-amylase is activated or synthesized is also unknown. One enzyme which does appear to be activated from a zymogen is a debranching enzyme, amylopectin-1,6-glucosidase [96]. This enzyme is absent from the ungerminated seed (smooth cv. Alaska) but increases in activity at about the same time as α- and β-amylase. It is claimed that the debranching enzyme is first released within the cotyledon cells (as a zymogen) from its stored form in a particulate body, and then activated by a proteinase.

Spatial separation of carbohydrases within the cell must also be considered if the contentions of Matile [95] are correct. Matile suggests that the pea protein bodies (aleurone vacuoles), which store protein and phytin in the dry seed, become lysosomal when their reserves are depleted, and contain hydrolases (including ribonucleases, phosphatases, proteinases). Furthermore, he contends that α-amylase (and presumably the amylopectin-1,6-glucosidase) act as free enzymes in the cytoplasm and that the products of α-amylolysis then pass into the vacuole/lysosome for completion of the digestion process. While such a suggestion is appealing, it is still far from proven.

The immediate fate of the products of amylolysis is still incompletely understood (and has received surprisingly little attention), although they must eventually be transported to the growing axis. Presumably the rate of transport in the smooth Early Alaska pea is more rapid than in the wrinkled Victory Freezer or Fullpod cultivars, since no sugars or dextrins accumulate in the cotyledons of the former. β-Amylase has been reported in the stem and root tissue of the Alaska pea, and it has been suggested that some of the oligosaccharides released by α-amylolysis in the cotyledons might be transported to the axis for further digestion [134]. There is no substantial evidence to support this suggestion. We note, though, that little or no α-glucosidase activity (for oligosaccharide catabolism) occurs in the cotyledons of either wrinkled or smooth peas, whereas low levels of the enzyme do occur in the axis. A debranching enzyme has been detected in the axis [154].

The fate of the products of phosphorolysis is a little better understood. Starch phosphorylase activity in the cotyledons produces glucose-1-phosphate. This is apparently not hydrolysed to free glucose but instead reacts with UTP to form UDPglucose which is then converted to sucrose by sucrose phosphate synthetase and sucrose phosphatase (Sect. 6.1.2). These two enzymes have also been reported to be present in cotyledons of germinated broad beans (*Vicia faba*) [65]. Enzymes capable of hydrolysing or transforming sucrose are absent from the cotyledons of broad beans and smooth peas, but they are present in the axis, to which the sucrose is presumably transported for utilization during seedling growth.

Cotyledons of many legumes contain cell walls which are rich in hemicelluloses. Mobilization of these components has been described in *Pisum*, *Phaseolus* and *Lupinus* spp., although the composition of the hemicelluloses and the products of hydrolysis are largely unknown. An illustration of the hydrolysis of the thick cell walls of *Lupinus angustifolius* is to be found toward the end of this chapter (Fig. 6.29).

6.3.2. Endospermic Legumes

Between the seed coat and cotyledons of legumes of the Trifolieae lies a well-developed endosperm, the tissue containing the reserve carbohydrate. The polysaccharide reserve material in the endosperm is not starch but a galactomannan, a large polymer of mannose residues linked (β-1,4) to form a linear backbone to which galactose residues are attached as single unit side chains by α-1,6 linkages. The galactose content of the galactomannan varies from 10–50% according to species, and may be of taxonomic significance (Sect. 2.6.4). Oligosaccharides are also present. The outermost layer of the endosperm is the aleurone layer which, unlike the rest of the endosperm is made up of living cells devoid of galactomannan reserves.

Isolated embryos of fenugreek grow if given suitable conditions of light and moisture, support for this process coming from the reserves of oil and proteins in the cotyledons. Thus the endosperm reserves of these leguminous seeds appear to be less necessary for the survival of the developing seedling than those of the cereal grain. Indeed, from an evolutionary point of view, the galactomannan may possibly have been retained because of its high capacity to hold water; interestingly, many plants of the tribe Trifolieae appear to have spread out from the dry regions of the Eastern Mediterranean. There is little or no control by the embryo of mobilization of the reserves in the endosperm, which might be expected if growth of the embryo is indeed not dependent upon these reserves.

As in the cereals and non-endospermic legumes, raffinose is one of the first carbohydrates to be mobilized by hydrolytic cleavage of the α-galactosidic linkage to yield sucrose and galactose. Hydrolysis of this trisaccharide in the fenugreek endosperm begins soon after imbibition and is probably accompanied by transport of the resulting sucrose and galactose moeities into the cotyledons [120]. In fenugreek, lucerne and carob, stachyose is also a major reserve oligomer, and this too is hydrolysed to sucrose and galactose.

Eighteen hours after the emergence of the radicle from the fenugreek seed, the galactomannan begins to be mobilized and this process continues rapidly over the next 24 h (Fig. 6.4 B). Electron microscope studies suggest that prior to enzyme release there is polysome formation and proliferation of rough endoplasmic reticulum in the aleurone cells, events which indicate enhanced protein synthesis. As yet, however, there is no substantive supporting biochemical evidence. The aleurone grains also disappear about this time as their protein reserves are mobilized, the amino acids probably being used for the synthesis of new proteins [121]. Such changes in the aleurone layer are reminiscent of some of those which are purported to occur in barley, although in the legume there is no control of enzyme production and release by plant hormones. The hydrolases which are released by the aleurone tissue, probably following de novo synthesis, include an α-galactosidase, a β-mannosidase and also β-mannanase [122].

α-Galactosidase is an exopolysaccharidase which hydrolyses the α-1,6 link between the unit galactose side chains and a mannose residue:

$$\text{α-galactosidase} \ \text{-----} \rightarrow \quad \begin{array}{cc} \text{Gal} & \text{Gal} \\ | & | \\ \end{array}$$
$$\text{Man-Man-Man-Man-Man-}$$

β-Mannanase is an endoenzyme which hydrolyses oligomers of mannose (tetrasaccharides or larger) to mannotriose and to mannobiose, while β-mannosidase hydrolyses the triose and biose to mannose, and might also hydrolyse single mannose residues from an oligomannan, acting as an exomannopolysaccharidase. There is no evidence for the breakdown of galactomannans by phosphorolysis. Mannose and galactose are absorbed into the cotyledons but they do not accumulate there.

Information on the mobilization of galactomannans in other legumes — carob, lucerne and soybean — has come from thorough studies by McCleary and Matheson (Figs. 6.5 and 6.6) [97–99]. Large amounts of galactomannan occur in carob endosperm, less in lucerne and very little in soybean (Fig. 6.5A). The amounts of β-mannanase activity reflect these different levels, (e.g. Fig. 6.6A), although in carob and lucerne the enzyme activity seems to increase well after hydrolysis of galactomannan is underway, an observation which still remains to be explained. α-Galactosidase activity (Fig. 6.6B) in lucerne and carob increases at the same time as that of β-mannanase and in lucerne the galactosidase may

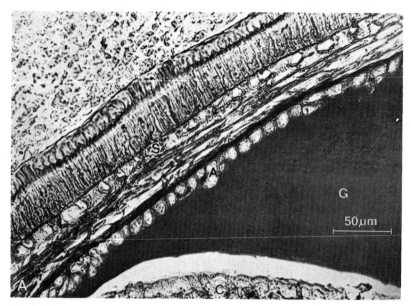

Fig. 6.4 A

Fig. 6.4A–C. Light of the outer part of the seed of *Trigonella foenum-graecum* (fenugreek).
(A) Before mobilization of the seed reserves. The three-layered seed coat (*S*), a small
part of the cotyledon (*C*) and the endosperm layer (*A* and *G*) are shown. The aleurone
layer (*A*) is the outer cell layer of the endosperm, the rest (*G*) being composed of large
cells with thin primary walls to the inside of which is deposited darkstaining galactomannan
which appears completely to fill the cell. (B) During galactomannan breakdown the reserves
in the endosperm (*G*) are being dissolved. The dissolution zone (*D*) begins at the aleurone
layer (*A*) and spreads towards the cotyledons (*C*). (C) The endosperm (*E*) is depleted
and only a remnant remains between the seed coat (*S*) and the cotyledon (*C*). The aleurone
layer has disintegrated. From Reid, 1971 [120]

remove some of the galactose side chains from the mannan backbone before
β-mannanase attack. The higher activity of α-galactosidase in lucerne is probably
related to the fact that this seed contains 45% galactose in its galactomannan,
whereas carob contains only 24%. β-Mannosidase activity is high in the endosperm of dry seeds of carob, declining rapidly from the third to eighth day
after imbibition, though a low level is maintained in the cotyledons. This enzyme
also remains at a low level in both the endosperm and cotyledons of lucerne
throughout the period when galactomannan is hydrolysed. Nevertheless it has
been calculated [98] that the level of enzyme activity in both seeds is just
sufficient to account for mannose production from manno-oligosaccharides.

Endosperms, when isolated, accumulate galactose and mannose, but do not
do so in intact seeds. This latter fact is illustrated in Figure 6.5 B, where the
differences between the amounts of galactose and mannose released on hydrolysis
of galactomannans (and the raffinose series oligosaccharides), and the accumulated levels of these monosaccharides in germinated carob seeds, can be seen
clearly.

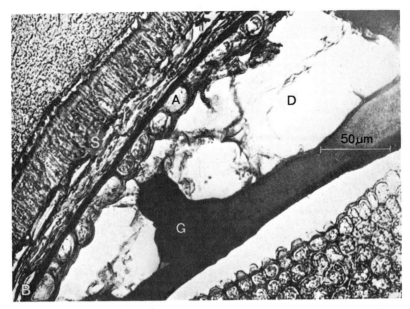

Fig. 6.4B

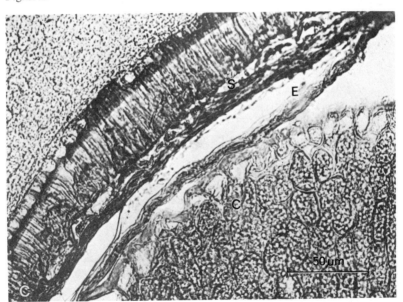

Fig. 6.4C

Most of the galactose and mannose produced in the endosperm of all these legumes is absorbed by the cotyledons and further metabolized. Several workers have proposed that these sugars are rapidly phosphorylated (to Gal-1-P and Man-6-P). If not directly used for energy metabolism they might be transformed to sucrose and even to starch (this accumulates in fenugreek cotyledons) which

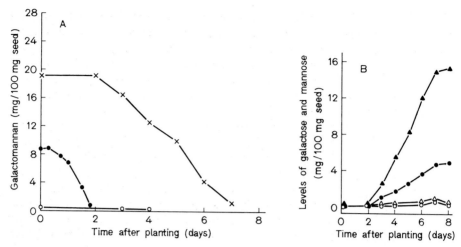

Fig. 6.5. (A) Depletion of galactomannan in carob (×), lucerne (●) and soybean (○). Redrawn from McCleary and Matheson, 1976 [99]. (B) Differences between the amounts of galactose and mannose released and accumulated during hydrolysis of galactomannan and raffinose series oligosaccharides in seeds of carob. ●——●: galactose released from stored carbohydrate, ▲——▲: mannose released from stored carbohydrate, ○——○: galactose accumulated in whole seed, △——△: mannose accumulated in whole seed. After McCleary and Matheson, 1976 [99]

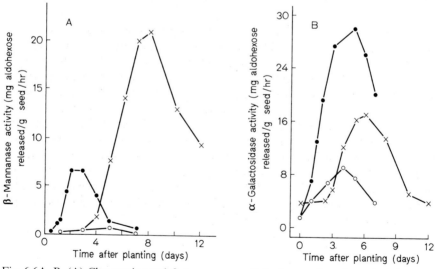

Fig. 6.6A, B. (A) Changes in total β-mannanase activity in germinated carob (×), lucerne (●) and soybean (○) seeds. After McCleary and Matheson, 1975 [98]. (B) Changes in total α-galactosidase activity in germinated carob (×), lucerne (●) and soybean (○) seeds. After McCleary and Matheson, 1974 [97]

could later be mobilized when the sucrose level in the cotyledons falls following its transport to the growing axis. Little is known about enzymes in the cotyledons capable of metabolizing galactose and mannose; but an enzyme (phosphomannoisomerase: PMI) capable of converting mannose-6-P to fructose-6-P has been isolated from the cotyledons of a number of endospermic legumes, and in honey locust cotyledons the level of PMI declines coincidentally with the depletion of galactomannan reserves [99].

Studies on the mobilization of reserves from the endosperm of legume seeds have only recently commenced, and although some interesting facts have already emerged, there is still much work to be done. Nevertheless it is pertinent to compare the pattern of mobilization in endospermic legumes, non-endospermic legumes and cereals. The endospermic legumes are unique in that the cotyledons do not synthesize reserve starch until after germination is completed. But concerning the pattern of stored reserve breakdown, there is a certain similarity between the endospermic legumes and some of the cereals. Although the endosperm reserves of these two kinds of seed are quite different, their hydrolysis is by enzymes released from the aleurone layer, and the products are absorbed and modified by the cotyledons (the scutellum being the reduced cotyledon in cereals) before being passed to the growing axis. In a sense, therefore, more contrast is shown between the patterns of breakdown in the two types of leguminous seed than between endospermic legumes and cereals.

6.3.3. Mannan-Containing Seeds Other Than Legumes

As noted in Chapter 2, a number of non-leguminous plants store mannans. Morphological and biochemical studies on date (*Phoenix dactylifera*) seeds have shown that hydrolysis of polysaccharides in the endosperm (89% mannose deposited in the secondary cell wall) is effected by exoenzymes which are probably released by the penetrating haustorium [81]. Mannose, the end product of mannan hydrolysis, is absorbed via the haustorium into the growing axis where it is converted to sucrose. Hydrolysis of mannans in the cell walls of the lettuce seed endosperm, on the other hand, involves synthesis of a mannanase within the endosperm itself [59]. Such a difference in the pattern of mannan hydrolysis between seeds might be expected, for the deposition of mannans within cell walls of the endosperm or perisperm of hard seeds like date, coffee, etc., results in destruction of the cell cytoplasm during seed development, but the lettuce seed endosperm is still comprised of living cells at maturity.

6.4. Stored Lipid Metabolism

6.4.1. General Metabolism

As indicated earlier, triglycerides are the major storage lipids in seeds (see Chap. 2). We will first consider the degradative and assimilative pathways of triglyceride metabolism in plants, and then consider modifications associated with these events in germinated seeds.

The initial degradation of triglycerides involves lipases, enzymes which catalyse the three-stage hydrolytic cleavage of the fatty acid ester bonds in triglycerides ultimately to produce glycerol and free fatty acids:

$$\text{Triglyceride} \rightarrow \text{Diglyceride} + \text{Fatty acid (FA)} \rightarrow$$
$$\rightarrow \text{Monoglyceride} + 2\,\text{FA} \rightarrow \text{Glycerol} + 3\,\text{FA}$$

Glycerol enters the glycolytic pathway following its phosphorylation and oxidation to the triose phosphates (dihydroxyacetone phosphate $\rightleftharpoons$ glyceraldehyde-3-phosphate), which can, in turn, be converted to pyruvate and then oxidized through the citric acid cycle. In tissues where triglyceride degradation predominates, however, the triose phosphate is more likely to be condensed by aldolase in the reversal of glycolysis to yield hexose units (reaction 21, Fig. 6.9).

The free fatty acids released by lipase may then be degraded by oxidation reactions to yield compounds containing fewer carbon atoms. The major oxidative pathway is that of β-oxidation, in which the fatty acid is first "activated" in a reaction requiring ATP and coenzyme A, and then, by a series of reactions involving the sequential removal of two carbon atoms, this "active fatty acid" is broken down to acetyl CoA (Fig. 6.9). Saturated fatty acids with an even number of carbon atoms yield only acetyl CoA. Chains containing an odd number of carbon atoms, if completely degraded by β-oxidation, will yield the 2-carbon acetyl moieties (as acetyl CoA) and one 3-carbon propionyl moiety (as propionyl CoA, $CH_3CH_2CO-S-CoA$). This, in turn, can be degraded in a multi-step process via malonic semialdehyde to acetyl CoA.

Alpha-oxidation of fatty acids has also been reported in plants, a process which involves sequential removal of one carbon at a time from free fatty acids of chain length ranging from C_{13} to C_{18}. It is unlikely that complete oxidation of fatty acids occurs in this manner, and the physiological significance of this pathway is obscure. It could well serve to shorten odd-chain fatty acids to even lengths and thus allow their degradation by β-oxidation.

Unsaturated fatty acids (e.g. oleic acid) are oxidized by the same general pathways, although certain extra steps need to be taken. The double bonds of naturally occurring unsaturates are in the *cis* configuration, but for step 4 (Fig. 6.9) of β-oxidation to be effected they must be in the *trans* position. A *cis-trans* isomerase ($\Delta^{3,4}$ *cis*-$\Delta^{2,3}$-*trans*-enoyl CoA isomerase) converts the fatty acid to its oxidizable form.

$\Delta^{3,4}$ *cis* enoyl CoA $\Delta^{2,3}$ *trans* enoyl CoA

Polyunsaturated fatty acids (e.g. linoleic and linolenic acids) containing a *cis, cis*-1,4-pentadiene system (i.e. with two double bonds spaced by two single bonds in a 5-carbon segment of a fatty acid chain) must be converted to a *cis, trans*-1,3-butadiene hydroperoxide system (i.e. with two double bonds spaced by one single bond in a 4-carbon segment) in order for them to undergo β-oxidation. An enzyme catalyzing this reaction has been extracted from seeds of legumes, cereal grains and high-oil-containing seeds. Lipoxygenase is the current name for this enzyme (formerly named lipoxidase) and its action is:

$$\text{RCH} \overset{cis}{=} \text{CH—CH}_2\text{—CH} \overset{cis}{=} \text{CH—R} + \text{O}_2 \rightarrow \text{RCH} \overset{cis}{=} \text{CH—CH} \overset{trans}{=} \text{CH—} \overset{\overset{\displaystyle \text{OOH}}{|}}{\text{CH}}\text{—R}$$

cis, cis-1,4-pentadiene   *cis, trans*-1,3-butadiene

Thus, all fatty acids are probably converted to a form from which they can ultimately be degraded to acetyl CoA by β-oxidation. This acetyl moiety may be completely oxidized in the citric acid cycle to CO_2 and H_2O, or utilized via the glyoxylate cycle for carbohydrate synthesis. During seedling growth the latter process is the most important.

The conversion of fat reserves to sugars was reported in castor beans in 1955 by Yamada [143], the pathway for this conversion was elucidated in peanut and sunflower by Bradbeer and Stumpf (1959) [32] and in castor bean by Canvin and Beevers (1961) [37]. Directly coupled to the β-oxidation system is the glyoxylate cycle, which takes in the acetyl CoA and, in a series of enzymatic reactions, links this to the glycolytic pathway, which then operates in reverse to produce hexose (Fig. 6.9). The key enzymes responsible for forging this link are malate synthetase and isocitrate lyase (isocitratase). Acetyl CoA is first converted to citrate (in the same manner as precedes its entry into the citric acid cycle) (Step 7, Fig. 6.9) then to isocitrate, but the steps in the citric acid cycle between isocitrate and succinate are avoided by the action of isocitrate lyase, which cleaves isocitrate directly to succinate and glyoxylate. In this way the two decarboxylating reactions of the citric acid cycle are bypassed. Another acetyl CoA is incorporated into the cycle (Step 10) when this is condensed with glyoxylate by malate synthetase to yield malate. At each turn of the cycle one molecule of succinate is released (Step 9) which is converted to oxaloacetate by citric acid cycle enzymes in the mitochondria (Steps 13 to 15) and thence into the glycolysis chain as phosphoenolpyruvate by phosphoenolpyruvate carboxykinase (Step 16).

6.4.2. Fat Mobilization in Seeds

Breakdown of the fat (oil) reserves involves three discrete bodies found within fat-storing cells. These are the fat-storing oil body, the mitochondrion and the glyoxysome. Briefly, these three organelles function as follows: (1) lipolysis to give fatty acids and glycerol occurs in the oil bodies; (2) oxidation of fatty acids and synthesis of succinate via the glyoxylate cycle takes place in the glyoxysomes; and (3) conversion of the succinate to oxaloacetate is performed by the mitochondria. The oxaloacetate is processed further in the cytoplasm ultimately to yield sucrose. We will now consider these stages in more detail.

Mobilization of the fat reserves from the oil bodies results in their gradual disappearance. In high-fat seeds, e.g. peanut (*Arachis hypogaea*) and castor bean (*Ricinus communis*) there is little change in the general appearance of these storage vesicles during digestion; they gradually decrease in size and disappear as the reserves are depleted [142]. At one time it was thought that the oil body became an autolytic vacuole and acted in a manner similar to a lysosome. Alternatively it was postulated that in pea (*Pisum sativum*) and bean (*Phaseolus vulgaris*) cotyledons the smooth membrane of the empty storage vesicles might be a precursor for smooth endoplasmic reticulum [103]. Nowadays it is generally held that in the above-mentioned seeds the oil body is either completely degraded, or else the residual pieces of broken membrane have no metabolic function.

Since fats are stored in oil bodies, it is reasonable to expect that the enzyme responsible for their degradation should be closely associated with these structures. During the first 11 days after imbibition a peanut cotyledon may decrease in dry weight from 345 mg to 143 mg, with a concomitant decrease in fat content of 55%. This represents hydrolysis of 9.4 μmoles of triglyceride per cotyledon per day. Even so, less than 1% of the total lipase activity of a peanut cotyledon has been found associated with the oil body and 99% is associated as an acid lipase (pH 4.6) with a particulate fraction [76]. This fraction has been claimed to be mitochondrial, but that is unlikely. The precise location of the acid lipase is still undetermined but it could be associated with the glyoxysomes.

In the castor bean, mobilization of the stored lipid begins on about the third day after imbibition, and digestion is completed within seven days (Fig. 6.7). By the end of this time the endosperm is liquefied and the contents are completely absorbed by the expanding cotyledons, the root-shoot axis having by now grown to about 15 cm in length. Within two days of the start of imbibition an acid lipase (optimum pH 5.0) is activated in the endosperm. This reaches a peak of activity and declines sharply to day 4, by which time an alkaline lipase (optimum pH 9.0) has appeared and reached maximal activity (Fig. 6.8A). Fractionation of the endosperm cell into fat (oil bodies), particulate (glyoxysomes, mitochondria), and supernatant (free of these organelles) fractions shows that the acid lipase remains associated with the fat (Fig. 6.8B) and the alkaline lipase with the particulate fraction (Fig. 6.8C).

The acid lipase associated with the oil bodies of the fat layer is capable of hydrolyzing tri-, di-, and monoglycerides and is most active (in extracts) over the first two days after imbibition, although strangely enough at this time the storage triglycerides are not being utilized (Fig. 6.7). The alkaline lipase is specific only for monoglycerides; it is found in the membrane of the glyoxysome, and its activity is highest when the triglyceride reserves are being mobilized. This paradox has not been resolved. We must assume that hydrolysis by acid lipase, the enzyme capable of degrading the storage triglycerides, is kept in check up to the third day after imbibition. Perhaps the acid lipase is bound to the outside of the oil body and the barrier separating it

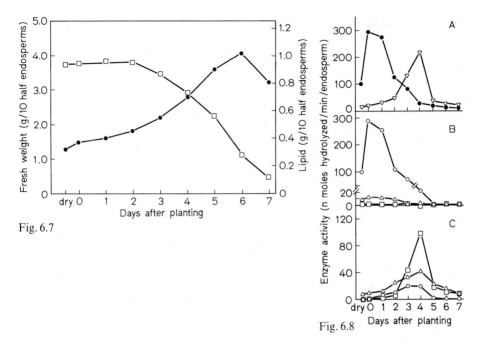

Fig. 6.7

Fig. 6.8

Fig. 6.7. Changes in the lipid (□) content and fresh weight (●) of the castor bean endosperm. After Muto and Beevers, 1974 [105]

Fig. 6.8A–C. Changes in lipase activity in the castor bean endosperm. (A) Total activity (fat and particulate and supernatant) at pH 5.0 (●) and pH 9.0 (▽). (B) Activity in the fat (○), particulate (□) and supernatant (△) fractions at pH 5.0. (C) As B), but at pH 9.0. After Muto and Beevers, 1974 [105]

from the lipid is the oil body membrane; a delay in the production of an enzyme to degrade the membrane could be responsible for keeping the lipase and its substrate spatially separated. Even so, the major triglyceride mobilization must then take place when the levels of acid lipase are declining. Another unresolved question is whether the acid lipase completely degrades the triglycerides, as it is capable of doing, or whether monoglycerides are produced for subsequent alkaline lipase action. Electron micrograph studies have shown a close juxtaposition of the oil body and the glyoxysome (Fig. 6.11) and so it is not unreasonable to assume that an alkaline lipase associated with the glyoxysome membrane could play some role in the final degradation of the triglyceride reserves and the entry of the fatty acid products into this organelle.

Lipases have been reported to be present in dry seeds of some species, e.g. Scots pine (*Pinus sylvestris*), Douglas fir (*Pseudotsuga menziesii*), castor bean, but at a low level, or absent in others e.g. apple (*Pyrus malus*). In most cases there appears to be a rise in lipase activity following imbibition, but whether this is due to de novo synthesis of the enzyme or activation of existing lipases has not been determined. A decline in lipase activity with declining triglyceride reserves has been widely reported.

6.4.3. The Fate of Glycerol and Fatty Acids

Seed lipases are generally non-specific and can hydrolyse a wide variety of triglycerides. The products of lipolysis often do not accumulate within the seeds but are rapidly metabolized. In hazel (*Corylus avellana*) seeds for example, even the most sensitive assays fail to detect any free glycerol at a time when lipase activity is high [131], and ^{14}C-glycerol fed to cotyledon slices from non-dormant seeds is utilized completely within 200 h. Some of this ^{14}C-glycerol is respired (perhaps after conversion to sucrose) and some incorporated into lipids in the cotyledons and embryonic axes. Most of the radioactive label in the lipids is still present in the glycerol itself. In castor bean, 25% of radio-labelled glycerol fed to endosperm slices is oxidized to CO_2 and much of the rest converted to sucrose [22], but there is no incorporation into lipids. The different patterns of utilization of glycerol in these seeds might be due to the fact that the castor bean endosperm is completely degraded within a few days during seedling growth, but the hazel cotyledons exist for a much longer time and turnover of the reserves occurs. Recent studies on the castor bean endosperm [68] have shown that the glycerol released by the lipases is phosphorylated by glycerol kinase in the cytosol (i.e. not within any organelle) to give α-glycerol phosphate; this in turn is oxidized by cytochrome-linked α-glycerol phosphate oxidoreductase to dihydroxyacetone phosphate in the mitochondria, which is released into the cytosol for conversion to hexose (Fig. 6.9).

In castor bean, as in many other fat-storing seeds, free fatty acids do not accumulate but are rapidly degraded and converted to carbohydrate within the endosperm. In other seeds, however, a different pattern of mobilization can be observed, e.g. in the germinated seeds of the West African oil palm (*Elaeis guineensis*). Here, a specialized structure called the haustorium invades the endosperm, and through its vascular system, which is connected to that of the shoot and root, the products of lipid catabolism are transported to the developing seedling. A build-up of free fatty acids in the endosperm occurs during its degradation and those which are not respired are absorbed directly by the haustorium without any prior conversion to carbohydrate [29]. Free fatty acids have also been found to accumulate in the haustorium and they might be reconverted therein to triglyceride for temporary storage until required by the growing axis [110].

6.4.4. The Glyoxysome

Pathways for the conversion of the liberated fatty acid to sugars via β-oxidation and the glyoxylate cycle are illustrated in Figure 6.9. At one time the enzymes

---

Fig. 6.9. Some pathways of triglyceride catabolism and hexose assimilation.   ▶
    Enzymes: *(1)*: Lipases; *(2)*: Fatty acid thiokinase; *(3)*: Acyl CoA dehydrogenase; *(4)*: Enoyl hydratase; *(5)*: β-Hydroxyacyl CoA dehydrogenase; *(6)*: β-Ketoacyl thiolase; *(7)*: Citrate synthetase; *(8)*: Aconitase; *(9)*: Isocitrate lyase; *(10)*: Malate synthetase; *(11)*: Malate dehydrogenase; *(12)*: Catalase; *(13)*: Succinate dehydrogenase; *(14)*: Fu-

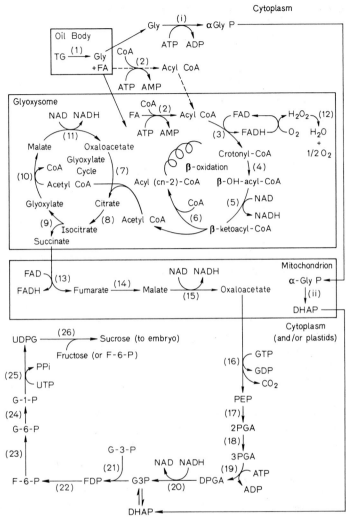

marase; *(15)*: Malate dehydrogenase; *(16)*: Phosphoenolpyruvate carboxykinase; *(17)*: Enolase; *(18)*: Phosphoglyceromutase; *(19)*: Phosphoglycerate kinase; *(20)*: Glyceraldehyde-3-phosphate dehydrogenase; *(21)*: Aldolase; *(22)*: Fructose-1,6-diphosphatase; *(23)*: Phosphohexoisomerase; *(24)*: Phosphoglucomutase; *(25)*: UDPG pyrophosphorylase; *(26)*: Sucrose synthetase (or Sucrose-6-P synthetase and Sucrose phosphatase)

(i): Glycerol kinase; *(ii)*: α-glycerol phosphate oxidoreductase.

Substrates: *TG*: triglyceride; *Gly*: glycerol; *FA*: fatty acids; *PEP*: phosphoenolpyruvate; *2PGA*: 2-phosphoglyceric acid; *3PGA*: 3-phosphoglyceric acid; *DPGA*: 1,3-diphosphoglyceric acid; *G-3-P*: glyceraldehyde-3-phosphate; *FDP*: fructose-1,6-diphosphate; *F-6-P*: fructose-6-phosphate; *G-6-P*: glucose-6-phosphate; *G-1-P*: glucose-1-phosphate; *UDPG*: uridine diphosphate glucose; α-*Gly P*: α-glycerol phosphate; *DHAP*: dihydroxyacetone phosphate

Coenzymes and energy suppliers: *FAD/(H)*: flavin adenine dinucleotide/(reduced); *NAD/(H)*: nicotinamide adenine dinucleotide/(reduced); *GTP*: guanosine triphosphate; *ATP*: adenosine triphosphate; *UTP*: uridine triphosphate; *GDP*: guanosine diphosphate; *ADP*: adenosine diphosphate; *AMP*: adenosine monophosphate; *CoA*: coenzyme A. Based on Ching, 1972 [5]

Fig. 6.9 (supplement). Proposed pathway for the catabolism of ricinoleic acid in castor bean. After Hutton and Stumpf, 1971 [73]

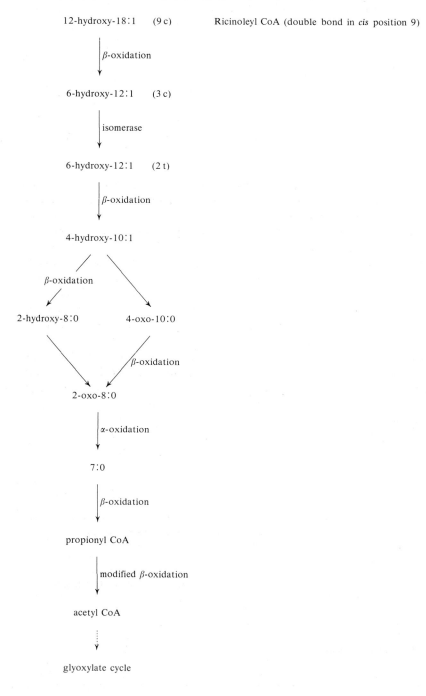

12-hydroxy-18:1   (9 c)   Ricinoleyl CoA (double bond in *cis* position 9)

β-oxidation

6-hydroxy-12:1   (3 c)

isomerase

6-hydroxy-12:1   (2 t)

β-oxidation

4-hydroxy-10:1

β-oxidation

2-hydroxy-8:0   4-oxo-10:0

β-oxidation

2-oxo-8:0

α-oxidation

7:0

β-oxidation

propionyl CoA

modified β-oxidation

acetyl CoA

glyoxylate cycle

Fig. 6.10. A typical separation on a sucrose density gradient
of the components of a crude particulate fraction from five-
day-old castor bean endosperm. After Breidenbach and
Beevers, 1967 [33]

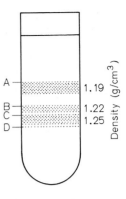

involved in these pathways were thought to be situated in either a particulate
fraction (presumed to be mitochondrial) or free in the cytoplasm as soluble
enzymes. In 1967 Breidenbach and Beevers [33] precisely determined the location
of two key glyoxylate cycle enzymes, isocitrate lyase and malate synthetase.
Following gentle disruption of the endosperm cells of 5-day-old castor beans,
these workers extracted a crude particulate fraction which they then placed
on a sucrose density gradient. On centrifugation the particulate fraction separated into three major bands of different densities (A, B and C), and one minor
band (D) (Fig. 6.10). Each of these bands was collected and assayed for enzyme
activity. Band B was identified as proplastids and band D was an unimportant,
unidentified band. The distribution of enzyme activities found in bands A and
C is shown in Table 6.3. The two enzymes exclusive to the glyoxylate cycle
(isocitrate lyase and malate synthetase) were located only in band C and enzymes
of the citric acid cycle (fumarase and succinate dehydrogenase) were associated
only with band A. The site of the glyoxylate cycle was thus shown to be in
a particle distinct from the mitochondrion, and this was named the glyoxysome.
The previous confusion over the site of the glyoxysome enzymes either resulted
from incomplete separation of this organelle from mitochondria (i.e. the enzymes

Table 6.3. Enzyme activities in subcellular fractions A and C separated from castor bean
endosperms

	Enzyme activity (micromoles substrate consumed/min/mg protein)	
	Band C (Density 1.25 g cm^{-3})	Band A (Density 1.19 g cm^{-3})
Isocitrate lyase	1.16	0
Malate synthetase	1.48	0
Citrate synthetase	1.16	0.72
Malate dehydrogenase (NAD)	22	50
Fumarase	0	1.4
Succinic dehydrogenase	0	0.55

Based on Breidenbach and Beevers, 1967 [33]

were correctly identified as being particulate, but were attributed to the wrong particle), or alternatively from disruption of the glyoxysome by harsh grinding techniques, thus liberating the enzymes (here the enzymes were incorrectly stated to be soluble, i.e. cytoplasmic).

Citrate synthetase and malate dehydrogenase are components of both the glyoxylate and citric acid cycles and it is not surprising to find these located in both the glyoxysomal and mitochondrial bands. Even so, evidence is accumulating that these enzymes extracted from the glyoxysomes have properties quite distinct from those in the mitochondria, although there are still some conflicting reports [18, 70].

At the time that the glyoxysome particle was discovered, it was assumed that β-oxidation of fatty acids was completed in the mitochondrion and the resultant acetyl CoA transported to the glyoxysome for conversion to succinate. But it was difficult to explain how this acetyl CoA avoided complete oxidation in the mitochondrion or even how it could be transferred from one organelle to another. The problem was resolved when it was shown that β-oxidation also occurs in the glyoxysome and that the acetyl CoA is produced and consumed in the same organelle [44, 72].

The first step in the conversion of succinate from the glyoxylate cycle to hexose is catalysed by succinic dehydrogenase (Step 13, Fig. 6.9), an enzyme which is not found in the glyoxysome, but is present as a membrane-bound component of the mitochondrion. This finding has led to an acceptance of the concept that mitochondria and glyoxysomes cooperate closely in the conversion of succinate to malate and that the succinate is transported from the glyoxysome to the mitochondrion for further conversion to oxaloacetate through a segment of the citric acid cycle. This oxaloacetate is then released from the mitochondrion into the cytoplasm for conversion to phosphoenolpyruvate (PEP) and then to hexose by reversal of the glycolytic pathway. Several workers have suggested that the reaction sequence from PEP to sucrose might also occur within the plastids, from which small amounts of enzymes of glycolysis have been recovered. It has also been pointed out that gluconeogenesis, the reversal of glycolysis to yield glucose (and then sucrose) takes place at the same time as glycolysis is occurring [138]: a spatial separation of these two events within the cell (gluconeogenesis in the plastid, and glycolysis in the cytoplasm) would conveniently explain how one pathway could operate in two directions at the same time in the same cell. The importance of the plastid in gluconeogenesis has not been satisfactorily determined as yet, however, and further studies need to be done.

In the reverse flow of glycolysis, fructose-1,6-diphosphatase (Step 22, Fig. 6.9) catalyses the hydrolysis of fructose-1,6-diphosphate to fructose-6-phosphate, thus by-passing the irreversible step of phosphorylation of fructose-6-phosphate by phosphofructokinase. The activity of fructose-1,6-diphosphatase increases 25-fold in the fatty endosperm of castor bean following imbibition, its developmental pattern following that of catalase, a marker enzyme for glyoxysome activity in this tissue [148] (see Fig. 6.9). There is little or no gluconeogenesis from lipid during the first two days after imbibition of marrow seeds but between days 2 and 5 this becomes extensive, and there is a large increase in fructose-1,6-diphosphatase activity (Table 6.4). The major products of gluconeogenesis are sucrose and stachyose which are presumably transported from the cotyledons for use in the growing regions of the seedling, although conversion of lipid catabolites to amino acids, nucleotides, protein and further lipids might occur to sustain the metabolism of the persistent cotyledons [138].

Activity of phosphofructokinase and fructose 1,6-diphosphatase in mammals can be finely controlled within the cell to regulate glycolysis and gluconeogenesis, and there is some evidence to suggest a similar control mechanism in seedling tissues [53a].

The complete process of degradation of saturated triglycerides and conversion to carbohydrate is summarized in Figure 6.9; enzymes for each step in this process have been located in their appropriate position in the cell. Whether free fatty acids or fatty acyl CoA moieties enter the glyoxysome is not completely resolved, and although acylation of CoA using free fatty acid is known to occur within glyoxysomes, the additional participation of the endoplasmic reticulum in this reaction has not been ruled out [43]. The fate of NADH produced in the glyoxysome has been questioned because there are no enzymes present in this organelle of sufficient activity to account for the re-oxidation of NADH, an event which is essential for the sustained operation of the glyoxylate cycle and β-oxidation. It is known that isolated glyoxysomes (but not mitochondria) can oxidize palmitoyl CoA and that NADH accumulates. When palmitoyl CoA is fed to a crude particulate fraction containing glyoxysomes and mitochondria, no NADH accumulates. This may be because NADH is being transferred from the glyoxysomes to the mitochondria and being re-oxidized by the electron transport chain [87]. Since NADH itself cannot pass through the mitochondrial membrane some shuttle system for the transfer of reducing equivalents must

Table 6.4. Lipid degradation and gluconeogenesis in the cotyledons of dark-grown marrow (*Cucurbita pepo*)

Days after imbibition	Stage of development	Fresh weight		Lipid (g/100 seedlings)	Sugar (mg/100 seedlings)	Fructose-1,6-diphosphatase activity (nmole substrate/cotyledon/min)
		(mg/cotyledon)	(mg/seedling)			
0	Dry seed	56	103	4.82	81	14.7
2	Radicle 1–5 mm long	70	137	4.66	45	40.5
4	Cotyledons within testa but above ground	–	254	4.34	227	–
5	Cotyledons free of testa	99	–	–	–	321
6	Cotyledons open	–	881	2.02	895	–
8	Cotyledons expanded, hypocotyl elongated, plant etiolated	121	1054	2.20	894	332

Based on Thomas and Ap Rees, 1972 [138]

exist. This, again, emphasizes the close working relationship between these two organelles. It also shows that fatty acid oxidation, by generating oxidizable NADH, can yield ATP.

One thing the scheme in Figure 6.9 does not take into account is the fact that storage triglycerides are usually unsaturated and, in the case of ricinoleic acid in castor bean, also hydroxylated. Complete oxidation of unsaturated and polyunsaturated fatty acids (as discussed earlier, Sect. 6.4.1) requires their modification by an isomerase or lipoxygenase: presumably these enzymes are present in all glyoxysomes.

β-Oxidation of ricinoleic acid (12-OH 18:1) presents two problems [73]:

1. Conversion of 12-OH 18:1 to 6-OH 12:1 by β-oxidation is possible, but this leaves the double bond in the *cis* position. A *cis-trans* isomerase (Sect. 6.4.1) must be present to convert the 6-OH 12:1 (3 *cis*) to 6-OH 12:1 (2 *trans*).

2. Further β-oxidation of 6-OH 12:1 (2 *trans*) would yield 2-OH 8:0, which cannot then be oxidized. To circumvent this the hydrogen is removed from the hydroxyl to yield an oxygenated (or oxo) derivative, which is converted to the 7:0 fatty acid by α-oxidation.

The proposed pathway for the complete degradation of ricinoleic acid to acetyl CoA in the glyoxysome of the castor bean is summarized in Figure 6.9 (supplement).

6.4.5. The Synthesis and Degradation of Glyoxysomes

The size and composition of these organelles varies little with species or tissue. They have equilibrium densities of 1.25 g/cm^3, are surrounded by a single unit membrane and may or may not have crystalline inclusions. Isolated glyoxysomes from different fatty seedlings have strikingly similar specific activities of individual enzymes, with the exception of alkaline lipase which might vary ten-fold in glyoxysomes from different sources [69]. When fat degradation is occurring most actively, glyoxysomes are often found closely associated with oil bodies (Fig. 6.11).

We will consider the synthesis and fate of the glyoxysome in two types of tissue, viz. that which is degraded as the triglyceride reserves are depleted (e.g. castor bean endosperm, Ponderosa pine megagametophyte, and maize scutellum) and that which persists (e.g. cotyledons of peanut, watermelon, cucumber, cotton and sunflower). Since information is not yet available on all stages of synthesis and disappearance of glyoxysomes in all tissues, we will highlight certain of these events in a few of these tissues. Whether such events are common to a larger number of fat-storing tissues still has to be resolved.

The production of malate synthetase and isocitrate lyase in peanut cotyledons and maize scutella is compared in Figure 6.12. In the isolated cotyledons of peanuts incubated in darkness, the enzyme levels associated with the glyoxysome continue to rise over the first eight days after imbibition. A similar increase occurs in cotyledons from intact dark-grown peanut seedlings [85]. In the scutellum, on the other hand, there is a sharp rise and fall of glyoxysomal activity, the fall being associated with a decline in triglyceride reserves. Catalase activity

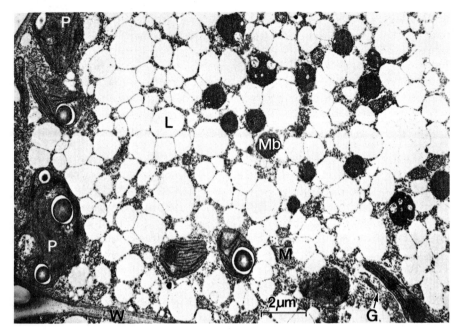

Fig. 6.11. Electron micrograph of a cucumber (*Cucumis sativus*) cotyledonary cell three days after germination in the light. Glyoxysomes (*Mb*) with their single limiting membrane are interspersed amongst the numerous oil bodies (*L*). Mitochondria (*M*) plastids (*P*) and dictyosomes (*G*) are also evident. From Trelease et al., 1971 [139]

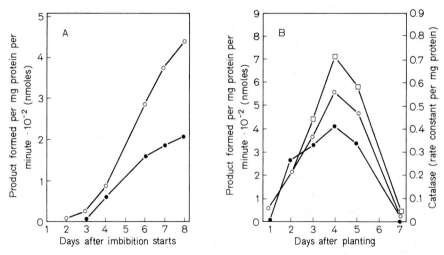

Fig. 6.12A and B. Variation in glyoxysomal isocitrate lyase (●) and malate synthetase (○) in: (A) isolated peanut cotyledons incubated in darkness; and (B) of these enzymes plus catalase (□) in the glyoxysomal fraction from the scutellum of germinated maize (*Zea mays*). After Longo and Longo, 1970 [85]

also rises and falls in the scutellum (it was not measured in the peanut cotyledon). This enzyme replenishes FAD for the acyl CoA dehydrogenase step of β-oxidation (Steps 3 and 12, Fig. 6.9) and it is now widely used as a marker enzyme for glyoxysomes. Ching [40] reported that a few small glyoxysomes can be detected in the megagametophyte of the dry seed of Ponderosa pine which enlarge two- to three-fold and increase exponentially in number during early seedling growth. Mitochondrial activity rises and falls in parallel to that of the glyoxysome, both in the pine megagametophyte and in the endosperm of the castor bean [53].

In castor bean the formation of isocitrate lyase does not require the presence of the embryonic axis, but the presence of the embryo is possibly required for at least the first two days after imbibition for this enzyme to develop in the pine megagametophyte (Chap. 7). Isocitrate lyase and malate synthetase are probably synthesized de novo during glyoxysome formation but whether their production is directed by conserved or de novo-synthesized messenger RNA is not clearly established. Little or nothing is known about the synthesis of other enzymes of the glyoxylate cycle or the β-oxidation pathway. Enzymes have been reported to be associated with the membrane of the castor bean glyoxysome but, with the exception of the alkaline lipase, it is probable that all of them are free within the organelle matrix, as in the maize scutellum [86]. The appearance and disappearance of the glyoxysomal enzymes in castor bean endosperm is accompanied by the synthesis and degradation of the glyoxysomal membrane [80]. It now seems likely that the major structural phospholipids of the glyoxysomal membrane are synthesized exclusively on the endoplasmic reticulum (ER) [30]. Furthermore, there is a marked similarity in the polypeptide composition of the ER and the glyoxysomal membranes, regardless of the developmental age of the tissue [31]. These observations are compatible with the suggestion that there is a common cellular site of origin of these membranes, and that glyoxysomes arise by vesiculation or invagination of the proliferating ER. The site of synthesis of most of the enzymes of the glyoxysome is still unknown. It is unlikely that the glyoxysomes themselves are self-sufficient for both RNA and protein synthesis because these organelles do not contain any characteristic or unique DNA species [46], or (in castor bean) any ribosomes. Enzyme synthesis, along with synthesis of the glyoxysomal membranes, might occur within the cytoplasm, on the ER, and the newly synthesized enzymes might be transported within the cisternae of the ER to a proliferation site where they are sequestered in newly forming glyoxysomes. Recent work has shown that both malate and citrate synthetase are found in relatively high concentrations in association with the endoplasmic reticulum at the time of glyoxysome formation [55].

The fate of glyoxysomes in greening tissue is still a subject for controversy. Extensive studies on cucumber (*Cucumis sativus*) by Newcomb's group [139] have shown that during greening of light-grown cotyledons there is a gradual loss of glyoxysomes, and the production of a similar microbody called the peroxisome. The latter contains a number of enzymes involved in photorespiration, including catalase, malate dehydrogenase and glutamate-glyoxylate aminotransferase: these enzymes are also present in the glyoxysome. However, peroxi-

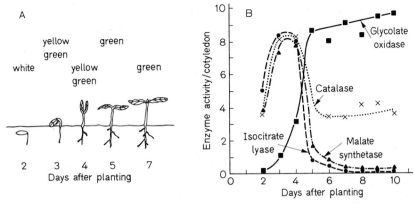

Fig. 6.13. (A) Diagrammatic representation of the appearance of cucumber seedlings grown under a 12–12 h light-dark cycle. (B) Changes in glyoxysomal and peroxisomal enzyme activities in homogenates of cucumber cotyledons grown as in (A). Enzyme activity: glycolate oxidase (■), nmol substrate consumed/min/cotyledon; isocitrate lyase (●), 0.1 × nmol substrate consumed/min/cotyledon; malate synthetase (▲), 0.04 × nmol substrate consumed/min/cotyledon; and catalase (×), 0.2 × units/cotyledon. After Trelease et al., 1971 [139]

somes do not contain either isocitrate lyase or malate synthetase and one of their unique marker enzymes is glycolate oxidase.

A correlation between the development of the glyoxysome and peroxisome in cucumber cotyledons is illustrated in Figure 6.13. Glycolate oxidase appears as the cotyledons start to turn green and there is a concomitant decrease in glyoxysomal enzymes. In seedlings grown in the dark for ten days, isocitrate lyase and malate synthetase maintain higher activities than those kept in light-dark cycles for only half that time. No glycolate oxidase forms in dark-grown cotyledons, indicating the persistence of the glyoxysome and failure of peroxisome formation. These observations show that cotyledons maintain their capacity to hydrolyse stored triglycerides until they become sufficiently chlorophyllous to support seedling growth through photosynthesis.

Three separate models could account for the reported organelle transitions: (1) the glyoxysomes and peroxisomes appear and disappear as totally independent bodies; (2) each type of microbody is present all the time, but its activation or deactivation occurs at different stages of development (perhaps due to enzyme formation or activation); or (3) there is a succession of glyoxysomal and peroxisomal enzymes in the same microbody. Discussion of these alternatives is beyond the intended scope of this book, but a glance through the plant physiology literature published over the past five years will soon reveal the intense interest there has been in elucidating the fate of the glyoxysome and the origin of the peroxisome.

6.4.6. Assimilation of the Breakdown Products

The initial products of triglyceride degradation, fatty acids and glycerol, may be utilized for further fat and membrane production, particularly in persistent cotyledons. A major proportion of the breakdown products, however, are converted to hexose and finally to sucrose by a sequence of reactions we have already discussed. Castor bean endosperms and maize scutella contain high levels of sucrose-6-P synthetase, sucrose phosphatase and also sucrose synthetase [65] (Sect. 6.1.2). In the endosperm of castor bean the predominant sugar present is sucrose which is taken up actively by the developing cotyledons. Little sucrose is cleaved to fructose and glucose before the transfer, but if this does occur the disaccharide is rapidly re-formed in the cotyledon and transported to the axis as such. Over 80% of the absorbed sucrose is transferred to the embryonic axis and the rate of this transfer exceeds 2 mg/h in five- to six-day-old seedlings [82, 83]. Removal of the embryonic axis drastically reduces uptake of sucrose by the cotyledons: thus removal of the sink alters replenishment of the source.

Acetyl CoA arising from β-oxidation of the fatty acids may also be utilized for amino acid synthesis via partial reactions of the glyoxylate and citric acid cycles [127]. Glyoxylate fed to cotyledons of pumpkin, linseed, watermelon or sunflower is incorporated into glycine and serine, and fed acetate is metabolised into those amino acids and amides which are readily formed from intermediates of the glyoxylate and citric acid cycles (e.g. glutamic acid, glutamine and γ-amino butyric acid — derived from α-ketoglutarate). In these tissues therefore, considerable depletion of the glyoxylate and citric acid cycle intermediates might occur for amino acid production, rather than for sucrose production. In castor bean endosperm, however, there is little drain of the intermediates and sucrose is the major product of acetate addition. This reflects the situation in the whole seed, where some 70% of the carbon from triglyceride is eventually recovered as sucrose.

For a discussion of the possible control processes involved in fat mobilization and of the role played by the axis turn to Chapter 7.

6.5. Stored Protein Metabolism

6.5.1. General Metabolism

Storage proteins are hydrolysed into their constituent amino acids by proteinases (proteases), enzymes which have been classified into four major groups based on their active site catalytic mechanisms. These are:

1. Serine proteinases: these have serine at the active site, e.g. chymotrypsin and trypsin.

2. Cysteine (sulphydryl or thiol) proteinases: these require a free SH group in their active centre, e.g. papain, ficin and bromelin.

3. Metalloproteinases: require metal ions as cofactors and are inhibited by metal chelators.

4. Acid proteinases: are active at an acidic pH, e.g. pepsin and rennin of animals.

Beyond this, these enzymes can be classified more specifically with regard to their substrates:

1. Endopeptidases: these cleave the internal bonds of polypeptides to yield smaller polypeptides.

2. Aminopeptidases: these sequentially cleave the terminal amino acid from the free amino end of the polypeptide chain.

3. Carboxypeptidases: as (2) but single amino acids are sequentially cleaved from the carboxyl end of the chain. Like aminopeptidases, carboxypeptidases are exopeptidases.

For an enzyme to be placed into one of these categories it must first be fully characterized. Since only a few plant proteinases have been characterized (see Ref. 2 for details) the general term "proteinase" will be used here for an enzyme with undefined proteolytic activity.

There is another class of hydrolysing enzymes to be considered, which hydrolyse various small peptides but not proteins: these are the peptide hydrolases.

Thus the hydrolysis of proteins to their component amino acids can be viewed as a consequence of two pathways, which may or may not be interacting:

$$\text{Polypeptides} \xrightarrow[\text{peptidases}]{\text{amino- or carboxy-}} \text{Amino Acids}$$

or

$$\text{Polypeptides} \xrightarrow{\text{endopeptidases}} \text{Small Poly-/peptides}$$

$$\Big\downarrow \text{peptide hydrolases}$$

$$\text{Amino Acids}$$

The liberated amino acids may be used for protein synthesis, or to provide energy by oxidation of the carbon skeleton after deamination. Ammonia produced in the latter reaction can be prevented from reaching toxic levels by fixation into glutamine and asparagine.

The mobilization of protein reserves in seeds will be considered under two headings: mobilization in cereals, and mobilization in dicotyledons. Our understanding of proteolysis following germination is still quite limited and less is known about the processes involved than, for example, those in carbohydrate or fat catabolism.

6.6. Protein Hydrolysis in Cereals

As outlined earlier (Chap. 2) reserve proteins are stored in two separate sites in the cereal grain: in the aleurone grains (bodies) of the aleurone layer, and in the protein bodies (sometimes disrupted) of the endosperm. Proteolysis within the cells of the aleurone layer will be considered in Chapter 7, but it is important to note here that mobilization of the aleurone grain proteins appears to be important for the provision of amino acids from which hydrolytic enzymes (e.g. α-amylase) are synthesized. In the barley cultivars Kristina and Iisvesi, aleurone layers release several different proteinases; the major one being a labile acid sulphydryl endopeptidase which is purported to play a central role in the mobilization of the reserve proteins of the starchy endosperm following germination [133]. In germinated grains of the Pirrka cultivar of barley there appear to be at least eight different proteinases: three carboxypeptidases, three aminopeptidases, and two peptidases which are capable of hydrolysing the dipeptides Leu-Tyr and Ala-Gly in vitro [101]. These proteinases are distributed unequally in the aleurone cells, scutellum, starchy endosperm and growing axis, but their importance in the mobilization and turnover of proteins at these sites is still undefined, and there appears to be some contradiction in the literature between the role of specific proteinases in the breakdown of endosperm proteins in this and in the aforementioned barley cultivars [101, 133]. More studies are required to clarify the specificity and the role of the various proteinases known to be present in germinated grains of barley (and other cereals for that matter), but on the basis of the limited studies completed so far we will speculate that in barley there could be three different proteolytic systems in operation:

1. A system to hydrolyse aleurone-layer proteins to provide amino acids for the synthesis of hydrolytic enzymes; this system is probably under the control of GA from the embryo.

2. A system for the mobilization of stored reserves in the starchy endosperm for use by the developing seedling. This might be comprised of two components: proteinase synthesized and secreted by the aleurone layer (and again controlled by GA), and proteinase pre-formed in the endosperm itself and activated therein.

3. A system in the axis which might be important for protein turnover in the growing embryo, and/or for hydrolysis of any small peptides transported into the scutellum from the starchy endosperm.

Within the endosperm of *Zea mays* are the storage proteins, zein (which is stored within protein bodies) and glutelin (stored in the cytoplasmic matrix of the cell). Mobilization of both of these reserves commences soon after germination (about 20 h after imbibition) and the "grand period" of hydrolysis occurs between the third and eighth day after imbibition. This coincides with the appearance in the endosperm of an acidic endopeptidase which is claimed to be synthesized de novo in the endosperm itself [63] independently of any control by the embryo, or by added gibberellin. No discrimination was made between the two endosperm components, i.e. the starchy endosperm and the aleurone layer, but we assume that any de novo enzyme synthesis must occur within the latter. Acid proteinase activity in the endosperm (again no discrimination made between the parts thereof) of sorghum (*Sorghum vulgare*) increases during seedling development. This enzyme is interesting because it has a narrow specificity and cleaves the peptide linkages between the α-carboxyl group of aspartate or glutamate and the amino group of the adjacent amino acid [52]. Prolamin and glutelin constitute a major proportion of the total protein of the sorghum endosperm, and are rich in aspartate and glutamate.

6.6.1. Fate of the Liberated Amino Acids

Relatively little change in the level of storage protein occurs in the endosperm of germinated Spratt Archer barley over the first two days after initial imbibition. Then occurs a period of intense activity during which hordein, followed shortly by hordenin, is utilized (Fig. 6.14). Protein breakdown is associated with a rise in the non-protein nitrogen fraction of the endosperm, which reaches a maximum at about day 4. The observed increase is far less, however, than the corresponding loss in protein nitrogen because translocation of simple peptides and amino acids into the growing axis commences after day 2 (Fig. 6.15B). The fate of the amino acids proline and glutamic acid (along with an amide fraction which is probably derived from glutamine) has been followed since these account for a considerable proportion of the nitrogen of hordein and hordenin (see Chap. 2) and thus are major contributors to the free amino acid pool resulting from proteolysis. A sharp decline in the amounts of proline, glutamic acid and amide occurs in the endosperm after the fourth day from the start of imbibition (Fig. 6.15A); it has been estimated that over 90% of the nitrogen required for new synthesis within the developing seedling can be contributed by these constituents. Proline might be converted to glutamic acid, thus increasing the pool size of this amino acid, which is very important as a donor of amino nitrogen. Glutamic acid can either be further aminated to

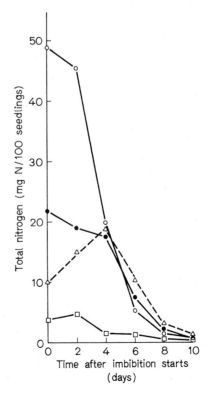

Fig. 6.14. Quantitative changes in the major nitrogenous fractions of the endosperm of barley following germination. o——o: hordein; ●——●: hordenin; □——□: albumin; △----△: non-protein nitrogen (amino acids and small peptides). Based on Folkes and Yemm, 1958 [51]

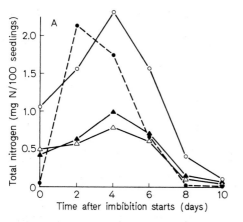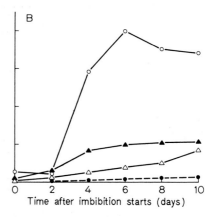

Fig. 6.15A and B. Changes in the contents of selected amino acids in the non-protein fractions of (A): barley endosperm; and (B): embryo. o——o: amide nitrogen; ▲——▲: glutamic acid; △——△: proline; ●----●: arginine. Based on Folkes and Yemm, 1958 [51]

form glutamine or be deaminated, transaminated or transported (along with glutamine) to the embryo. Increases in the endosperm pools of alanine, glycine, lysine, aspartic acid and, in particular, arginine (Fig. 6.15A) have been noted, all of which could be additional transportation forms of nitrogen from the endosperm to the developing seedling. These amino acids do not accumulate in the embryo to any extent (e.g. arginine, Fig. 6.15B).

One of the most abundant free amino acids in the partially digested endosperm of *Z. mays* is glutamine. Radioactive glutamine applied to the epithelial layer of the scutellum can be recovered in the growing points of the embryonic axis within a few hours. Thus, it is known that glutamine is both available for uptake by the scutellum, and that it can be transported to the growing axis [129]. Other amino acids might also be taken up by the scutellum, although there is limited competition between some amino acids and glutamine. This observation is consistent with the suggestion that glutamine uptake by the scutellum is achieved by a carrier or transport protein. During the growth of the maize seedling uptake of sugars by the scutellum coincides with the uptake of glutamine: even at 0.5 M concentration sucrose does not interfere with glutamine influx, suggesting that their respective mechanisms for uptake are completely independent.

6.7. Protein Hydrolysis in Dicots

The utilization of protein reserves has been studied mostly in the non-endospermic legumes on which we will concentrate here.

 Dry seeds of *Phaseolus vulgaris* have low activity of proteinase but a marked burst occurs about five days after the start of imbibition (Fig. 6.16A), which

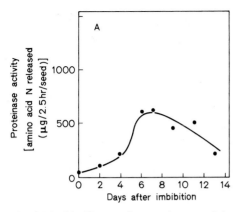
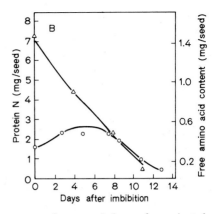

Fig. 6.16. (A) Changes in proteinase activity in extracts from cotyledons of germinated *Phaseolus vulgaris*. Enzyme activity measured by amino acid release from a casein solution incubated with cotyledon extract at pH 5.5. (B) Changes in protein nitrogen level (△) and free amino acid level (○) in cotyledons of germinated *Phaseolus vulgaris*. Based on Yomo and Srinivasen, 1973 [145]

is probably due to enzyme activation rather than de novo synthesis [145]. A decrease in protein nitrogen accompanies the increase in activity of this enzyme (a sulphydryl proteinase) although there is no large accumulation of free amino acids in the cotyledons (Fig. 6.16 B). The digestion of the protein bodies commences towards the centre of the cotyledons and spreads towards the periphery (Fig. 6.24) but, for unexplained reasons, there is no digestion in the two or three cell layers below the cotyledon surface, or in several cell layers around the vascular bundles [146]. As the protein (and at the same time, starch) bodies disappear from the cells the resultant vesicles coalesce to form larger vacuoles, and ultimately the cotyledon cells degenerate and die.

Pea seeds (*Pisum sativum* cv. Alaska), when dry, also contain low levels of soluble proteinase, which increase approximately 4-fold from about the fifth day after imbibition until about the tenth day [147]. Similarly in the Burpeeana cultivar the capacity of extracts from pea cotyledons to hydrolyse the natural storage proteins legumin and vicilin increases from the fifth day after planting (Fig. 6.17A and B). This increase in proteolytic activity coincides well with the sharp decrease in most of the various protein fractions stored within the cotyledons (Fig. 6.17C). Prior to this major mobilization, the stored legumin and vicilin undergo some alterations in composition, although this does not appear to involve deamidation or decarboxylation [20]: in contrast, an early feature of protein utilization in peanuts and soybeans is thought to be the provision of ammonia to the developing seedling by the removal of amido groups from cotyledon storage proteins by the action of deaminases [38, 45]. Removal of the sugar components from pea vicilin and legumin appears to occur after cleavage of the peptide links has commenced, and the glycosyl units are then released as complete oligosaccharides [21].

As can be seen by comparing Figure 6.17C with Figures 6.17A and B, some protein hydrolysis commences before there is any detectable activity of the

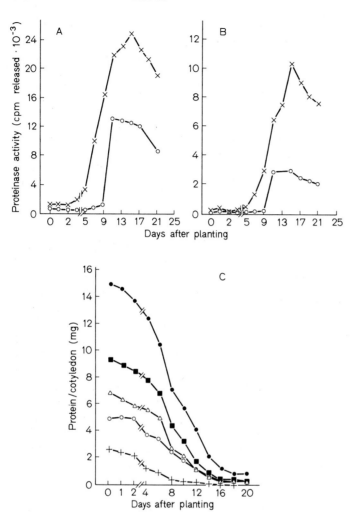

Fig. 6.17. (A) and (B) Changes in the capacity of extracts from pea cotyledons (cv. Bur-peeana) to hydrolyse radioactively labelled legumin (*A*) and vicilin (*B*) at pH 5.0 (×) and pH 7.5 (o). (C) Changes in the content of various protein fractions in pea cotyledons. ●——●: total protein; o——o: albumin; ■——■: globulin; △——△: legumin; +—+: vicilin. After Basha and Beevers, 1975 [20]

major proteinase [attributed to acid sulphydryl proteinase(s)]; a similar observation has been made for the Alaska pea cultivar. Early protein degradation (and perhaps also modification) might be due to the combined activity of pre-formed soluble peptidases (perhaps aminopeptidases capable of degrading small polypeptides released by endopeptidase activity) present in the dry seed, whose activity declines rapidly over the first few days after imbibition [23], and an insoluble, membrane-bound proteinase (an endopeptidase) which appears to be synthesized de novo within 48 h of initial imbibition [106].

The association between proteolytic enzymes and protein storage bodies appears to be variable. In *P. vulgaris* cotyledons the enzymes are probably cytoplasmic but, as outlined above, in pea at least some proteolytic activity is membrane associated and therefore insoluble. Matile [95] has further claimed that an acid proteinase in pea (like various carbohydrases, Sect. 6.3.1) is associated with the protein bodies of the cotyledons and this body becomes lysosomal in nature following digestion of the enclosed protein. The location of proteinase enzymes within protein-storing bodies has been reported for other seeds, e.g. hemp, cotton, vetch, sunflower and in the monocot sorghum, and in at least some of these species the protein bodies assume an autolytic role during reserve mobilization [14, 111, 144]. More recent studies on proteolysis in mung bean (*Phaseolus aureus*) cotyledons have shown that this process depends upon the appearance of a sulphydryl type endopeptidase(s) [41]. Furthermore, there is a close temporal relationship between the appearance of this enzyme and the catabolism of stored proteins [61]. Protein bodies isolated from mature, unimbibed cotyledons contain several hydrolytic enzymes including carboxypeptidase, α-mannosidase (for hydrolysis of mannose from storage glycoprotein), N-acetyl-β-glucosaminidase (for hydrolysis of glucosamine residues from storage glycoprotein) and other proteolytic activity (defined as caseolytic, since casein was used as the in vitro substrate). Despite the presence of these hydrolytic enzymes within the protein body there is no evidence for autolysis of storage proteins until several days after imbibition. Then there is rapid hydrolysis of the storage proteins and release of amino acids from the protein bodies, which remain as membrane-bounded organelles though the matrix becomes increasingly eroded [40a]. This increased autolytic activity of the protein bodies results from the accumulation of an endopeptidase possibly acquired from vesicles which seem, from electron-microscopic evidence, to coalesce with the storage bodies. The endopeptidase itself is not derived from a pre-formed zymogen but is almost certainly synthesized de novo [40a]. Thus, the endopeptidase appears to be the limiting enzyme for proteolysis and only after its synthesis has occurred does reserve breakdown commence.

Biochemical and histochemical studies on proteinase activity and reserve protein mobilization in cotyledons of germinated cowpeas (*Vigna unguiculata*) suggest that the major enzyme(s) involved in protein breakdown is also of the endopeptidase type and that there is a close relationship in timing and location between enhanced enzyme activity and protein mobilization [62]. Electron microscope studies also suggest a possible mechanism whereby a proteolytic enzyme might be introduced to the protein body to initiate hydrolysis. Such an enzyme could be synthesized on the rough ER within the cytoplasm, packaged into vesicles which then migrate to the protein bodies, fuse with them and release the proteinase within. While such a scheme still requires confirmation it does fit in with findings for mung bean [40a] and with previous observations made in the cotyledons of other legumes that protein synthesis increases early after imbibition, as apparently does the abundance of ribosomes associated with the endoplasmic reticulum [109].

The eventual fate of the protein bodies of many species (dicots and monocots) following internal digestion of their contents is that they fuse to form vacuoles.

In different species and tissues thereof this fusion may occur between more or less empty bodies, between bodies with incompletely degraded contents or, in some cases, between swollen but seemingly intact protein bodies (see Ref. 2 for examples). After depletion of the reserves the fused protein bodies may from the large central vacuole of the cell (e.g. Fig. 6.28).

6.7.1. Fate of the Liberated Amino Acids

Dry pea cotyledons contain few free amino acids but their content increases following germination (Fig. 6.18), at the time when the protein reserves are being mobilized. Some of these amino acids are re-utilized in protein synthesis in situ, and protein turnover has also been demonstrated in cotyledons even when they are senescing. However, the majority of the amino acids are translocated, via the phloem, to the growing axis, and the amino acid pool in the cotyledons is gradually depleted. There is evidence that vascular bundle parenchyma cells bordering the xylem and phloem of *Vicia faba* develop into transfer cells, which are specialized to aid the transport of solutes over short distances [34]. Movement of amino acids out of the cotyledons of pea starts about the second or third day after imbibition commences and may continue for a further seven to ten days until the cotyledons disintegrate. This is accompanied by a ten-fold increase in the level of the free amino acid pool in the growing shoot between the third and the tenth day (Fig. 6.18). The most dramatic changes

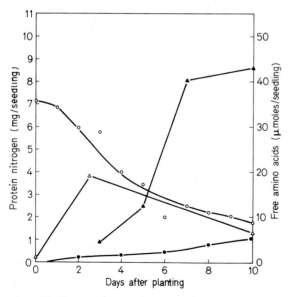

Fig. 6.18. Changes in protein and amino acid content in the cotyledons and axis of dark-grown Alaska pea seedlings. o——o: protein content of cotyledons; ●——●: protein content of root plus shoot; △——△: free amino acid content of cotyledons; ▲——▲: free amino acid content of shoots. Based on Larson and Beevers, 1965 [84]

during proteolysis are associated with only three amino acids however: glutamic acid/glutamine (glu/glu NH_2) and aspartic acid/asparagine (asp/asp NH_2), which are the major amino acids of vicilin and legumin, and particularly homoserine, which is not incorporated in any protein (Table 6.5). Two of these amino acids (glu/glu NH_2 and homoserine) increase first in the cotyledons as protein stores are broken down before being transported into the growing seedling; asp/asp NH_2 does not accumulate to any great extent in the cotyledons (Table 6.5).

Simply analysing levels of various amino acids and finding that an increase in the cotyledons is followed by an increase in the growing axis cannot be regarded as sufficient evidence that amino acids are being transported from the former to the latter, a criticism which is also relevant to the experiments illustrated in Figure 6.15. The growing axis of light-grown legume seedlings is itself capable of amino acid biosynthesis by amination of carbon skeletons produced by photosynthesis, and this process could conceivably account for increases in amino acid levels in the seedling. However, in tracer experiments, where radioactively-labelled amino acids were applied to cotyledons of dark-grown peas it was found that these were subsequently transported to the growing parts. In particular, it was found that when ^{14}C-glutamate and ^{14}C-aspartate were injected into the cotyledons of etiolated pea seedlings the radioactivity was detected in various components of the roots and shoots. Considerable quantities of the radioactive label originally in aspartate and glutamate (36% and 48% respectively) were recovered as respired CO_2 and, while no label was incorporated into either sugars or lipids, the largest part of the remaining label was found to be within homoserine [84], indicating a significant conversion of aspartate and glutamate to this non-protein amino acid.

The conversion of aspartate to homoserine occurs first within the cotyledons, and then the homoserine is transported to the roots and the shoot (Fig. 6.19). Very little aspartate or glutamate is transported to the axis, so in pea, unlike in cereals and castor bean, these are not the major transport forms of amino acids: this role is attributed to homoserine. Homoserine itself is utilized only slowly in etiolated peas, for when ^{14}C-homoserine is injected into cotyledons only 6% is respired over the subsequent ten days while 80% remains unchanged. At this time there is an almost equal distribution of this unmodified homoserine between the cotyledons and the growing axis [84].

Table 6.5. Changes in the content of glutamic acid/glutamine, aspartic acid/asparagine and homoserine in the cotyledons and shoots of light-grown peas cv. Alaska

	Cotyledons μ moles amino acid/seedling			Shoots μ moles amino acid/seedling		
Age (Days)	0	2.5	10	3	7	10
Glu/GluNH₂	0.69	5.76	1.86	0.49	0.90	3.58
Asp/AspNH₂	0.08	0.47	1.33	0.12	5.06	11.24
Homoserine	—	2.71	0.59	3.40	28.50	24.20

Based on Larson and Beevers, 1965 [84]

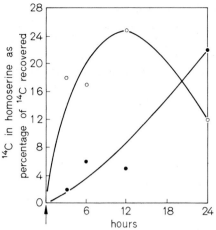

Fig. 6.19. Changes in the homoserine content of cotyledons (○) and root and shoot (●) after injecting L-aspartic acid-U-^{14}C into cotyledons of two-day-old Alaska pea seedlings. *Arrow* indicates the time of injection of radioactive label into the cotyledons as ^{14}C-aspartic acid. After Larson and Beevers, 1965 [84]

The observed slow utilization of homoserine in the growing shoots and roots raises the question of whether it can really be an effective nitrogen donor to these organs which are undergoing extensive growth obviously involving amino acid and protein synthesis. Could other amino acids be translocated from the cotyledons and be supplied in sufficient amounts to the growing roots and shoots? When ^{14}C-leucine was injected into pea cotyledons, only 3% of the radioactivity was recovered from the seedling axis 24 h later: most had been incorporated into proteins in the cotyledons or respired (6% had been converted to homoserine). While these results tell us that transport of leucine (or the ^{14}C label therefrom) is very limited, they do not tell us whether the leucine detected in the growing seedling was translocated there without modification, or whether the low level of transport observed was sufficient to satisfy the amino acid supply requirements for protein synthesis.

It is interesting to consider the fate of the translocated homoserine, for it is not a compound which is readily converted to other amino acids, nor does it donate its amino group in transamination reactions (hence, presumably, its low turnover rate). The three most significant products of homoserine metabolism appear to be threonine, an unknown organic acid (or acids), and respired CO_2, although only 2, 3, and 4% respectively were formed from labelled homoserine fed for 16 h to seven-day-old pea seedlings [56]. Further studies on the kinetics of homoserine utilization and the formation and translocation of other amino acids in relation to the kinetics of protein synthesis in the growing seedling of pea (and other legumes) would be of interest, and also of relevance in determining the major source of nitrogen for the developing seedling.

The situation in castor beans is different, for here the site of protein storage is the endosperm. The predominant form of transported nitrogen from this region is glutamine [130]. When ^{14}C-labelled aspartate, glutamate, alanine, glycine and serine were fed to the endosperm they were all found to be converted to sucrose in situ and then transported to the seedling axis. This was accompanied by the production of glutamine, which appeared to act as the sink for amide nitrogen derived from the deamination of the gluconeogenic amino acids. The role of sucrose as the translocatable product of fat catabolism has been outlined earlier in this chapter. In contrast, amino acids such as valine, phenylalanine, proline and arginine, which are not gluconeogenic, were found to be only slowly metabolized by the endosperm and were recovered unchanged in the seedling axis. Arginine, and to a lesser extent glutamine and asparagine, accumulate in the female gametophyte of the gymnosperm *Pinus banksiana* (Jackpine) during breakdown of the proteins stored therein [47]. Whether only these, or all amino acids are transported to the growing embryo is not known.

Similarities between the mechanisms by which the products of proteolysis of the endosperms of castor bean and of maize are processed and transported are evident. The involvement of homoserine may be peculiar to mobilization from the cotyledons of *Pisum* (or perhaps legumes) since little or no homoserine has been reported in other plants, and the full significance of this interesting digression from the norm is not yet understood.

6.8. Proteinase Inhibitors

There are numerous reports in the literature documenting the existence within monocot and dicot seeds of proteins which specifically inhibit the action of proteinases, particularly those of animal origin [125]. The function of plant proteinase inhibitors is a subject of considerable interest and debate, but it could include one or more of the following:

1. Storage. Trypsin inhibitors constitute as much as 5–10% of the water-soluble proteins in the embryos and endosperms of barley, wheat, and rye [100].

2. Control of endogenous proteinases. Proteinases could be inactivated by proteinase inhibitors. Removal of these inhibitors in the developing seedling could then account for activation of these hydrolytic enzymes. Evidence for such a role of these inhibitors in seeds is still not completely convincing, although the inhibitor in rice grains appears to reduce the activity of the native proteinase [66]. Mung bean cotyledons also contain an inhibitor which acts against endogenous endopeptidase but the discrepancy between the kinetics of inhibitory activity and endopeptidase activity argues against a regulatory relationship. Perhaps the inhibitors could function to protect the cytoplasm from accidental rupturing of proteinase-containing bodies [21a].

3. Protection or dissuasion. Proteinase inhibitors might function to inhibit the intestinal proteolytic digestive enzymes of invading insects or the extracellular proteinases of invading microorganisms.

6.9. Stored Phosphate Metabolism

6.9.1. General Metabolism

Phytic acid (myo-inositol hexaphosphate) has been clearly established as a major phosphate reserve in many seeds, containing more than 50% of the total stored phosphate. At one time it was thought that an important event during germination was the transfer of phosphate from phytic acid to ADP to produce ATP. There are three major arguments against this hypothesis however: (1) ATP formation is an early event in germination (see Chap. 5), but phytic acid breakdown is a late event during growth; (2) the total amount of $\sim P$ which could be produced by phytic acid utilized in a growing seedling over the first 6–10 days after imbibition has been calculated to be equivalent to that amount produced by respiratory metabolism in about 10 min [27]; and (3) phytic acid does not possess a phosphate ester bond of sufficient "high energy" to act as donor of a phosphoryl group for nucleoside triphosphate formation [119]. Thus, the role of phytic acid as a phosphagen is doubtful, and its role as a phosphate reserve appears to be the most important. Since the storage form of phytic acid is the mixed potassium, magnesium and calcium salt (and as such is called phytin or phytate), it is also the major source of these macronutrient mineral elements in the seed.

Phytase hydrolyses the phytin to release phosphate (and its associated cations) and myo-inositol. Myo-inositol derived from phytin may be of importance to the growing seedling, since it is a known precursor of all pentosyl and uronosyl sugars units normally associated with pectin and certain other polysaccharides present in the cell wall [123].

Other forms of phosphate found in smaller amounts in seeds include lipid phosphate, protein phosphate and nucleic acid phosphate. Phospholipids and phosphoproteins are most probably dephosphorylated during their hydrolysis, the free phosphate being utilized by the growing axis, and the lipid and protein moieties catabolized in the manner outlined in earlier sections of this chapter. The fate of nucleic acids and nucleotides will be discussed briefly at the end of this section.

6.9.2. Phosphate Metabolism in Cereals

In one of the most complete studies of phosphate metabolism in seeds, Hall and Hodges [58] have described the changes which occur in various parts of the oat (*Avena sativa*) grain during the first eight days after imbibition. In the dry grain phytic acid accounts for 53% of the total phosphate, with 27% being made up of lipid, nucleic acid, and protein phosphate. The fate of these phosphate fractions can be followed from Figure 6.20 B–F. Radicle emergence occurs around the second day after imbibition starts (Fig. 6.20A), and subsequent growth is marked by the increasing fresh weight of the shoots and roots. The rise in total phosphate levels in the growing shoot/root axis is accompanied by a fall in the non-axial parts of the seed (Fig. 6.20 B). In their original paper, Hall and Hodges [58] described what we have termed the non-axial parts (NAP) as the "endosperm", which they stated included the scutellum and aleurone layer, as well as the starchy endosperm per se. Since it is now known that the aleurone layer is a rich source of phytic acid in cereals, we have effected a change in terminology in Figure 6.20 from endosperm to NAP to avoid any misconception that the starchy endosperm in the sole source of this storage

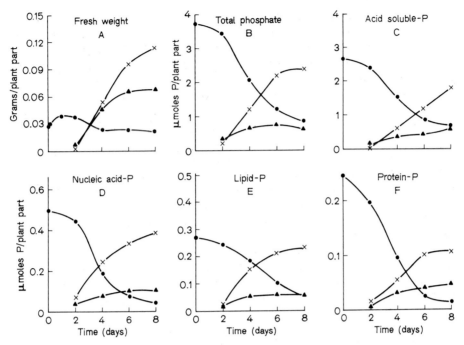

Fig. 6.20A–F. Changes in various phosphate components in the roots (▲), shoots (×) and non-axial parts—NAP (●) of germinated oat seedlings. (A) Fresh weight, (B) Total phosphate, (C) Acid soluble phosphate. In the NAP this is comprised mainly of phytic acid, and in the shoot and root of inorganic phosphate and some acid soluble organic phosphate (but no phytic acid), (D) Nucleic acid phosphate, (E) Lipid phosphate, (F) Protein phosphate. Based on Hall and Hodges, 1966 [58]

phosphate. Since phytic acid itself cannot be transported to the developing axis, the rise in inorganic (and to a lesser extent, organic) phosphate in the axis, accompanying the near stoichiometric decline in phytic acid in the non-axial regions (Fig. 6.20C), suggests a release of phosphate from its storage form and its transport as free, inorganic phosphate to the growing regions. Of the remaining classes of phosphate-containing compounds, nucleic acid phosphate is the most abundant in the dry grain, and this, along with the lipid and protein phosphate, decreases in the non-growing regions of the grain and increases in the shoot and root regions soon after germination (Fig. 6.20D, E, F).

While the fate of the various phosphate fractions is best understood in germinated oats, it is perhaps unfortunate that most studies on the fate of the macronutrient components of phytin, and the role of its hydrolytic enzyme (phytase), have been carried out using other cereals.

A phytase has been isolated from the protein bodies of whole, milled, ungerminated barley, and this enzyme appears to be associated with the protein bodies of the aleurone layer (aleurone grains), a known storage site of phytin [141]. There is strong evidence (e.g. in wheat and rice) against the association of phytin and phytase with the protein bodies of the starchy endosperm of cereals [48, 137]. Phytase activity in wheat begins to increase during the first day after the start of imbibition of the intact grain (see also Chap. 7). Part of this increase is apparently due to increased (four-fold) activation of the enzyme in the aleurone layer: the reason for this increase is unknown, because it has been estimated that there is sufficient enzyme present in the dry grain to account for the observed rates of phytin hydrolysis in vivo following germination [48]. A similar paradox is apparent in rice, where the phytase activity reaches its maximum after the phytin reserves have been hydrolysed [104]. Perhaps phytase has an additional, undefined role in metabolism. It is worth noting, however, that in some studies the assays used for "phytase" activity also detect activity of other phosphatases whose role, although unknown, might be completely unrelated to phytin hydrolysis.

Phytase activity increases markedly in the scutellum and growing axis of germinated, isolated wheat embryos [26]. Since the scutellum and the embryo are only minor sites of phytin storage and since, as we have already seen, the aleurone layer itself contains phytase in excess of its requirements, the significance of this observation is not clear. The development of phytase activity in the axis and scutellum can be prevented by inorganic phosphate, so it is possible that in the intact grain synthesis of this enzyme in the embryo of the grain is permanently suppressed by the incoming products of phytin hydrolysis in the aleurone layer. Thus observations made on isolated embryos must be treated with caution, for they may have little or no bearing on events occurring within the intact grain.

The catabolites of phytase activity in wheat are myo-inositol, phosphate, and the macroelements potassium, magnesium and calcium. Myo-inositol phosphate ester intermediates with fewer than six phosphate groups do not accumulate within the grain during hydrolysis of the phytin molecule [94], which is indicative of its rapid and complete breakdown by phytases. Movement of the catabolites from the aleurone layer and their redistribution into the developing seedling appears to be by simple diffusion through the endosperm, without any apparent competition for uptake by the scutellum.

6.9.3. Phosphate Mobilization in Dicots

Changes in the distribution of phosphate in the cotyledons and seedling axis of the asparagus (or yard-long) bean, *Vigna sesquipedalis* are shown in Figure 6.21A and B. Depletion of the reserve phosphate in the cotyledons is completed by the fifth day after imbibition (Fig. 6.21A). Most of the phosphate is released from phytin associated with the protein bodies in the cotyledons, and from the acid-insoluble phosphate (see legend Fig. 6.21). Depletion of the lesser amounts of alcohol-soluble organic phosphate (see legend Fig. 6.21) also occurs. There is no increase in free inorganic phosphate in the cotyledons during this massive depletion of the phosphate reserves (Fig. 6.21A). In the seedling axis, however, there is a build-up of both inorganic and organic phosphate fractions over the same time period as the loss from the cotyledons, and (as

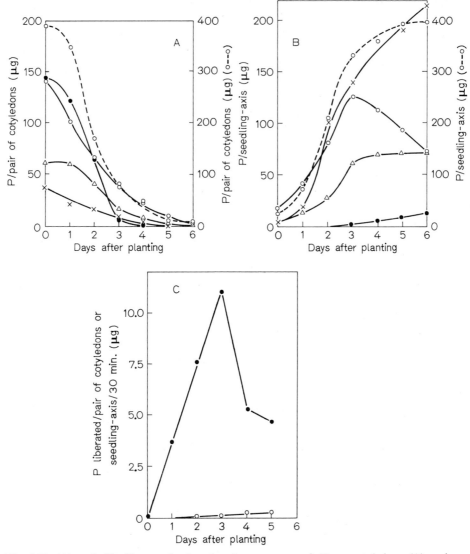

Fig. 6.21. (A) and (B) Changes in the phosphate pattern of *Vigna* cotyledons (A) and seedling axis (B). o----o: total phosphate (right ordinate only); ●——●: phytin phosphate; o——o: acid-insoluble phosphate; △——△: alcohol-soluble organic phosphate; ×——×: inorganic phosphate. (C) Changes in phytase activity in the cotyledons (●) and seedling axis (o). After Sugiura and Sunobe, 1962 [132]

Fraction	Probable content
Total phosphate	Sum of the following four fractions
Phytin phosphate	As labelled
Inorganic phosphate	As labelled
Acid-insoluble phosphate	Phospholipids, phosphoproteins, nucleic acids (DNA and RNA)
Alcohol-soluble organic phosphate	Unknown

Taken from Sugiura and Sunobe, 1962 [132], and Umbreit, W.W., Burris, R.H. and Stauffer, J.F., *Manometric Techniques and Related Methods for the Study of Tissue Metabolism*, Burgess Publ. Co., Minneapolis, 1945 noted for cereals) there is barely any accumulation of phytin in the axis tissue (Fig. 6.21 B). The full extent of the transport of the phosphate reserves from the cotyledons to the axis can be judged by following the changes in total phosphate levels. The level of reserves in the cotyledons of the dry seeds is approximately 400 μg phosphate/cotyledon pair: by day 6 after imbibition this level has almost decreased to zero in the cotyledons, but has increased to 400 μg phosphate/seedling axis. As shown in Figure 6.21 C, phytase activity increases in the cotyledons from the time of imbibition, and reaches a peak of activity on day 3, but no significant phytase activity can be detected in the seedling axis.

Although the above studies on the changes in the phytase and phosphate levels in *Vigna* were not related to the time course of germination, it is widely recognized in other dicotyledonous plants (e.g. lettuce, cotton, dwarf and mung beans, peas) that phytase increases in activity following germination. The pattern of mobilization of the phosphate reserves in peas (cv. Alaska) is, in many respects, remarkably like the pattern described above for *Vigna,* except that the process is very slow until after the fifth day following imbibition, at which time the sugar phosphates, free nucleotide phosphates and phytin reserves (i.e. all the acid-soluble phosphates) start to decline rapidly. Inorganic phosphate accumulates in the cotyledons and by day 10 up to 50% of the total released is still present [57]. This accumulation of phosphate in the cotyledons does not appear to slow down or suppress phytase activity, although the actual synthesis (or activation) of the enzyme occurs between the second and fifth day after imbibition, before the phosphate has accumulated.

6.9.4. Mobilization of Nucleic Acids from the Storage Regions of the Seed

Changes in the nucleic acid phosphate fraction in the non-axial parts and growing regions of the developing oat seedling are outlined in Figure 6.20 D. In this plant, nucleic acid phosphate accounts for less than 10% of the stored phosphate in the non-axial parts, and a minute fraction of all the storage material available in this region. Similarly, little nucleotide or nucleic acid is stored in the kernel of *Z. mays,* where it appears that neither the scutellum, nor the endosperm (including the aleurone layer) contain enough of these materials to account for the observed increases in the growing embryo [74]. Also, the endosperm of rice cannot supply more than a small percentage of the precursors necessary for the RNA synthesis observed in the growing axis. Thus, while ribonucleases are produced in the aleurone layer of germinated cereals, secreted into the endosperm to degrade the RNA therein, and the resultant nucleotides (probably) transported to the axis, these are not in sufficient quantity to maintain growth and there must be net synthesis of nucleotides in the growing axis itself. The nitrogen source required for such synthesis could be provided by amino acids present in, or transported to the young embryo. Prior to these events, during the first one and two days after imbibition, the growing leaf and root tissues might be provided with nucleotides resulting from hydrolysis of nucleic acids in the coleoptile and coleorhiza: such a pattern of events appears to occur in wheat, for example [117].

Following the germination of garden pea (*P. sativum*) there is a depletion of RNA in the cotyledons but, as in cereals, the increase in axis RNA is greater than the decline in cotyledonary RNA, indicating a net nucleotide synthesis, which possibly utilises nitrogen initially in reserve protein. The nucleotide products of ribonuclease activity do not accumulate in the cotyledons but may be translocated to the growing axis to support, in part, nucleic acid synthesis therein [24]. Ribonuclease activity is initially low in both the cotyledons and axis of field pea (*Pisum arvense*), but increases rapidly as seedling growth progresses. An initial increase in enzyme activity in the cotyledons appears to be due to its activation, with a second increase occurring about eight days after imbibition, due to de novo synthesis [19]. The significance of this second increase in enzyme activity is not clear, however, because it occurs when nearly all the RNA has gone from the cotyledons. In an experiment where ^{14}C-adenine was fed to the cotyledons of germinated garden peas, it was found that considerable quantities (50%) were incorporated into RNA in the cotyledons of seeds germinated for up to seven days, suggesting that RNA synthesis (and turnover) occurs in the cotyledons even at a time when the major reserves are being degraded and transported [25] (a similar observation was made with respect to protein synthesis). Less than 10% of the adenine was transported from the cotyledons to the developing axis, and only 20% of that was incorporated into nucleic acids.

In summary, studies on the role of storage nucleic acids in seed germination and seedling growth are extremely limited, but on the basis of some fragmentary evidence it is apparent that storage tissues are unable to supply the nucleotide requirements of the growing axis. On the other hand, the axis has the capacity to synthesize its own supply, nitrogen for this process perhaps being provided by amino acids produced by hydrolysis of reserve proteins. See also Chapter 5 for further consideration of RNA and protein synthesis in storage organs.

6.10. Patterns of Reserve Mobilization in Seeds — Examples

We have so far discussed details concerning mobilization of the major reserves — carbohydrate, fat, protein, etc. — concentrating largely on the metabolic changes experienced by any one reserve. Virtually all seeds contain more than one type of food store so that we see the occurrence of more than one of the processes discussed above. It is important now to redirect our attention and to consider the mobilization processes in toto taking place in any one seed. We will therefore conclude this chapter with a presentation of some patterns of reserve mobilization from the storage organs of selected seeds, as an illustration of events occurring within individual species. Surprisingly few comprehensive studies have been made of the breakdown of all the reserves from a single storage tissue, and so no attempt has been made here to compare

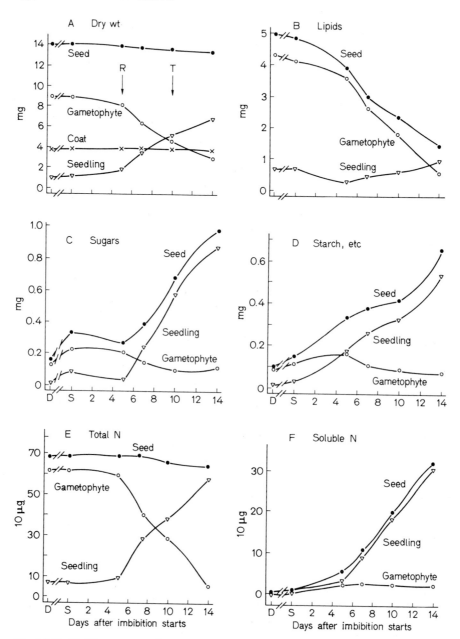

Fig. 6.22A–H. The pattern of reserve mobilization in stratified seeds of Douglas fir. *R*: time of radicle emergence; *T*: time of cotyledon emergence; *D*: dry seed; *S*: end of stratification period. After Ching, 1966 [39]

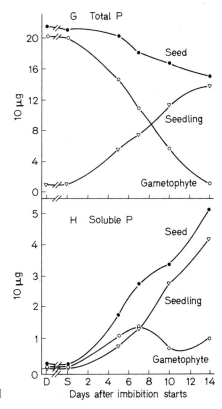

Fig. 6.22 G–H

or contrast the sequence of events in different species. Presumably there is a set pattern, or sequence of events controlling the hydrolysis of reserves in every species, but whether or not this is similar in related species cannot be determined at present due to the paucity of available information. Many more studies are needed. The examples presented here represent some of the most comprehensive studies carried out to date but, as will be evident, even some of these are incomplete. One study which is not detailed here, but is also worthy of the reader's attention is the work on *Yucca* seeds [67].

Changes occurring during the hydrolysis of reserves from the (mega)gametophyte of Douglas fir are shown in Fig. 6.22. Dry seeds (*D*) used for these analyses were soaked for four weeks at 3°C and after this stratification period (*S*) (during which there were no changes in the stored reserves) seeds were germinated in an 8-h light (30°C) and 16-h dark (20°C) photoperiodic regime. The radicle emerged on day 5 (*R*, Fig. 6.22A), which was followed by an increase in dry weight of the seedling and loss of weight (reserves) of the gametophyte. Cotyledons emerged on day 10 (*T*, Fig. 6.22A) and were elongating by day 14, at which time the plumule also emerged. Reserves in the gametophyte were found to be more or less depleted by day 14, but by then photosynthesis in the cotyledons would be responsible for the provision of carbon to the growing seedling. Lipids (the major storage reserve) mobilized from the gametophyte

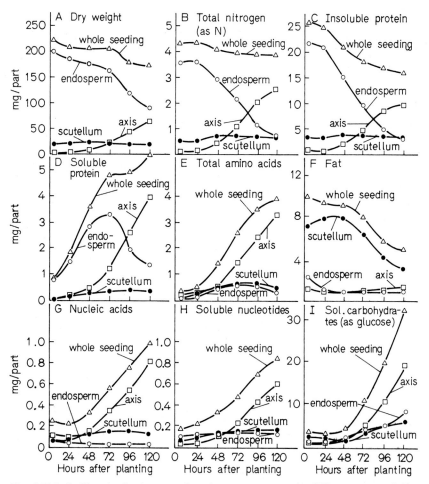

Fig. 6.23A–I. Changes in content of various components in different parts of *Zea mays* during germination and growth. After Ingle et al., 1964 [75]

did not accumulate in the growing seedling as such (Fig. 6.22 B), but free sugars (Fig. 6.22 C), and starch (and other insoluble carbohydrates) did (Fig. 6.22 D), although there was little previous build-up of these in the gametophyte. Total nitrogen, including storage protein, was lost from the gametophyte at about the same time as lipids (Fig. 6.22 E), and there was an increase in nitrogen in the growing seedling as soluble nitrogen (Fig. 6.22 F). Only 10–15% of the total nitrogen in the seedlings was found to accumulate as free amino acids and amides [39]. There was no accumulation of amino acids or soluble nitrogen in the gametophyte. Prior to radicle emergence the total phosphate in the gameto-phyte apparently decreased (mobilization of phytin) accompanied by an increase in the soluble phosphate in both the gametophyte and the growing seedling (Fig. 6.22 G and H). The study outlined in Figure 6.22 (one of the most complete in any seed) thus gives some insight into the chain of events involving degradation

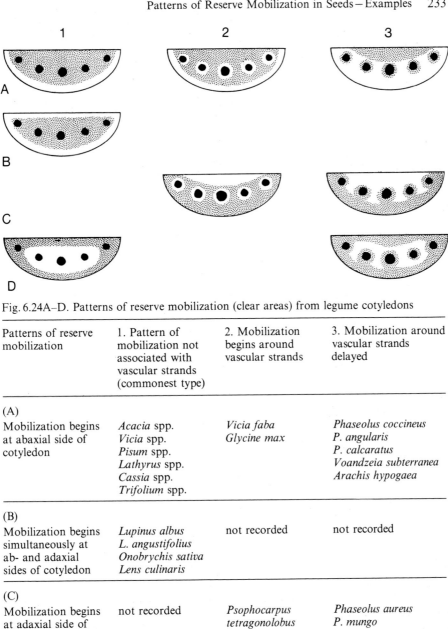

Fig. 6.24A–D. Patterns of reserve mobilization (clear areas) from legume cotyledons

Patterns of reserve mobilization	1. Pattern of mobilization not associated with vascular strands (commonest type)	2. Mobilization begins around vascular strands	3. Mobilization around vascular strands delayed
(A) Mobilization begins at abaxial side of cotyledon	*Acacia* spp. *Vicia* spp. *Pisum* spp. *Lathyrus* spp. *Cassia* spp. *Trifolium* spp.	*Vicia faba* *Glycine max*	*Phaseolus coccineus* *P. angularis* *P. calcaratus* *Voandzeia subterranea* *Arachis hypogaea*
(B) Mobilization begins simultaneously at ab- and adaxial sides of cotyledon	*Lupinus albus* *L. angustifolius* *Onobrychis sativa* *Lens culinaris*	not recorded	not recorded
(C) Mobilization begins at adaxial side of cotyledon	not recorded	*Psophocarpus tetragonolobus* (the only example so far)	*Phaseolus aureus* *P. mungo*
(D) Mobilization begins at centre of cotyledon	*Piptanthus laburnifolius* *Hovea trisperma* + other species	not recorded	*Phaseolus vulgaris* *P. lunatus* *Vigna sinensis* *Dolichos lablab*

Kindly provided by Dr. D.L. Smith

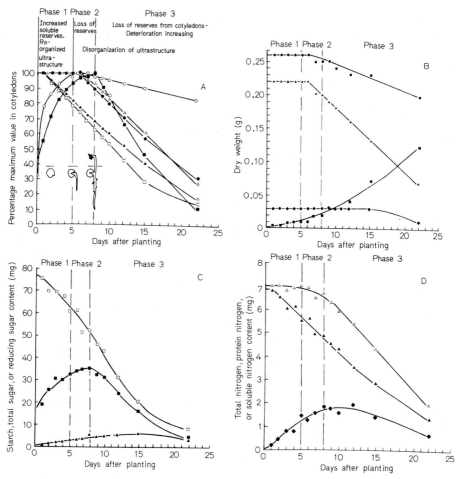

Fig. 6.25. (A) Changes in fresh weight (○), dry weight (●), total nitrogen (△), protein nitrogen (▲), starch (□), and sugar (■) in the cotyledons of peas (Victory Freezer) during germination and establishment of the seedling. (B) Changes in average dry weight of the seedling (●), axis (■), pair of cotyledons (▲), and the testa (◆) during the three morphological phases in development of the pea seedling. (C) Changes in the starch (□), total sugar (■), and reducing sugar (▲) content per pair of cotyledons during pea seedling development. (D) Changes in total nitrogen (△), protein nitrogen (▲), and soluble nitrogen (◆) content per pair of cotyledons during pea seedling development. Germination is hypogeal. After Bain and Mercer, 1966 [16]

of reserve lipids, proteins and phosphate-containing compounds from the storage tissue of Douglas fir seeds, and the transport of the breakdown products to the developing seedling for synthesis of essential cellular components. While we might expect the same general pattern of events to be involved in the mobilization of reserves in gymnosperms, we might also expect variations, and the contention that proteolysis in *Picea abies* is more rapid than lipolysis [126] appears to represent but one of these.

Metabolic changes associated with germination and growth (in darkness) of *Zea mays* grains are shown in Figure 6.23 although, sadly, in this study changes in the major endosperm reserve material—starch—were not followed. The Figure is more or less self-explanatory, but it is worth calling to attention that the major fat reserve lies within the scutellum (Fig. 6.23 F), and the protein reserves within the endosperm (Fig. 6.23 C). There are published reports that protein hydrolysis in the endosperm occurs more rapidly than that of starch, e.g. in rice [114]. More comprehensive, comparative studies are still needed on the order of mobilization of starch, phosphate and protein (including the different storage types) within intact, germinated cereal grains.

Some interesting work on reserve hydrolysis in legumes has recently been carried out by D.L. Smith, whose exhaustive studies of 500 legume species have revealed that there are eight basic patterns of hydrolysis of reserves from the cotyledons. These patterns are presented in Figure 6.24, along with a list of representative species for each type. This work is important in that it is the first to concentrate on the similarities and differences in a large number of related species, and this approach could profitably be extended to other groups of species as well as supplemented with studies to determine the order of hydrolysis of the stored reserves in representative species of each type. With this sort of basic information we might be better equipped to understand the complex metabolic controls which must be associated with the breakdown and utilization of stored reserves.

Development of the seedling of glasshouse-grown wrinkled *Pisum sativum* cv. Victory Freezer has been divided into three morphological phases (Fig. 6.25A). Mobilization of protein and starch reserves from the cotyledons commences between day 1 and 2, (just after radicle emergence) and the losses of these reserves more or less parallel each other. Such is also the case in the smooth pea, cv. Early Alaska (Fig. 6.3A). Changes in the dry weight of the seed parts, and changes in carbohydrate and protein content of the cotyledon are also emphasized (Fig. 6.25 B–D). Note that Figure 6.25 C shows some accumulation of sugar (or perhaps dextrin) within the cotyledons: this does not occur in the Early Alaska cultivar (Fig. 6.3A). Protein bodies commence to be hydrolysed before starch in *Phaseolus vulgaris* cotyledons, and finally cell wall material is hydrolysed and the cells collapse [128]. On the other hand, starch decreases before protein in *Vicia faba* cotyledons, and free sugars accumulate therein also [34].

Protein, nitrogen and starch reserves in the cotyledons of *Vigna sesquipedalis* also appear to be hydrolysed at the same time (Fig. 6.26B and D), although in *Vigna unguiculata* cotyledons starch remains in cells from which all protein (and fat) has been mobilized [34]. Changes in the dry weight of *V. sesquipedalis* (Fig. 6.26A) show that both radicle and hypocotyl emerge within the first 24 h of the beginning of imbibition. Starch reserves are mobilized from the cotyledons (Fig. 6.26B) and there is an increase in insoluble polysaccharides (probably new cell wall material, not starch) in the tissues of the growing axis. Soluble sugars do not accumulate in the cotyledons but are transported to the seedling axis, where they appear to aggregate in the hypocotyl region (Fig. 6.26 C). Protein nitrogen hydrolysis in the cotyledons is accompanied by an increase in proteins and soluble nitrogen (amino acids) in the axis (Fig. 6.26 D and E). Changes in nucleic acid phosphate are also shown (Fig. 6.26 F).

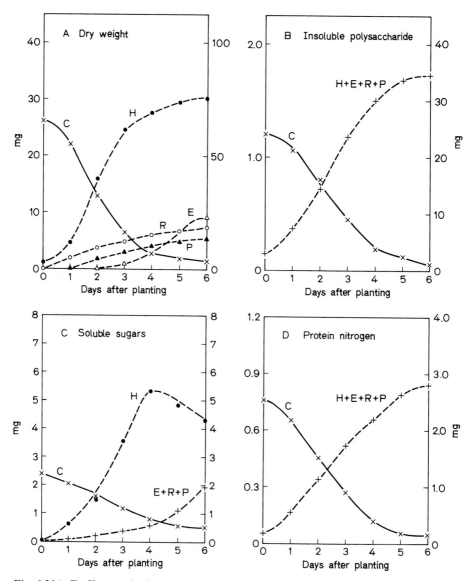

Fig. 6.26A–F. Changes in the cotyledons and axes of *Vigna sesquipedalis* during seedling development. *C*: cotyledons; *H*: hypocotyl; *E*: epicotyl; *R*: radicle; *P*: plumule. Scale on the right-hand ordinate is for *C*, and on the left-hand ordinate for *H*, *E*, *R* and *P*. Germination is epigeal. Based on Oota et al., 1953 [108]

Hydrolysis of proteins in the cotyledons of germinated *Lupinus albus* commences several days prior to that of lipids (Fig. 6.27), with the degradation of the cotyledon cell wall cellulose and hemicelluloses occurring after about 12 days, when the organs are senescing and losing their chlorophyll. The pattern of degradation of the storage proteins within the protein bodies is shown in

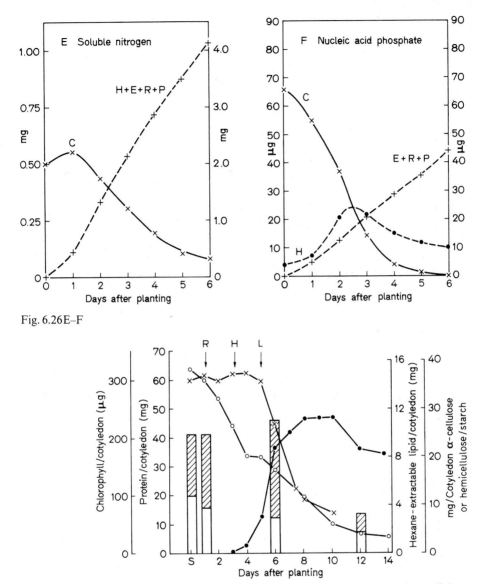

Fig. 6.26E–F

Fig. 6.27. Changes in protein (○), lipid (×), total chlorophyll a and b (●), α-cellulose (hatched bar) and hemicellulose/starch (unhatched bar) content per cotyledon during germination and seedling development in *Lupinus albus*. *R*: time of radicle elongation; *H*: time of hypocotyl expansion; *L*: time of expansion of first leaves and plumule. Germination is epigeal. Based on Parker, 1975 [115]

Figure 6.28A–F. Digestion of the swollen protein bodies commences after 24 h with extensive internal cavitation (Fig. 6.28 B), the cavities eventually coalescing to form a large central vacuole (Fig. 6.28 C), a few starch grains now becoming visible. Further vacuolation results in the swelling of the protein bodies, and expansion and rounding of the cells (Fig. 6.28 D). Later, about day 5, the protein

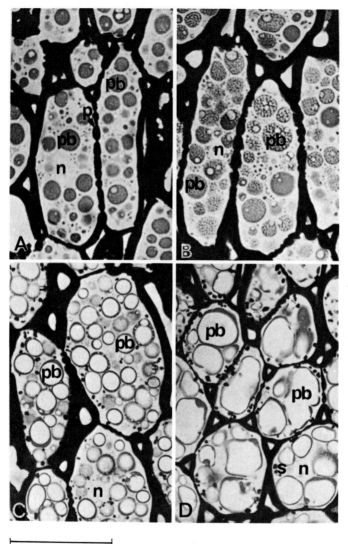

100μm

Fig. 6.28 A–H. Changes in the protein reserves within cotyledon mesophyll tissue of seedlings of *Lupinus albus* L. cv. Agricultural White. Protein digestion occurs in a wave from the epidermis inwards so several stages of digestion can be seen in any cotyledon section. It is not possible to equate stage of digestion exactly with days after imbibition. (A) Is usually found in day 1 cotyledons; (B) (C) and (D) in day 1–4. Plate (E) is characteristic of day 5 and (F) and (G) are found in day 6–8 cotyledons. In (H) can be seen the specially persistent type of protein body which remains in the mid-region of the cotyledon after other reserves have been withdrawn. *n*: nucleus; *p*: pitfield; *pb*: protein body; *pbf*: protein body fragment; *ppf*: persistent protein fragment; *s*: starch; *v*: vacuole. From Parker, 1975 [115]

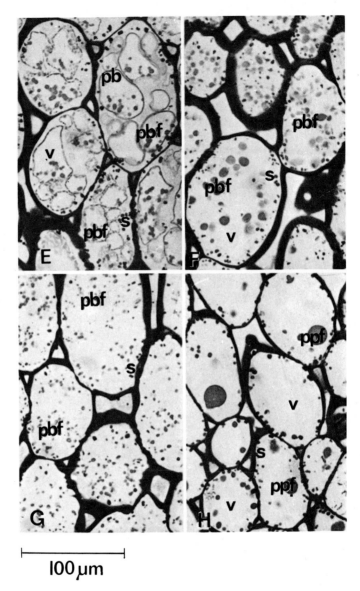

Fig. 6.28 E–H

fragments within the protein bodies (Fig. 6.28 E) and finally the cell walls become thinner; the cells become highly vacuolated and by day 8 only a protein body fragment persists (Fig. 6.28 F–H). Alpha- and β-conglutin are the two major storage proteins in the protein bodies, accompanied by the minor globulin γ-conglutin. During protein hydrolysis the β-conglutin moeity is apparently mobilized first, its hydrolysis being complete within six days: α- and γ-conglutin are broken down later, and some α-conglutin might persist undegraded.

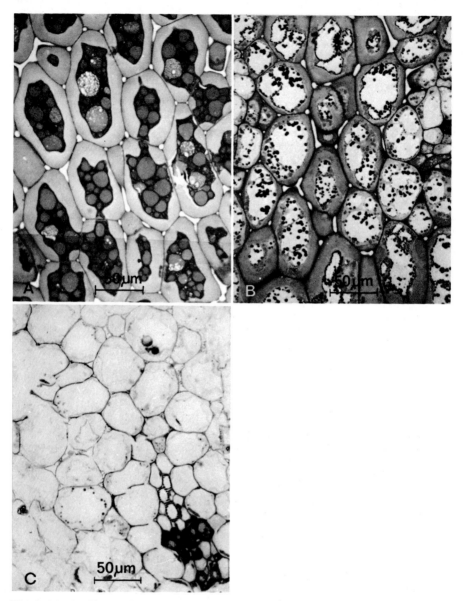

Fig. 6.29 A–C. Changes in the protein reserves and cell wall material within cotyledon mesophyll tissue of *Lupinus angustifolius* L. cv. New Zealand Bitter Blue. (A) Cotyledon material from dry seed imbibed in fixative (2 h, glutaraldehyde; 4 h, osmium tetroxide). (B) 5-day cotyledons. (C) 9-day cotyledons. Kindly provided by Dr. M. Parker

Mature dry seeds of *Lupinus luteus* contain no starch, but starch grains appear during mobilization of the protein and lipid reserves [102]. The source of material for starch synthesis is not photosynthesis (for starch grains also form in dark-germinated seeds), nor the proteins, nor the small quantities of lipids stored in the cotyledons. It is apparently the products of hydrolysis of the cell wall hemicelluloses which accumulate into starch grains because they are not transported to the growing axis. Hydrolysis of protein bodies and subsequent cell wall degradation is illustrated clearly in Figure 6.29 A–C for *Lupinus angustifolius.*

Finally a word about the endospermic legumes. As we have mentioned previously, the galactomannan reserves of the endosperm of these seeds might be more important for retention of water within the seed to allow early embryo growth than for the provision of nutrients. The cotyledons of these legumes are usually rich in fats and protein but their mobilization does not appear to have received any attention.

Some Articles of General Interest

1. Appelqvist, L.-Å: Biochemical and structural aspects of storage and membrane lipids in developing oil seeds. In: Recent Advances in the Chemistry and Biochemistry of Plant Lipids. Galliard, T., Mercer, E.I. (eds.). London: Academic Press, 1975, Proc. Phytochem. Soc. **12**, pp. 247–286

2. Ashton, F.M.: Mobilization of storage proteins of seeds. Ann. Rev. Pl. Physiol. **27**, 95–117 (1976)

3. Beevers, H.: Organelles from castor bean seedlings: biochemical roles in gluconeogenesis and phospholipid biosynthesis. In: Recent Advances in the Chemistry and Biochemistry of Plant Lipids. Galliard, T., Mercer, E.I. (eds.). London: Academic Press, 1975, Proc. Phytochem. Soc. **12**, pp. 287–299

4. Briggs, D.E.: Hormones and carbohydrate metabolism in germinating cereal grains. In: Biosynthesis and Its Control in Plants. Milborrow, B.V. (ed.). London: Academic Press, 1973, pp. 219–277

5. Ching, T.M.: Metabolism of germinating seeds. In: Seed Biology. Kozlowski, T.T. (ed.). New York: Academic Press, 1973, Vol. II, pp. 103–218

6. French, D.: Chemistry and biochemistry of starch. In: Biochemistry of Carbohydrates. Whelan, W.J. (ed.). London: Butterworths, 1975, Vol. V, pp. 267–335

7. Galliard, D.T.: Degradation of plant lipids by hydrolytic and oxidative enzymes. In: Recent Advances in the Chemistry and Biochemistry of Plant Lipids. Galliard, T., Mercer, E.I. (eds.). London: Academic Press, 1975, Proc. Phytochem. Soc. **12**, pp. 319–357

8. Macleod, A.M.: The utilization of cereal reserves. Sci. Prog. (Oxf.) **57**, 99–112 (1969)

9. Manners, D.J.: Some aspects of the enzymic degradation of starch. In: Plant Carbohydrate Biochemistry. Pridham, J.B. (ed.). London: Academic Press, 1973, pp. 109–125

10. Marzliak, P.: Lipid metabolism in plants. Ann. Rev. Pl. Physiol. **24**, 287–310 (1973)

10a. Palmer, G.H., Bathgate, G.N.: Malting and Brewing. In: Advances in Cereal Science and Technology. Pomeranz, Y. (ed.). St. Paul, Minn.: Am. Assoc. Cereal Chem. Inc., 1976, pp. 237–324

11. Richardson, M.: Microbodies (glyoxysomes and peroxisomes) in plants. Sci. Prog. (Oxf.) **61**, 41–61 (1974)

12. Tolbert, N.E.: Microbodies—peroxisomes and glyoxysomes. Ann. Rev. Pl. Physiol. **22**, 45–74 (1971)

References

13. Abbott, I.R., Matheson, N.K.: Phytochemistry **11**, 1261–1272 (1972)

14. Adams, C.A., Novellie, L.: Pl. Physiol. **55**, 7–11 (1975)

15. Adams, C.A., Watson, T.G., Novellie, L.: Phytochemistry **14**, 953–956 (1975)

16. Bain, J.M., Mercer, F.V.: Australian J. Biol. Sci. **19**, 69–84 (1966)

17. Bain, J.M., Mercer, F.V.: Australian J. Biol. Sci. **19**, 85–96 (1966)

18. Barbareschi, D., Longo, G.P., Servattez, O., Zulian, T., Longo, C.P.: Pl. Physiol. **53**, 802–807 (1974)

19. Barker, G.R., Bray, C.M., Walter, T.J.: Biochem. J. **142**, 211–219 (1974)

20. Basha, S.M.M., Beevers, L.: Planta (Berl.) **124**, 77–87 (1975)
21. Basha, S.M.M., Beevers, L.: Pl. Physiol. **57**, 93–97 (1976)
21a. Baumgartner, B., Chrispeels, M.J.: Pl. Physiol. **58**, 1–6 (1976)
22. Beevers, H.: Pl. Physiol. **31**, 440–445 (1956)
23. Beevers, L.: Phytochemistry **7**, 1837–1844 (1968)
24. Beevers, L., Guernsey, F.S.: Pl. Physiol. **41**, 1455–1458 (1966)
25. Beevers, L., Splittstoesser, W.E.: J. Exp. Botany **19**, 698–711 (1968)
26. Bianchetti, R., Sartirana, M.L.: Biochim. Biophys. Acta **145**, 485–490 (1967)
27. Bieleski, R.L.: Ann. Rev. Pl. Physiol. **24**, 225–252 (1973)
28. Bilderback, D.E.: Pl. Physiol. **53**, 480–484 (1974)
29. Boatman, S.G., Crombie, W.M.: J. Exp. Botany **9**, 52–74 (1958)
30. Bowden, L., Lord, J.M.: FEBS Lett. **49**, 369–371 (1975)
31. Bowden, L., Lord, J.M.: Biochem. J. **154**, 491–499 (1976)
32. Bradbeer, C., Stumpf, P.K.: J. Biol. Chem. **234**, 498–501 (1959)
33. Breidenbach, R.W., Beevers, H.: Biochem. Biophys. Res. Commun. **27**, 462–469 (1967)
34. Briarty, L.G., Coult, D.A., Boulter, D.: J. Exp. Botany **21**, 513–524 (1970)
35. Briggs, D.E.: Planta (Berl.) **108**, 351–358 (1972)
36. Brown, H.T., Morris, G.H.: J. Chem. Soc. **57**, 458–528 (1890)
37. Canvin, D.T., Beevers, H.: J. Biol. Chem. **236**, 988–995 (1961)
38. Catsimpoolas, N., Campbell, T.G., Mayer, E.W.: Pl. Physiol. **43**, 799–805 (1968)
39. Ching, T.M.: Pl. Physiol. **41**, 1313–1319 (1966)
40. Ching, T.M.: Pl. Physiol. **46**, 475–482 (1970)
40a. Chrispeels, M.J., Baumgartner, B., Harris, N.: Proc. Natl. Acad. Sci. **73**, 3168–3172 (1976)
41. Chrispeels, M.J., Boulter, D.: Pl. Physiol. **55**, 1031–1037 (1975)
42. Chrispeels, M.J., Tenner, A.J., Johnson, K.D.: Planta (Berl.) **113**, 35–46 (1973)
43. Cooper, T.G.: J. Biol. Chem. **246**, 3451–3455 (1971)
44. Cooper, T.G., Beevers, H.: J. Biol. Chem. **244**, 3514–3520 (1969)
45. Daussant, J., Neucere, N.J., Conkerton, E.J.: Pl. Physiol. **44**, 480–489 (1969)
46. Douglass, S.A., Criddle, R.S., Breidenbach, R.W.: Pl. Physiol. **51**, 902–906 (1973)
47. Durzan, D.J., Chalupa, V.: Can. J. Botany **46**, 417–428 (1968)
48. Eastwood, D., Laidman, D.L.: Phytochemistry **10**, 1275–1284 (1971)
49. Edelman, J., Shibko, S.I., Keys, A.J.: J. Exp. Botany **10**, 178–189 (1959)
50. Fincher, G.B., Stone, B.A.: Australian J. Pl. Physiol. **1**, 297–311 (1974)
51. Folkes, B.F., Yemm, E.W.: New Phytologist **57**, 106–131 (1958)
52. Garg, G.K., Virupaksha, T.K.: Europ. J. Biochem. **17**, 13–18 (1970)
53. Gerhardt, B.P., Beevers, H.: J. Cell Biol. **44**, 94–102 (1970)
53a. Givan, C.V.: Planta (Berl.) **108**, 29–38 (1972)
54. Goldstein, L.D., Jennings, P.H.: Pl. Physiol. **55**, 893–898 (1975)
55. Gonzalez, E., Beevers, H.: Pl. Physiol. **57**, 406–409 (1976)
56. Grant, D.R., Voelkert, E.: Phytochemistry **11**, 911–916 (1972)
57. Guardiola, J.L., Sutcliffe, J.F.: Ann. Botany (London) **35**, 809–823 (1971)
58. Hall, J.R., Hodges, T.K.: Pl. Physiol. **41**, 1459–1464 (1966)
59. Halmer, P., Bewley, J.D., Thorpe, T.A.: Planta (Berl.) **130**, 189–196 (1976)
60. Hardie, D.G.: Phytochemistry **14**, 1719–1722 (1975)
61. Harris, N., Chrispeels, M.J.: Pl. Physiol. **56**, 292–299 (1975)
62. Harris, N., Chrispeels, M.J., Boulter, D.: J. Exp. Botany **26**, 544–554 (1975)
63. Harvey, B.M.R., Oaks, A.: Pl. Physiol. **53**, 453–457 (1974)
64. Harvey, B.M.R., Oaks, A.: Planta (Berl.) **121**, 67–74 (1974)
65. Hawker, J.S.: Phytochemistry **10**, 2313–2322 (1971)
66. Horiguchi, T., Kitagishi, K.: Pl. Cell Physiol. (Tokyo) **12**, 907–915 (1971)
67. Horner, H.T., Jr., Arnott, H.J.: Botan. Gaz. **127**, 48–64 (1966)
68. Huang, A.H.C.: Pl. Physiol. **55**, 555–558 (1975)
69. Huang, A.H.C.: Pl. Physiol. **55**, 870–874 (1975)
70. Huang, A.H.C., Bowman, P.D., Beevers, H.: Pl. Physiol. **54**, 364–368 (1974)
71. Humphreys, T.E.: Phytochemistry **14**, 333–340 (1975)

72. Hutton, D., Stumpf, P.K.: Pl. Physiol. **44**, 508–516 (1969)
73. Hutton, D., Stumpf, P.K.: Arch. Biochem. Biophys. **142**, 48–60 (1971)
74. Ingle, J., Hageman, R.H.: Pl. Physiol. **40**, 48–53 (1965)
75. Ingle, J., Beevers, L., Hageman, R.H.: Pl. Physiol. **39**, 735–740 (1964)
76. Jacks, T.J., Yatsu, L.Y., Altschul, A.M.: Pl. Physiol. **42**, 585–597 (1967)
77. James, A.L.: New Phytologist **39**, 133–144 (1940)
78. Jørgensen, O.B.: Acta Chem. Scand. **19**, 1014–1015 (1965)
79. Juliano, B.O., Varner, J.E.: Pl. Physiol. **44**, 886–892 (1969)
80. Kagawa, T., Lord, J.M., Beevers, H.: Pl. Physiol. **51**, 61–65 (1973)
81. Keusch, L.: Planta (Berl.) **78**, 321–350 (1968)
82. Kriedemann, P., Beevers, H.: Pl. Physiol. **42**, 161–173 (1967)
83. Kriedemann, P., Beevers, H.: Pl. Physiol. **42**, 174–180 (1967)
84. Larson, L.A., Beevers, H.: Pl. Physiol. **40**, 424–432 (1965)
85. Longo, C.P., Longo, G.P.: Pl. Physiol. **45**, 249–254 (1970)
86. Longo, G.P., Bernasconi, E., Longo, C.P.: Pl. Physiol. **55**, 1115–1119 (1975)
87. Lord, J.M., Beevers, H.: Pl. Physiol. **49**, 249–251 (1972)
88. Macleod, A.M., Palmer, G.H.: J. Inst. Brew. **72**, 580–589 (1966)
89. Macleod, A.M., Palmer, G.H.: Nature (London) **216**, 1342–1343 (1968)
90. Macleod, A.M., Palmer, G.H.: New Phytologist **68**, 295–304 (1969)
91. MacWilliam, I.C., Harris, G.: Arch. Biochem. Biophys. **84**, 442–454 (1959)
92. Manners, D.J., Marshall, J.J., Yellowlees, D.: Biochem. J. **116**, 539–541 (1970)
93. Mares, D.J., Stone, B.A.: Australian J. Biol. Sci. **26**, 813–830 (1973)
94. Matheson, N.K., Strother, S.: Phytochemistry **8**, 1349–1356 (1969)
95. Matile, P.: Z. Pflanzenphysiol. **58**, 365–368 (1968)
96. Mayer, A.M., Shain, Y.: Science **162**, 1283–1284 (1968)
97. McCleary, B.V., Matheson, N.K.: Phytochemistry **13**, 1747–1757 (1974)
98. McCleary, B.V., Matheson, N.K.: Phytochemistry **14**, 1187–1194 (1975)
99. McCleary, B.V., Matheson, N.K.: Phytochemistry **15**, 43–47 (1976)

100. Mikola, J., Kirsi, M.: Acta Chem. Scand. **26**, 787–795 (1972)
101. Mikola, J., Kolehmainen, L.: Planta (Berl.) **104**, 167–177 (1972)
102. Młodzianowski, F., Wesłowska, M.: Acta Soc. Botan. Polon. **44**, 529–536 (1975)
103. Mollenhauer, H.H., Totten, C.: J. Cell Biol. **48**, 395–405 (1971)
104. Mukherji, S., Dey, B., Paul, A.K., Sircar, S.M.: Physiol. Plantarum **25**, 94–97 (1971)
105. Muto, S., Beevers, H.: Pl. Physiol. **54**, 23–28 (1974)
106. Nakano, M., Asahi, T.: Pl. Cell Physiol. (Tokyo) **14**, 1205–1208 (1973)
107. Nomura, T., Akazawa, T.: Pl. Physiol. **51**, 979–981 (1973)
108. Oota, Y., Fujii, R., Osawa, S.: J. Biochem. (Tokyo) **40**, 649–661 (1953)
109. Öpik, H.: J. Exp. Botany **17**, 427–439 (1966)
110. Opute, F.I.: Ann. Botany (London) **39**, 1057–1061 (1975)
111. Ory, R.L.: In: Symposium: Seed Proteins. Inglett, G.E. (ed.). Westport, Conn.: AVI Publ., 1972, pp. 86–98
112. Palmer, G.H.: J. Inst. Brew. **78**, 470–472 (1972)
113. Palmer, G.H.: J. Inst. Brew. **79**, 513–518 (1973)
113a. Palmer, G.H.: Am. Soc. Brew. Chem. Proc. 1975, 174–180 (1975)
114. Palmiano, E.P., Juliano, B.O.: Pl. Physiol. **49**, 751–756 (1972)
115. Parker, M.L.: PhD thesis. Univ. Wales (Bangor) (1975)
116. Pollard, C.J.: Pl. Physiol. **44**, 1227–1232 (1969)
117. Price, C.E., Murray, A.W.: Biochem. J. **115**, 129–133 (1969)
118. Radley, M.: Planta (Berl.) **86**, 218–223 (1969)
119. Raison, J.K., Evans, W.J.: Biochim. Biophys. Acta **170**, 448–451 (1968)
120. Reid, J.S.G.: Planta (Berl.) **100**, 131–142 (1971)
121. Reid, J.S.G., Meier, H.: Planta (Berl.) **106**, 44–60 (1972)
122. Reid, J.S.G., Meier, H.: Planta (Berl.) **112**, 301–308 (1973)
123. Roberts, R.M., Deshusses, J., Loewus, F.: Pl. Physiol. **43**, 979–989 (1968)
124. Rowsell, E.V., Goad, L.J.: Biochem. J. **90**, 12p (1964)
125. Ryan, C.J.: Ann. Rev. Pl. Physiol. **24**, 173–196 (1973)

126. Simola, L.K.: Z. Pflanzenphysiol. **78**, 41–51 (1976)
127. Sinha, S.K., Cossins, E.A.: Can. J. Biochem. **43**, 1531–1541 (1965)
128. Smith, D.L.: Protoplasma **79**, 41–57 (1974)
129. Stewart, C.R.: Pl. Physiol. **47**, 157–161 (1971)
130. Stewart, C.R., Beevers, H.: Pl. Physiol. **42**, 1587–1595 (1967)
131. Stobart, A.K., Pinfield, N.J.: New Phytologist **69**, 939–949 (1970)
132. Sugiura, M., Sunobe, Y.: Botan. Mag. (Tokyo) **75**, 63–71 (1962)
133. Sundblom, N.-O., Mikola, J.: Physiol. Plantarum **27**, 281–284 (1972)
134. Swain, R.R., Dekker, E.E.: Biochim. Biophys. Acta **122**, 87–100 (1966)
135. Swain, R.R., Dekker, E.E.: Pl. Physiol. **44**, 319–325 (1969)
136. Taiz, L., Jones, R.L.: Planta (Berl.) **92**, 73–84 (1970)
137. Tanaka, K., Yoshida, T., Asada, K., Kasai, Z.: Arch. Biochem. Biophys. **155**, 136–143 (1973)
138. Thomas, S.M., Ap Rees, T.: Phytochemistry **11**, 2177–2185 (1972)
139. Trelease, R.N., Becker, W.M., Gruber, P.J., Newcomb, E.H.: Pl. Physiol. **48**, 461–475 (1971)
140. Tronier, B., Ory, R.L.: Cereal Chem. **47**, 464–471 (1970)
141. Tronier, B., Ory, R.L., Henningsen, K.W.: Phytochemistry **10**, 1207–1211 (1971)
142. Vigil, E.L.: J. Cell Biol. **46**, 435–454 (1970)
143. Yamada, M.: Sci. Pap. Coll. Gen. Educ., Univ. Tokyo. **5**, 161–169 (1955)
144. Yatsu, L.Y., Jacks, T.J.: Arch. Biochem. Biophys. **124**, 466–471 (1968)
145. Yomo, H., Srinivasen, K.: Plant Physiol. **52**, 671–673 (1973)
146. Yomo, H., Taylor, M.P.: Planta (Berl.) **112**, 35–43 (1973)
147. Yomo, H., Varner, J.E.: Pl. Physiol. **51**, 708–713 (1973)
148. Youle, R.J., Huang, A.H.C.: Biochem. J. **154**, 647–652 (1976)

Chapter 7. Control Processes in the Mobilization of Stored Reserves

By far the most work on the control of enzymes responsible for the mobilization of stored reserves has been carried out on cereal grains and some of the possible reasons for the preferential use of this material should become clear in this chapter. Discussion of this topic inevitably involves a consideration of the effects of plant hormones on cell metabolism and, more specifically, of their actions in non-growing storage tissue. The reason for this is that in the best-understood system — the cereal grain — mobilization of food reserves is quite clearly under hormonal control. The major part of this chapter is therefore devoted to regulation in these grains, especially barley and wheat, but we conclude with an account of control processes in other seeds.

7.1. Control Processes in Cereals

We saw in Chapter 6 how the reserves in the endosperm of cereals are mobilized largely as a result of the activity of enzymes secreted by the aleurone layer. We also referred to the stimulation of aleurone layer activity by gibberellin coming from the embryo. Herein lies the basis of the control exerted by the embryo over food mobilization, which we will now examine more closely.

It has been known since the last century [15] that degradation of the endosperm of barley is markedly slowed down or prevented by removal of the embryo. At the time when these first experiments were done it was not realized that the aleurone layer in fact secreted most of the enzymes acting upon the endosperm, and so the real nature of the link between the embryo and the control of endosperm dissolution was missed. Even when it was shown by Haberlandt [46] that the aleurone layer produced mobilizing enzymes the full role of the embryo remained unappreciated. Nevertheless, it became clear that the rate of degradation of the endosperm depended upon the embryo and, in the case of barley at least, some factor(s) produced during the first three days of embryo growth seemed to be required [75]. Surprisingly, it was not until more than half a century had passed that a diffusible promotive factor from the embryo was experimentally demonstrated. Evidence for this came from the finding that embryoless barley grains produced more α-amylase when incubated together with isolated embryos [88, 117] than they did when incubated alone. The promotive factor was collected, partially purified and identified as a gibberellin. Moreover, it was then shown that chemically characterized gibberellic acid stimulated degradation of the endosperm in embryoless grains, liquefaction occurring virtually as in the intact kernel. The control of food mobilization in barley was thus established: the embryo produces gibberellin which induces the aleurone layer to secrete the food-mobilizing enzymes. This therefore explained why applied gibberellin accelerated the malting process in barley [49].

Gibberellins A_1 and A_3 (gibberellic acid) are the major gibberellins produced by the germinating embryo though there are rival claims as to which is the predominant one [1]. GA_4 and GA_7 have also been detected and it has even been suggested that these various gibberellins might serve different roles — GA_3 and GA_7 to activate the aleurone cells and GA_1 and GA_4 to control embryo growth [1]. In most in vitro experiments GA_3 is used but GA_1 is somewhat more active. Other effective gibberellins are GA_2, GA_4, GA_7 and GA_{22} but some, such as GA_{12}, GA_{17}, GA_{26} are without any promotive action [28].

7.1.1. Gibberellin and the Barley Aleurone Layer

Since the major reserve of the barley endosperm is starch it is not surprising that the most studied enzyme is the one which is largely responsible for its degradation, i.e. α-amylase. This enzyme is synthesized and secreted by the aleurone cells under the influence of GA. But a number of other enzymes are similarly affected, namely proteinase, pentosanase, limit dextrinase and α-glucosidase. In addition, some enzymes are secreted by GA-treated aleurone cells, but do not depend on the gibberellin for their synthesis; these include ribonuclease, β-1,3-glucanase and phosphatase. We will begin the detailed discussion of the nature of gibberellin control by considering the events associated with the synthesis and secretion of α-amylase. This is followed by a discussion of some of the other enzymes and events induced by GA and their possible role in reserve mobilization.

7.1.2. Gibberellins and α-Amylase — "The α-Amylase Story"

Much of our understanding of the action of GA has come from work using the isolated aleurone layers, i.e. those which have been dissected out of the imbibed grain and then incubated in a sterile medium, to which appropriate additions (e.g. GA) have been made. The induced hydrolases are secreted into the medium which can then be assayed for their activity. Not only has this aleurone layer system been studied by seed physiologists interested in the control of mobilization of reserves after germination, but also by developmental physiologists and biochemists to follow the mode of action of GA at the molecular level. As has already been pointed out [2], the aleurone tissue offers several unique features which make it attractive to work with: (1) GAs are the natural trigger of enzyme production in vivo; (2) its response is virtually confined to this one regulator; (3) the cells do not undergo division; (4) the target tissue consists of only one cell type; (5) the cells can be isolated free of other tissues and still respond to GAs; and (6) substrate does not influence the response of isolated aleurone tissue. It is not surprising, therefore, that there is a wealth of literature on the subject, some of which has been incorporated into reviews [1–6].

For many years the favourite experimental material has been the isolated three-cell-thick aleurone layer of the barley (*Hordeum vulgare*) cultivar, Himalaya. This cultivar is conveniently huskless, but it is not used agriculturally and therefore not widely available. Unfortunately, not all harvests of Himalaya give grains that are equally responsive to GA and neither do those of some other, more abundant cultivars of barley; aleurone layers of some cultivars may produce α-amylase without any requirement for GA (i.e. produce a high "background" level of enzyme) while others are completely unresponsive. The reasons for this are not known. There now appear to be certain experimental advantages to using the aleurone layer from wheat grains, not least of which is that this tissue can be isolated by mechanical means in gram quantities, and relatively free of the starchy endosperm [93]. These mechanically isolated aleurone layers still respond to GA, whereas mechanically isolated barley aleurone layers have lost their sensitivity.

Dissection of the aleurone tissue out of intact grains means, of necessity, that it is now a wounded tissue and it may well undergo some metabolic changes to reflect this. Furthermore, the pre-treatment to which barley grains are subjected prior to dissection probably provokes certain modifications in the metabolism of aleurone cells which those in intact, germinated grains are unlikely to undergo, and vice versa [see discussion on changes in wheat grains; Sect. 7.1.10]. Thus, while GA-induced changes observed in isolated aleurone layers may provide a good insight into events taking place in the aleurone cells of the intact grain, additional changes might be occurring which are unique to the isolated tissue.

Aleurone tissue, when isolated from 3-day-old water-imbibed, embryoless half-grains (cv. Himalaya) and incubated in 1 μM GA_3 solution, begins to secrete α-amylase about 8 h after being introduced to the hormone. This release proceeds in a linear fashion over the next 16 h (Fig. 7.1A). The aleurone layer, on the other hand, accumulates only a little α-amylase which reaches a threshold level after 12 h. Even some of this apparent accumulation might really be exterior to the cell, in the cell wall. Several factors influence the continued production of α-amylase: (1) calcium chloride in the incubation medium (10–100 mM) stimulates a four- to five-fold increase in α-amylase; (2) removal of GA from the medium after 7 h incubation (or longer) causes a marked decrease in α-amylase production, restorable by subsequent re-addition of GA; (3) abscisic acid stops enzyme production within a few hours of application (Fig. 7.1B); and (4) introduction of cycloheximide (a protein synthesis inhibitor) to the medium results in almost immediate cessation of further α-amylase production, the enzyme level remaining steady (Fig. 7.1B). This latter observation suggests that α-amylase might be synthesized de novo, and that the released enzyme is stable, surviving undegraded during the subsequent 10 h of incubation [despite the fact that a proteinase is released into the medium along with α-amylase (Fig. 7.5A)].

It is worth noting that it is often assumed that all cells of an isolated aleurone layer are more or less synchronous in their response to GA and that each can continue to produce this enzyme for many hours as long as GA is present. This might not be so, however, and the increase in α-amylase from isolated aleurone layers over 16 h (Fig. 7.1A) might simply reflect an increasing number of cells responding to GA, rather than an increased production of enzyme by cells which are simultaneously responsive after about 8 h. Certainly, observations by Jacobsen and Knox [56] suggest that the middle of the three layers of aleurone cells produces α-amylase before the outer layers. Thus any assay for α-amylase (or any events leading up to its synthesis) probably represents the sum of events of individual cells at different stages of induction, rather than a simultaneous response.

Definitive proof for the de novo synthesis of α-amylase by GA-stimulated barley aleurone tissue was obtained by Filner and Varner [36] in 1967. They used a density-labelling technique which involved incubation of aleurone layers in $H_2{}^{18}O$ (plus GA). The ^{18}O in the water becomes introduced as a density-label in the carboxyl oxygen of amino acids formed during breakdown of protein in the aleurone cells.

Stored protein (aleurone grain)
(for simplicity, a dipeptide is shown)

Amino acids

$$(R_2CH(NH_2)C^{18}OOH)_n \xrightarrow[\text{synthesis}]{\text{protein}} \alpha\text{-Amylase } (^{18}O)$$

They argued that if the extracted α-amylase contained ^{18}O (and was consequently heavier) it must have been synthesized from the ^{18}O amino acids. This would be proof of de novo synthesis of the enzyme which could be separated from light (^{16}O) enzyme on the basis of its density. In order to obtain a convenient label for the light enzyme, Filner and Varner first incubated aleurone layers in GA_3, plus ^3H-lysine. The ^3H-labelled α-amylase-^{16}O produced was collected and placed on an equilibrium density gradient (cesium chloride). An initial experiment showed a coincidence between the radioactive (^3H) peak and that of assayable α-amylase activity (Fig. 7.2A). The fact that α-amylase could incorporate ^3H-lysine was, of course, good evidence for its de novo synthesis, but the final proof came when it was shown that assayed α-amylase synthesized in GA_3-treated aleurone layers incubated in $H_2{}^{18}O$ was heavier than the radioactive marker (^3H) peak (Fig. 7.2 B).

In Chapter 6 it was pointed out that there are several isozymes of α-amylase — according to recent work, as many as seven. In earlier studies, where only four isozymes were isolated [61], all four were found to be synthesized de novo in GA-treated aleurone layers. Whether the other three isozymes arise de novo

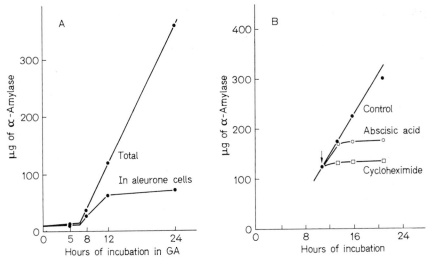

Fig. 7.1. (A) Time course of α-amylase release by 10 aleurone layers of Himalaya barley incubated with 1 μM GA. Enzyme activity measured in the medium surrounding the aleurone layers and in the supernatant of a 0.2 M NaCl extract of the aleurone layers. Total refers to the sum of these two activities. After Chrispeels and Varner, 1967[22]. (B) Midcourse inhibition of α-amylase production and release by abscisic acid and cycloheximide. Aleurone layers were incubated in 0.1 μM GA for 11 h and then 5 μM abscisic acid or 10 μg/ml cycloheximide was added (arrowed). After Chrispeels and Varner, 1967 [23]

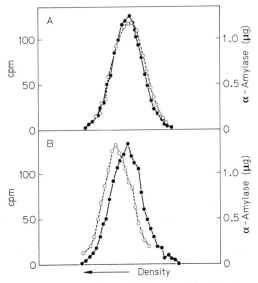

Fig. 7.2. (A) Equilibrium distribution of radioactivity (●) and α-amylase activity (○) after centrifugation of a mixture of purified α-amylase-^{16}O-^{3}H and crude α-amylase induced in $H_2^{16}O$ (i.e. α-amylase-^{16}O). Density of the CsCl gradient increases to the left. Radioactivity measures α-amylase-^{16}O-^{3}H, and enzyme assay α-amylase-^{16}O. The enzyme activity contributed by the radioactive α-amylase-^{16}O-^{3}H is negligible with respect to that contributed by the crude α-amylase-^{16}O. (B) Equilibrium distribution of radioactivity (●) and α-amylase activity (○) of a mixture of purified α-amylase-^{16}O-^{3}H and crude α-amylase induced in $H_2^{18}O$ (i.e. α-amylase-^{18}O). Density of the CsCl gradient increases to the left. Radioactivity measures α-amylase-^{16}O-^{3}H and enzyme assay α-amylase-^{18}O. After Filner and Varner, 1967[36]

has yet to be determined, but there seems to be no reason to doubt that they do.

One criticism levelled at the above studies has come from Clutterbuck and Briggs [25]. They direct attention to the fact that the α-amylase assay used by many workers, including Filner and Varner, is not necessarily specific for this enzyme, but also detects activity of additional carbohydrases, including α-glucosidase, another enzyme which increases in activity in aleurone layers following GA treatment (see later). However, recent studies using a different density-labelling technique, but a specific α-amylase assay, have confirmed that this enzyme is indeed synthesized de novo [47]. For the sake of convenience and brevity in our discussions of α-amylase synthesis and activity we have assumed that all the enzyme assays have α-amylase specificity. Nevertheless, it is likely that some results need confirming and the reader should keep watch on the published literature for any indications that our assumption has been too bold.

7.1.3. Events During the Lag Period

Isolated barley aleurone tissue requires about 8 h of exposure to GA before secretion of α-amylase commences (Fig. 7.1A). The obvious question which arises is: what events must occur in this lag period to culminate in enzyme secretion? There is good evidence that there is no large accumulation of α-amylase within aleurone cells prior to its secretion (e.g. Fig. 7.1A) so the lag period must consist of events which are preparative for synthesis of this enzyme, one of which might involve the transcription of the appropriate mRNA. This particular event could be the rate-limiting process.

It has been claimed, on the basis of electron microscopy, that marked changes occur in the internal organization of aleurone layers (dissected out from three-day-imbibed half grains), during the lag phase between the time of GA addition and the secretion of α-amylase. For example, following GA addition the protein-containing aleurone grains lose their spherical appearance and within 6 h they undergo distinct volume and shape changes [66]. This is probably because of hydrolysis of the proteins within these grains to provide amino acids for the de novo synthesis of α-amylase and other enzymes. Furthermore, electron micrographs appear to show that by the 10th h after GA addition [close to the time that α-amylase is first observed in the surrounding medium (Fig. 7.1A)] the rough endoplasmic reticulum (RER) is more prolific and becomes increasingly stacked [65] in the region of the nucleus [71] an event indicative of increased protein synthesis in that area (but see later).

Some workers have claimed that the number of ribosomes present in isolated barley aleurone cells increases in response to GA [35, 65] although others have shown no GA-stimulated synthesis of ribosomal RNA (or transfer RNA) [59, 109]. Since rRNA synthesis can be stopped in GA-treated aleurone layers without significant effect on either α-amylase synthesis or cellular rRNA levels [59] this can be taken as evidence that synthesis of new ribosomes is not a pre-requisite for enzyme synthesis. An increase in the number of poly-somes in GA-stimulated aleurone cells, commencing some 3–4 h after GA addition has been reported [35]. This may indicate that aleurone cells are becoming increasingly active at synthesizing proteins, although the polysome extraction technique used was one which could have led to spurious increases. GA might act through effecting a qualitative change in protein synthesis, rather than a quantitative one. There is evidence that GA can redirect protein synthesis in aleurone cells. Ten hours after treatment with GA, aleurone tissue synthesizes little of those proteins characteristic of the control tissue incubated in water [110], and between 50 and 60% of the GA-directed protein synthesis is for α-amylase.

7.1.4. Membranes, Polysomes, and α-Amylase

Based on the claims that GA induces polysome and RER proliferation, a hypo-thesis was put forward that an important event in the lag phase between GA addition and α-amylase production is the association of newly-formed poly-somes, bearing mRNA for α-amylase, with newly-formed membranes. Several studies have therefore been made to determine the effects of GA on membrane synthesis, and early results were indeed positive. It appeared, for example, that CDP[1] choline: 1,2-diglyceride phosphorylcholine transferase (which con-verts CDP choline to lecithin) increases in activity in membrane fractions of barley aleurone cells in response to GA [12, 63]. This enzyme was extracted and assayed in vitro using the aleurone cells' own 1,2-diglyceride as substrate, which was bound to the membrane along with the enzyme. Unfortunately, therefore, the concentration of diglyceride was not under experimental control and was probably higher at the time of extraction in membranes from GA-treated aleurone layers, perhaps due to GA-stimulation of its synthesis. Thus the real effect of GA may not have been to stimulate membrane synthesis, but to enhance an unrelated reaction, the product of which was unfortunately available as a substrate, leading to spurious results in the in vitro enzyme assay. A rigorous approach has failed to find any evidence that GA enhances the incorpora-tion of radioactive precursors such as glycerol and acetate into phospholipids and membranes. Over a 12-h period of GA treatment total lipids, total lipid phosphorus and membrane phospholipid remain unchanged compared with the water control [38, 76]. We can therefore conclude that during the lag period there is no net synthesis of membranes. This, in turn, must cast some doubt on the suggestion that there is GA-induced proliferation of RER [66], a sugges-tion founded on a number of electron micrographs which have not been subjected to quantitative analysis. Ideally, for such analysis of structural changes within a cell, electron micrographs of serial sections right across that cell are required. Logistically this is more or less impossible. Thus observations on a limited number of sections can sometimes be misleading, giving rise to conclusions about differences which do not really exist for the cell or tissue as a whole.

---

[1] CDP=cytidine diphosphate

But if future studies confirm that RER assembly indeed takes place in response to GA, then this must involve relocation of existing membrane/lipid units already present in the aleurone cell prior to GA addition, without any new membrane synthesis.

7.1.5. Synthesis and Release of α-Amylase

The 8-h lag period after GA addition is followed by a 16-(or more) h period of rapid α-amylase synthesis (Fig. 7.1A). More ultrastructural changes are said to occur in the aleurone cells during this time [66]. These include further proliferation of the RER, distention of the RER cisternae, continued reduction in the size of the aleurone grains, decreases in the number of oil bodies, an increase in the number of plastids, and loss of the phytin globoid.

There is little doubt that the secretion of α-amylase from the barley aleurone cell is an energy-dependent process [111] although there is considerable disagreement as to the mode of release of this enzyme. Jones and co-workers favour a "soluble" mode of release, i.e. they find no evidence from electron microscope studies for the accumulation of discrete secretory vesicles prior to α-amylase secretion nor, from autoradiographic studies, for the accumulation of newly-synthesized proteins in distinct organelles within the cell [19]. Any α-amylase accumulation which might occur is claimed to be in the perinuclear region, the proposed site of enzyme production on the RER [71]. Using fractionation techniques also, Jones [68] was unable to isolate any cell particle containing α-amylase; more than 95% of the extracted activity was present in the supernatant (i.e. soluble) fraction. According to another ultrastructural study, however, RER vesicles aggregate along the outer cell membrane particularly towards the basal end of the cell (i.e. toward the starchy endosperm) [114], where some α-amylase is also claimed to accumulate prior to release [56]. Moreover, by using improved techniques two distinct fractions of α-amylase have been obtained — one associated with the cell wall (i.e. already secreted) and one associated with, and apparently enclosed by membranes (the latter could be dispersed by addition of a detergent [37]). Failure to remove the cell wall-associated α-amylase and/or disperse the membrane, to make the enclosed enzyme available for assay, substantially reduces the ability to detect this particulate enzyme; this might explain why Jones failed to detect enzyme contained within vesicles. The evidence published to date, therefore, though sometimes conflicting, tends to favour the packaging of α-amylase in vesicles prior to release from the aleurone cells.

7.1.6. α-Amylase and Its Messenger RNA — Site of Action of GA?

The barley aleurone system has attracted biochemists interested in determining if GA acts at the nuclear level by inducing transcription of the gene for α-amylase. Recent studies have taken advantage of the observation that many messenger RNAs in eukaryotic cells contain a covalently linked poly(adenylic acid) seg-

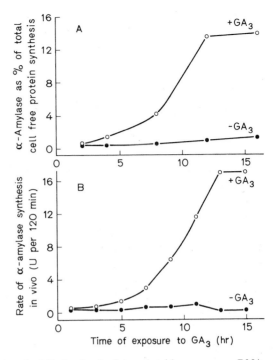

Fig. 7.3A and B. The increase with time in (A) the level of translatable messenger RNA for α-amylase, and (B) the increase in the rate of synthesis of the enzyme in vivo in response to GA treatment. U: units of α-amylase activity. For (A) poly-(A) RNA was extracted from aleurone layers treated with GA for different time periods, and used to support α-amylase synthesis in vitro. After Higgins et al., 1976[50]

ment—poly(A). This allows RNA fractions containing such a segment to be isolated by their affinity for an inert matrix containing their complementary nucleotides: oligo(deoxythymidylic acid)—oligo(dT), or poly(uridylic acid)—poly(U).

It has now been shown that GA induces an increase in poly(A)-containing RNA within isolated aleurone layers [50, 51, 60] and that this RNA when placed in an in vitro protein-synthesizing system catalyses the synthesis of several proteins, including α-amylase (Fig. 7.3A). On the other hand, aleurone layers incubated in water produce an insignificant amount of mRNA for this enzyme (Fig. 7.3A), although messages for other (undefined) proteins are made [50]. The increase in in vitro activity of the mRNA fraction extracted from aleurone cells after increasing times of incubation in GA coincides well with the in vivo production of α-amylase, which occurs over the same time period (Fig. 7.3B). These observations are suggestive of α-amylase messages starting to be transcribed within 3–4 h of the addition of GA, which are translated over a subsequent 10-h period. Other possible explanations exist however. For example, GA (or a product of GA action) might protect mRNA from degradation, so that the mRNA for α-amylase could be destroyed in the aleurone cells incubated in water but not in those incubated in GA. Alternatively, GA might be responsible for enhancing the translational capacity of pre-formed mRNA for α-amylase by effecting its processing, activation or release from an inactive form. If future studies do confirm that GA can activate the gene for α-amylase the next question to be resolved is, of course, — how? Whatever its final action, GA must presumably first bind to some kind of receptor site and the GA-receptor complex must then initiate one or more primary events which eventually lead to the synthesis of α-amylase (and other hydrolases).

The possibility of post-transcriptional control of enzyme synthesis has been raised by several workers, and it has been noted that methylation of purine residues of tRNA and heavy rRNA is enhanced in GA-treated aleurone tissue [18]. It is not clear how GA might influence this process, or why methylation of certain types of RNA should be essential for α-amylase synthesis. Other studies claiming to show post-transcriptional control by GA [16] have received no support [17].

The synthesis and secretion of α-amylase by aleurone cells is sharply reduced when GA is withdrawn, say after a 12-h treatment [22, 67]. If the presence of GA is required all the time for the continued production of enzyme, then it is necessary to determine if there is continued synthesis of α-amylase mRNA after 12 h of incubation on GA. If not (and results in Fig. 7.3A suggest not) then the requirement for the continued presence of GA must be to control some event in the synthesis of α-amylase other than the production of mRNA — perhaps the translation process.

Finally, a brief word about cyclic AMP. Activation of the enzyme adenyl cyclase which catalyses cyclic 3′,5′-adenosine monophophate synthesis constitutes the initial molecular event in the target cells of several mammalian hormones. The nucleotide then initiates a series of events leading to a final response characteristic of the hormone. Despite numerous studies on the barley aleurone layer there is no convincing evidence that GA action is mediated via cyclic AMP [73]; in fact the weight of evidence is against this nucleotide playing an important role in any plant tissue [7].

7.1.7. GA and α-Amylase: Regulation in the Intact Grain

Formation of GA and α-amylase in intact grains of barley (*H. distichon,* cv. Proctor) has been compared between those germinated at 14.4°C under malting conditions (see Chap. 5) and those placed on moist filter paper at 25°C. In the whole, malted grains (Fig. 7.4A) the content of gibberellin-like material (called such because the material extracted gives positive responses in a bioassay for gibberellins, but is not positively identified as a gibberellin) rises to a peak by the second day after imbibition starts, a time when α-amylase synthesis commences. The enzyme continues to be produced as the content of GA-like material declines. The rate of synthesis of the enzyme does decline, however, lagging behind the decline in GA levels by about one day. Removal of the embryo from malted grains 30 or more h after the start of imbibition does not prevent subsequent α-amylase synthesis (Fig. 7.4C). This is probably because

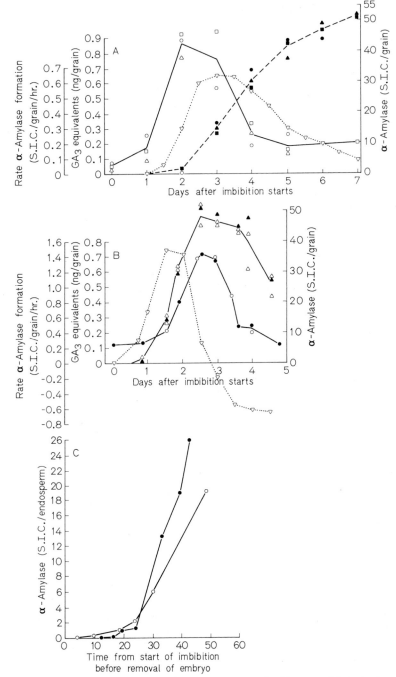

Fig. 7.4. (A) and (B) Changes in levels of α-amylase and gibberellin-like materials in barley (*Hordeum distichon* cv. Proctor) (A) malted at 14.4°C, and (B) germinated on moistened filter paper at 25°C. Symbols in (A) (represent values from three separate experiments): □──○──△: gibberellin-like material; ■----●----▲: α-amylase; ···▽···▽···: rate of formation of α-amylase. Symbols in (B) (values from two experiments): ──○──●──: gibberellin-like material; ──△──▲──: α-amylase; ···▽···▽···: rate of formation of α-amylase. (C) α-amylase development in malted grain de-embryonated at various times from 5 to 50h after the start of imbibition and then incubated further for 3 days (○) or 4¹/₂ days (●). After Groat and Briggs, 1969[44]

gibberellins are produced by the embryos over the first 30 h and released into the endosperm, where they accumulate. From there they can continue to diffuse into the aleurone tissue for several days, promoting synthesis of α-amylase.

In barley germinated on moist filter paper at 25° (Fig. 7.4B) an increase in gibberellin-like compounds actually follows the increase in α-amylase synthesis (which itself is produced sooner than under malting conditions). Moreover, the rate of α-amylase synthesis starts to decline at a time when the synthesis of GA-like material commences. It could be argued that the low amount of gibberellin present in the grains over the first one and a half days after imbibition commences (Fig. 7.4B) is sufficient to trigger α-amylase synthesis: if this is so, one wonders what is the function of the subsequent massive production of gibberellin. It could, of course, be concerned mainly with the regulation of embryo growth. A similar non-coincidence between GA-levels and α-amylase synthesis has also been reported in *H. vulgare* cv. Himalaya [70]. The reasons for these interesting discrepancies are not known, but they serve to illustrate that in the intact grain, as in the isolated aleurone layer, there are still some fundamental aspects of the relationship between GA activity and α-amylase synthesis which are incompletely understood.

7.1.8. Control of α-Amylase Synthesis by the Products of Enzyme Hydrolysis

Attempts have been made to determine why α-amylase synthesis eventually declines in the intact grain (e.g. see the decline in rate of α-amylase synthesis in Figure 7.4A and B after three and two days respectively). One possible reason is that production of the stimulus (i.e. GA) for α-amylase synthesis decreases, perhaps because hormonal synthesis is itself inhibited by the products of starch degradation. Unequivocal evidence supporting this possibility is lacking. In any case, sufficient gibberellin is released into the endosperm during the first one or two days after the start of imbibition to support α-amylase synthesis for at least four to five days thereafter (Fig. 7.4C). Thus any limitation in GA synthesis after one or two days is unlikely to affect enzyme production.

Intact grains of Himalaya barley germinated on sterile sand initiate α-amylase synthesis on the second day after imbibition, but this synthesis ceases by day 4 [70]. Addition of GA to intact grains at this latter time does not induce more enzyme production, indicating that cessation of α-amylase synthesis is unlikely to be a consequence of limiting endogenous GA. Neither has the aleurone layer lost its ability to make α-amylase, for it does so when dissected out from the intact grain and incubated with GA. Thus it appears that something in the intact grain is suppressing continued α-amylase production by the aleurone layer. Jones and Armstrong [70] postulate that the inhibition is caused by an accumulation of the products of hydrolysis of starch (glucose and maltose) in the endosperm, which reach osmotic concentrations as high as 400–500 milliosmolar by the third day after imbibition starts. This increased osmotic concentration in the endosperm might cause the aleurone cells to undergo osmotic stress, thus reducing their metabolism. Certainly, α-amylase synthesis by isolated aleurone layers is increasingly inhibited when they are incubated in 0.1–0.4 M glucose or maltose, and osmotic agents which are not products of starch hydrolysis, e.g. polyethylene glycol and mannitol, also inhibit enzyme synthesis [8].

Thus, the control mechanism for α-amylase synthesis in the intact barley grain might operate as follows. Following imbibition there is synthesis of GA within the embryo which then diffuses into the starchy endosperm and aleurone layer (see Chap. 6). During this time the aleurone cells must undergo some metabolic changes to make them "ripe" for stimulation by GA [120] which then provokes synthesis of hydrolases, particularly of α-amylase. This is secreted into the endosperm and there commences hydrolysis of the stored starch grains. Some of the products of amylolysis (maltose and glucose) may be converted to sucrose by the aleurone cells and transported to the growing embryo [24] but most are absorbed directly through the scutellum (Chap. 6). Since production of these sugars exceeds their rate of transport and utilization, they accumulate and act as an effective switch to stop synthesis of further α-amylase, an enzyme whose activity in the endosperm is already in excess of the requirement for the products of its hydrolytic activity.

The effective level of control in the aleurone cells elicited by the osmotica is unknown, although polyethylene glycol can cause a general reduction in the metabolic activity of these cells, including protein synthesis [21]. The possibility that osmotica can also reduce α-amylase release cannot yet be discounted.

7.1.9. Regulation of Other Hydrolases in the Barley Aleurone Layer

Studies on the early responses of aleurone layers to GA have shown that α-amylase is not the first enzyme to be secreted. Within 5 h of the addition of GA to isolated aleurone layers of barley (cv. Betzes) there is increased secretion into the surrounding medium of several hydrolytic enzymes, including an ATPase, phytase, β-glucosidase and phosphomonoesterase, as well as some soluble carbohydrate [96] (probably sucrose). Similar observations were made using GA-treated half-grains of both barley and wheat (the distal embryoless half), although in barley additional enzymes, e.g. an esterase, phosphodiesterase and α-galactosidase were also found to be secreted early [96]. From the limited information available to date it appears that some enzymes which are secreted early after GA addition (e.g. phosphomonoesterase and phosphodiesterase) do so in response to lower concentrations of GA than do the later ones (e.g. α-amylase) [97]. Furthermore, α-amylase and proteinase, which are secreted simultaneously, have identical GA dose-response curves (Fig. 7.5A and B). There are exceptions, however, and further work on more enzymes is required to determine if the sequence of enzyme production by aleurone tissue is related to GA concentration in the cells.

Although GA-induced de novo synthesis and release of proteinase shows characteristics which are remarkably similar to those of α-amylase (Fig. 7.5A and B), the level of proteinase is some 30-fold less than that of α-amylase. The role of this proteinase in the hydrolysis of protein reserves within the aleurone grains and endosperm is still not fully understood. As we have already

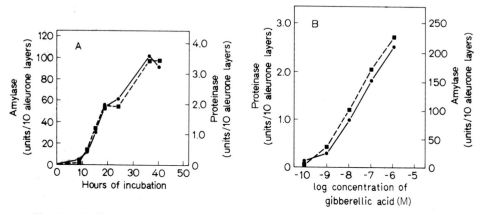

Fig. 7.5. (A) Time course of release of amylase and proteinase by aleurone layers of Himalaya barley in the presence of 1 μM GA. (B) Release of these two enzymes by aleurone
layers in response to various GA_3 concentrations. ●——●: amylase; ■----■: proteinase.
After Jacobsen and Varner, 1967[58]

pointed out, hydrolysis of protein reserves in the aleurone cells takes place
prior to α-amylase synthesis to provide amino acids for this process. Likewise,
some of these amino acids can be used for the de novo synthesis of the proteinase
itself [58]. Some proteinase activity is occurring within the aleurone cells to
provide precursors for these de novo syntheses, but whether this is due to
the activity of the same proteinase which is secreted along with α-amylase
still remains to be determined. Figure 7.5A shows only the pattern of secretion
of the proteinase enzyme: it is possible that some enzyme is synthesized and
is active in the aleurone cell prior to its secretion and that some of the synthesized
enzyme is not released. Other minor proteinases also exist within aleurone
cells, and these might also play a role in the provision of amino acids. However,
there do not appear to be any unique proteinase types which are synthesized
in the aleurone layer but not released [104], although there are some unique
proteinases in the starchy endosperm which are synthesized or stored there
during maturation — see Chapter 6.

At least a dozen proteins are released into the surrounding medium by
aleurone layers treated with GA. Of these, ten appear to be de novo-synthesized
proteins, and two do not [57]. The latter are pre-formed proteins which might
be released from the cells in response to added GA, but they might also be
proteins bound outside the cell within the cell wall and released therefrom
by secreted proteinase(s). Thus the secreted proteinase(s) might have several
roles in the intact germinated grain: to release proteins from the aleurone
cell wall; to hydrolyse these and stored endosperm proteins in order to provide
amino acids for the embryo; and to hydrolyse proteins within the aleurone
cells to provide amino acids for synthetic processes therein. The possibility
cannot be ignored that some amino acids might be resorbed into, and used
within the aleurone layers following hydrolysis of those proteins bound to the
wall and of those in the starchy endosperm.

While α-amylase and proteinase are synthesized de novo and secreted from aleurone cells in response to GA, β-1,3-glucanase is synthesized de novo equally well by aleurone cells incubated in water, although it is only released when GA is present [11, 106]. The significance of this enzyme to the germinated grain is obscure. At one time it was thought that its function is to digest the aleurone cell wall, thus facilitating the passage of other released hydrolases into the endosperm. Cell wall degradation in the basal area of the aleurone cells (i.e. on the side towards the endosperm) does occur, but this cannot be due to β-1,3-glucanase because the cell wall contains an unappreciable amount of β-1,3-glucans. In fact the cell wall has a high arabinoxylan content (85%), with little cellulose (8%) [83]. Recently, it has been shown that three pentosanases capable of degrading this polymer (endo-β-1,4-xylanase, β-xylanopyranosidase and α-arabinofuranosidase) increase in activity in germinated barley grains and are stimulated by GA in isolated aleurone layers [53, 105a]. It is presumably these enzymes which are really responsible for the observed cell wall degradation. It should be remembered, furthermore, that β-1,3-glucanase, other β-glucanases and pentosanases all play an extremely important role in degrading the walls of the starchy endosperm, thus rendering the starch and protein more accessible for enzymatic attack (Chap. 6 and Ref. [4a]).

α-Amylase and acid phosphatase are both known to accumulate in the cell walls of the aleurone layers [9, 111], the latter enzyme even being present at low levels in the aleurone cell walls in the mature dry grain (Fig. 7.6A). Isolated aleurone layers incubated in the absence of GA accumulate acid phosphatase in the inner region of the wall (Fig. 7.6B), indicating that synthesis and release of this enzyme from the cytoplasm is not a GA-dependent process. In the absence of GA, however, no enzyme is released from the whole tissue into the surrounding medium. The accumulation of enzyme within the inner wall could be because it is selectively bound to the wall in this region and/or because the wall outside this area is impermeable to the enzyme. Introduction of GA to the aleurone cells results in digestion of the cell walls (presumably due to synthesis and release of pentosanases) and, after several hours, phosphatase can be detected in localized pockets of wall digestion (Fig. 7.6C), some of which extend to the surface of the layer, forming wall channels through which this (and other) enzymes can diffuse to the outside. Release of acid phosphatase from aleurone layers commences some 6–12 h after the introduction of GA [9]. Aleurone cell walls are rich in plasmodesmata [107], and the detection of acid phosphatase activity in digested areas around these structures suggests that wall-digesting enzymes might widen them to provide the channels for enzyme release. Acid phosphatase thus represents an example of an enzyme whose synthesis within the aleurone cell is not greatly stimulated by GA, but whose release from the aleurone layer is. Nevertheless, it should be noted carefully in this case that it is not release from the cytoplasm (i.e. through the plasmalemma) which is stimulated by GA, but release from the cell wall, indirectly, as a consequence of the prior induction of cell wall-degrading enzymes.

Ribonuclease synthesis by isolated aleurone layers is enhanced only little by applied GA, although, like acid phosphatase, its release is massively stimulated (Fig. 7.7A). This release occurs considerably later than that of α-amylase,

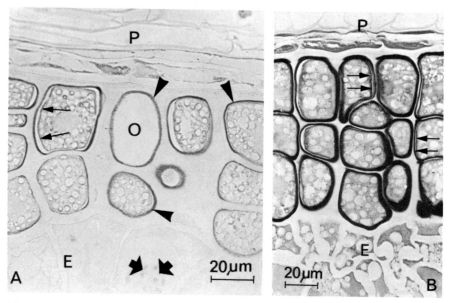

Fig. 7.6. (A) Transverse section of dry barley grain showing acid phosphatase activity
in the inner region of the wall around each aleurone cell *(large black arrows)* and at
the periphery of the cell cytoplasm *(small black arrows)*. Large wide arrows indicate
weak enzyme activity in deposits in the sub-aleurone cells (*E*). Cell *O* has lost its protoplast
in sectioning. *P*: seed coat. (B) TS of isolated aleurone layer incubated for 16 h in the
absence of GA. Note the increase in intensity of acid phosphatase activity in the wall
enzyme-band compared with (A). No enzyme activity remains in the sub-aleurone cells (*E*).
Arrowed is the acid phosphatase activity in the periphery of the cytoplasm; there is little
activity elsewhere in the cytoplasm. *P*: seed coat. (C) Section of aleurone layer incubated
16 h + GA to demonstrate changes in enzyme distribution in the wall. Wall enzyme is
found in fine plasmodesmata-like strands *(small arrows)* and in much broader strands
(white arrow) either crossing the wall to adjacent cells or stopping a short distance
out in the wall from the cell where they originate *(large black arrows)*. Large irregular
patches of enzyme activity (*D*) are also present. From Ashford and Jacobsen, 1974 [9]

proteinase (Fig. 7.5A) and acid phosphatase [9] and the enzyme builds up within
the aleurone layers for at least a day after GA application (Fig. 7.7B). The
block to ribonuclease release is not known, but it probably does not reside
within the cell wall since other enzymes are released within 6–8 h of GA application. It is likely that the cell wall channels formed by this time provide a
common path for the release of all enzymes into the surrounding medium.
The role of the released ribonuclease could be to degrade residual RNA present
in the endosperm, and perhaps provide nitrogenous bases for the growing seedling. It is known, however, that many plant tissues produce ribonuclease in
response to wounding, and the production of this enzyme by isolated aleurone
layers incubated on water might, in part, reflect the wounded status of this
tissue. How ribonuclease present within the aleurone cells themselves is
prevented from destroying the RNA therein (e.g. the putative newly synthesized
mRNA for α-amylase) is not known. The simplest explanation is that it is
somehow sequestered, perhaps in vesicles [72].

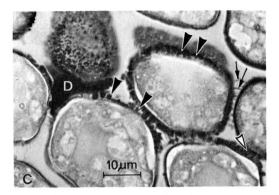

Fig. 7.6C

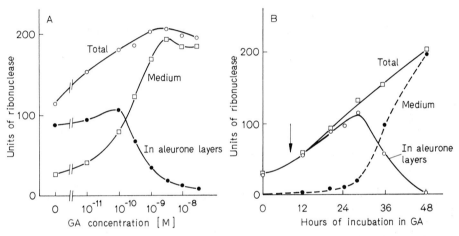

Fig. 7.7. (A) Dose-response curve of the production of ribonuclease by aleurone layers at different concentrations of GA. Enzyme activity was measured in the medium and an extract of the tissue after 48 h incubation of the isolated aleurone layers. (B) Time course of ribonuclease production by 10 aleurone layers incubated with 5 nM GA. *Arrowed* is the time of α-amylase release (see Figure 7.1A). After Chrispeels and Varner, 1967 [22]

A rapid efflux of potassium, magnesium and phosphate ions occurs some 6–8 h after addition of GA to isolated aleurone layers [69]. These ions might be released by acid phosphatase from the phytin complex associated with the aleurone grains. It has been suggested that GA might bring about the formation of an energy-linked ion-carrying system within the cell membrane, thus facilitating ion release from the aleurone cells [69]. On the basis of the available evidence though, it is entirely possible that release of ions from the aleurone cells, like that of acid phosphatase, is retarded merely because of the lack of available channels for the release through the walls, and must await GA-enhanced induction of the cell wall-degrading pentosanases. In turn, the ions themselves might facilitate passage of enzymes through the cell walls because it has been observed

Table 7.1. What is known (or we think is known) about GA-induced hydrolase synthesis in isolated barley aleurone layers

Event	Status of knowledge
GA induces de novo α-amylase(s) synthesis	Undisputed
During lag phase	
Amino acids must be provided for de novo hydrolase synthesis	Undisputed
GA induces ribosome and membrane synthesis and RER proliferation	Disputed
GA enhances polysome formation and protein synthesis	Disputed
GA causes qualitative changes in protein synthesis	Undisputed
GA enhances synthesis of poly(A)-containing RNA, including the mRNA for α-amylase	Possible, but GA-enhanced activation or protection of mRNA not eliminated
GA acts at the translational level of control	No firm evidence
GA action is mediated via cyclic AMP	Highly unlikely
α-Amylase release	
Mode of enzyme release is soluble	Possible
Mode of enzyme release is vesicular	More likely, but requires confirmation
Release is an energy-dependent process	Undisputed
Other hydrolases etc.	
GA induces de novo synthesis of proteinase, limit dextrinase, pentosanases and α-glucosidase	Undisputed
Acid Phosphatase, ribonuclease, β-1,3-glucanase and peroxidase synthesized in the absence of GA: GA enhances release	Undisputed
Major enzyme release follows formation of release channels through aleurone cell walls	Undisputed
Pentosan biosynthesis ceases when cell walls are degraded	Undisputed
GA induces ion release	Undisputed, although how GA does this is debatable

See text for further details

that at low ionic strengths α-amylase release is considerably retarded [111]. The release of these, and other ions (e.g. Mn^{2+}, Ca^{2+} and Fe^{2+}) is possibly also important to the nutrition of the seedling.

Two other enzymes associated with the process of starch degradation are synthesized de novo in the aleurone cells of the barley grain in response to

GA — these are limit dextrinase and α-glucosidase [47]. The significance of these enzymes in the hydrolysis of starch reserves is discussed in Chapter 6. β-Amylase is not synthesized de novo in the presence or absence of GA, but is carried over in an inactive form from the developing grain. It may be activated indirectly by GA through induction of a proteinase which releases it from its inactive bound form (Chap. 6).

Besides stimulating the production and release of certain hydrolases, GA can also retard enzyme activity. Consistent with the observation that GA induces degradation of aleurone cell walls is the observation that GA inhibits the synthesis of pentosan (arabinose and xylose) components of the wall. Between 4 and 16 h after GA addition to isolated aleurone tissue there is a marked decline in pentosan biosynthesis, in part due to the reduced activity of a membrane-bound arabinosyl transferase, and perhaps also of xylosyl transferase [62]. Changes in such membrane-bound enzymes could reflect a shift in the role of membranes from the production of cell wall components to the production, or packaging of hydrolases.

Table 7.1 is a summary of some of the events which, over the years, have been reported to take place in isolated barley aleurone layers in response to GA. The present status of knowledge for each event is also noted, although readers are encouraged to keep an eye on the literature and note future amendments.

7.1.10. GA-Induced Enzymes in Other Cereals

Gibberellin-induced synthesis and secretion of hydrolases has also been studied in a number of other cereals particularly wheat (*Triticum aestivum*) and wild oat (*Avena fatua*).

A considerable contribution to our understanding of the metabolic changes which the aleurone cells undergo within the cereal grain during the early stages after the start of imbibition has come from work by Laidman and coworkers. They have found that some changes which occur in the aleurone layer of wheat grains are stimulated by the embryo, and others occur independently. These findings are summarized in Table 7.2. Dependency upon the embryo is not necessarily just for GA, since, as shown in Table 7.3, induction of triglyceride metabolism (and neutral lipase activity) cannot be promoted by GA alone, whereas induction of metabolism of fatty acids via the glyoxylate cycle can. Other factors, e.g. IAA plus glutamine, can induce triglyceride metabolism in the aleurone tissues of wheat half-grains (Table 7.3) although it is not yet known whether these are the controlling factors normally transmitted from the embryo. α-Amylase induction in the aleurone layer appears to be enhanced by the presence of either cytokinin or the attached starchy endosperm [34] and it has been suggested that GA-stimulated α-amylase induction in the aleurone layer of the intact wheat grain follows its prior "sensitization" by cytokinin released from the endosperm.

That phospholipid synthesis and proliferation of the endoplasmic reticulum occur within wheat aleurone cells independently of the embryo (Table 7.2) or

Table 7.2. Embryo-dependent and embryo-independent changes in the aleurone layer of imbibed wheat grains

Embryo-independent
1. Development of activity of enzymes of the glycolysis-gluconeogenesis pathway, pentose phosphate pathway, citric acid cycle and the electron transport chain [31]
2. Development of phytase activity and initial mobilization of phytin (but see embryo-dependent No. 3) [32]
3. Induction of total nucleic acid synthesis [78]
4. Induction of phospholipid synthesis and proliferation of the endoplasmic reticulum [26, 113]

Embryo-dependent
1. Induction of neutral lipase activity [108], i.e. triglyceride metabolism
2. Development of activity of enzymes of β-oxidation, and isocitrate lyase and malate synthetase of the glyoxylate cycle [30], i.e. activity of glyoxysomes
3. Development of phytase activity above levels present in dry seed and initially present after imbibition [32]
4. Release of inorganic products of phytin hydrolysis (K^+, Ca^{2+}, Mg^{2+} and phosphate ions) [33]
5. Induction of α-amylase synthesis [89]

Table 7.3. Sensitivity of embryo-dependent responses of wheat aleurone tissue to GA and other promoters

Event	Response to GA	Alternative promoting substances
Induction of triglyceride metabolism	None	Starchy endosperm factor (cytokinin?) induces ca. 20% activity Cytokinin (100 nM) induces ca. 10% activity Indole acetic acid (10 μM) induces ca. 10% activity Indole acetic acid (10 nM) + 1 mM glutamine induces full activity
Neutral lipase activity	None	Embryo factor (including GA?) alone induces ca. 15% activity Indole acetic acid (10 μM) + 1 mM glutamine induces ca. 70% activity Nitrogenous compounds induce 0–55% activity [108]
Isocitrate lyase and malate synthetase activity	Stimulated	
Release of inorganic phosphate and other ions	Stimulated	Starchy endosperm factor (cytokinin?) might reduce ion release
Induction of α-amylase synthesis	Stimulated	Pre-incubation of aleurone tissue with cytokinin enhances subsequent GA stimulation

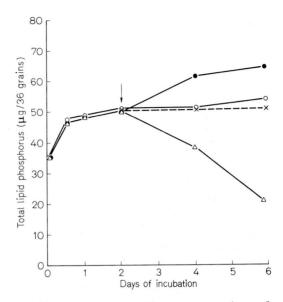

Fig. 7.8. The levels of total phospholipids in wheat aleurone tissues. o——o: tissue of germinating and growing grain; ●——●: tissue of de-embryonated grain incubated without GA_3; △——△: tissue of de-embryonated grain incubated $+1\,\mu M$ GA_3; ×----×: tissue of de-embryonated grain incubated with embryo diffusate$+1\,\mu M$ GA_3. *Arrow* indicates time when α-amylase activity is first detectable. Based on Varty and Laidman, 1976 [113]

GA [26, 113] is an interesting observation, particularly in the light of the controversy over the putative role of GA in controlling these events in barley aleurone tissue (see Table 7.1 and associated text). In aleurone layers of intact wheat grains the synthesis of phospholipid increases over the 1st day after imbibition starts, a time before GA induces α-amylase synthesis in the whole grain (Fig. 7.8). Aleurone layers of embryoless grains also show an increase in phospholipid levels, an event uninfluenced by added GA. After 48 h of incubation, however, phospholipid levels decline in GA-treated tissue (Fig. 7.8) which indicates that GA-enhanced phospholipid breakdown occurs. On the other hand, in the presence of the embryo, or of an embryo diffusate, this phospholipid breakdown is prevented (Fig. 7.8). Such results serve to illustrate the complex role of the embryo in modifying the metabolism of the aleurone layer. Furthermore, they show that GA does not increase or regulate phospholipid biosynthesis prior to the time of enhancement of α-amylase activity (Fig. 7.8), an observation which is in agreement with the current view that GA is without effect on membrane biosynthesis in barley aleurone layers (see earlier).

Ultrastructural studies on wheat aleurone cells over the same time period show that proliferation of long profiles of endoplasmic reticulum (ER) occurs during the first two days of incubation independently of the embryo and therefore of control by gibberellic acid (Fig. 7.9). The aleurone cells of ungerminated wheat grains (like those of barley [64]) contain numerous oil bodies around the periphery of the aleurone grains, with mitochondria packed within the cytoplasm, and sparse, short profiles of ER (Fig. 7.9A). By day 2 after the commencement of imbibition, after germination of the intact grain, the ER has developed into long profiles (Fig. 7.9B) and is often associated with the plastids. This type of pattern persists for a further two to four days. Similar proliferation of ER occurs in the aleurone cells of de-embryonated grain, as shown in Figure 7.9C. By comparison, the same cells of de-embryonated grains incubated for four days on GA show few long profiles of ER, but shorter distended profiles and many vesicles (Fig. 7.9D). Such micrographs agree with some of those of barley aleurone tissue [114], where similar vesiculation has been suggested to result from GA action. Since wheat aleurone cells, like those of barley, produce α-amylase in response to GA, it might be that the vesiculation of ER represents packaging of this hydrolase for secretion following induction of its synthesis. We might note that the mode of secretion of α-amylase from the isolated wheat aleurone layer is probably vesicular, for over 80% of the GA-induced enzyme in this tissue appears to be membrane enclosed [40, 41]. Disagreement with the barley-aleurone studies stems from the observation that here in wheat there is ER (mainly RER) proliferation independent of the presence of GA, whereas in barley this event has been ascribed to an effect of GA [66]. Further studies are obviously necessary to determine if there are any differences in the control of RER proliferation in the aleurone cells of these two cereals.

Cells of the isolated single aleurone layer of wheat will not respond to GA immediately after mechanical isolation, but will do so after about 12 h, although for the first 6 h of this time GA does not have to be present. Thus, aleurone cells incubated on water for 6 h and then exposed to GA start secreting α-amylase 6–8 h later [27].

No studies have yet been carried out on the intact wheat grain to determine how α-amylase production might be controlled once starch hydrolysis has commenced (see Sect. 7.1.8 for a discussion of the control mechanism in barley). The naturally embryoless wheat grains which are found in low percentages in certain cultivars have intact aleurone layers which respond to exogenous GA [74]. These seem to us to be an interesting grain in which to study how α-amylase production is regulated by the products of its hydrolytic action. This grain has the inherent advantage that there is no natural sink to which sugars might be transported, no embryo dissection is required, and there is presumably no endogenous GA production to contend with. Before leaving the case of wheat we should note

▶

Fig. 7.9A–D. Ultrastructural changes in the aleurone cells of wheat. (A) Aleurone cell from ungerminated wheat grain. Note short profiles of endoplasmic reticulum (*ER*), mitochondria (*M*), oil bodies (*S*) and large aleurone grains (*AG*). (B) Aleurone cell from grain two days after the start of imbibition. Long profiles of *ER* are present associated with plastids (*P*). Thylakoid structures are visible within the plastids. (C) Aleurone cell from de-embryonated grain incubated for six days in the absence of gibberellin. *ER* profiles are ramified through spaces between the aleurone grains and are associated with plastids (*P*). (D) Aleurone cell from de-embryonated grain incubated for four days in the presence of 10 μM gibberellin. Long profiles of *ER* are rarely seen, but shorter profiles of distended endoplasmic reticulum (*DER*) are found, and many vesicles (*V*) are present. Plastids (*P*) in association with some short profiles of *ER* are visible, and dictyosomes (*D*) with vesicles budding off from them are often seen. From Colborne et al., 1976[26]

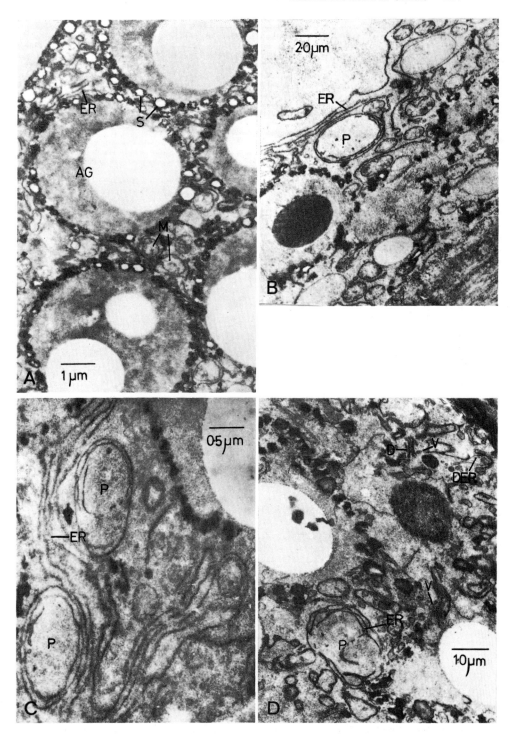

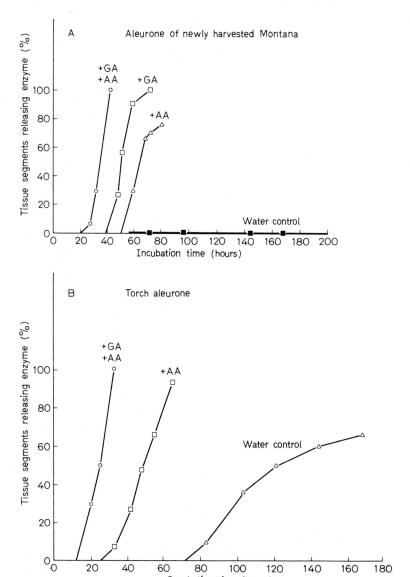

Fig. 7.10. (A) Interaction of GA and a complete amino acid mixture (*AA*) on α-amylase production by "Montana" aleurone strips. (B) Response of "Torch" aleurone strips to GA and a complete amino acid mixture (*AA*). The response curve for GA alone was not published. After Naylor, 1966 [85]

that there are quantitative differences in the responses of various cultivars to gibberellin. Of particular interest is the observation that aleurone cells of dwarf wheats have a greatly diminished response to the regulator [38a].

Avena aleurone layers featured in some of the earliest experiments utilizing isolated tissue to study GA effects on enzyme production. Naylor (1966) [85] dissected aleurone layers from 30 min-imbibed grains of a dormant inbred line of wild oats (*A. fatua*, "Montana") and a non-dormant domestic variety of oats (*Avena sativa*, "Torch"), placing these on an incubation medium with and without GA. The non-GA-treated aleurone layers of "Montana" produced no α-amylase (Fig. 7.10A), whereas the treated (+GA+AA) "Montana" and "Torch" aleurone tissue synthesized this enzyme within 20 and 40 h respectively (Fig. 7.10A and B). Both RNA and protein synthesis were observed to occur in "Montana" during the long lag phase prior to the GA-induced α-amylase synthesis, but the lag could be shortened by 20 h by the inclusion of a complete amino acid mixture with the GA (Fig. 7.10A). The implication is that GA-induced amino acid production (presumably via a GA-induced proteinase) within the aleurone layers is the limiting factor for enzyme synthesis. This is unlikely to be the only limiting factor, however, for while addition of amino acids alone was found to induce α-amylase synthesis, this occurred 15 to 20 h later than in the GA-treatment (Fig. 7.10A). The lag phase in "Torch" aleurone layers was likewise shortened by amino acids added with GA, although "Torch" aleurone layers did eventually produce α-amylase even without treatment (Fig. 7.10B). Either "Torch" aleurone layers had a low endogenous level of GA which itself activated amino acid and α-amylase production after longer incubation periods or (though less likely) the block to α-amylase synthesis was relieved by the slow production of a proteinase and/or a subsequent increase in the endogenous amino acid pool.

Unlike those of barley, wheat and oats, de-embryonated kernels of some *Zea mays*, (line Seneca Chief sweet corn, and hybrids Wf9 × 38–11, Wf9–39–11, 1097 × 1126, and Pride-5) do not require added GA for the development of α-amylase activity [43, 48]. Enzyme synthesis might of course be stimulated by any GA in the aleurone cells and/or starchy endosperm deposited there during seed development. Alternatively (and as discussed in Chap. 6) hydrolases might be pre-formed in the endosperm during maturation and then released, rather than synthesized, following germination. Other hybrids or lines of *Z. mays*, however, seem to require GA to induce catabolism of carbohydrate and proteins, e.g. Wf9 × M14 [55] and dwarf (d_5) mutant [48]. Mature grains of these might have a low endogenous GA level [94].

Transversely cut apical (embryoless) half-grains of rice respond to GA and produce α-amylase and proteinase: water-imbibed controls produce neither enzyme [87, 90]. In embryoless grains of IR8 rice, proteinase production precedes that of the amylase by 2.5 days (cf. barley). The embryos of normal varieties of rice appear to produce sufficient GA to induce the production of hydrolases in the germinated grain, but certain dwarf rice lines have a low endogenous GA level, and hence lower α-amylase and proteinase production [90].

7.2. Control Processes in Other Seeds

Knowledge concerning the control of reserve food mobilization in seeds other than cereals is limited. The reason for this may be because no system as

experimentally manipulatable as the cereal aleurone layer has yet been discovered. We have seen in Chapters 3 and 6 that in dicot seeds reserves
are stored in the endosperm or in the cotyledons. As in cereal grains, food
mobilization must be linked to embryo growth because it does not occur
until after germination has taken place. Thus, we might expect the embryonic
axis of the growing dicot seedling to regulate breakdown of reserves in the
storage tissues. Now since the endosperm might completely enclose the embryo
it is impossible to remove the latter (as was done in cereals) without seriously
damaging the endosperm and thereby interfering with its normal functioning.
Similarly, to investigate the effect of the axis upon the cotyledons the experimenter may again resort to surgical methods, excising the axis from the rest of
the embryo — a treatment which is less than ideal. We should also note that
one great advantage shown by cereals is found in only a minority of other
seeds; this is the possession of a discrete group of cells (the aleurone layer)
which is responsible for the production of virtually all of the enzymes concerned
with the initial breakdown of food stores. This layer of cells, as we have seen,
can be easily isolated and used for experimental investigations.

Several studies have nevertheless been made on the control of mobilization
in non-cereal seeds, most of the investigations falling into two groups: (1) Those
which examine the effect of the embryo or axis upon reserve breakdown and
enzyme synthesis; (2) Those which investigate the influence of exogenous plant
growth regulators (hormones) on the storage tissues.

But before discussing these aspects some questions should be considered:
are the enzymes which participate in food mobilization present in the dry seed,
requiring only hydration and possibly activation, or alternatively, are they synthesized during germination and growth?

Most of the evidence suggests that the enzymes are newly synthesized, though
in the majority of cases rigorous proof is unavailable. Enzyme development
in a number of seeds is prevented by inhibitors of protein and/or RNA synthesis.
For example, dipeptidase and isocitrate lyase development in *Cucurbita maxima*
[92, 105] are suppressed by protein synthesis inhibitors, while actinomycin D,
which inhibits some DNA-dependent RNA synthesis, prevents the increase in
lipase and isocitrate lyase of castor beans [14, 77]. Studies with inhibitors can
be criticized on several grounds especially since these chemicals may have previously unsuspected side effects. But more satisfactory evidence of the kind known
for barley aleurone layers has been found in some seeds. De novo synthesis
of isocitrate lyase in cotyledons of *Citrullus vulgaris* (watermelon) [52] and of
endopeptidase in mung bean [21a] have been shown to occur by means of
density-labelling experiments with D_2O.

Nevertheless, there are also reports, some of which are again based on
the use of inhibitors, that certain enzymes involved in mobilization are not
newly synthesized. The compound azetidine-2-carboxylic acid, an analogue of
proline, does not stop isocitrate lyase increase in *Cucurbita pepo* or *Cucumis sativus*
and it is therefore presumed that the enzyme existed in an inactive form before
the chemical was applied [98]. Secondly, in some cases mobilization enzymes
can be extracted from dry seeds (i.e. before germination) such as acid lipase
in castor bean (see Fig. 6.8 [84]). Finally, it is thought that in some seeds trypsin-

like proteolytic enzymes are laid down during development but are prevented from acting by trypsin (proteinase) inhibitors. These inhibitors occur in many seeds and it has been suggested that they play an important regulatory role in connection with protein mobilization. This has been disputed, however, and their true function is unclear (see Chap. 6).

7.2.1. Control by the Embryo and Embryonic Axis

It will be recalled that food mobilization in barley, rice and wheat does not occur in endosperms from which the embryo has been removed. This is an easy way to demonstrate that the embryo is responsible for the initiation (and subsequent control) of reserve breakdown. The same simple method has been applied for other seeds, with variable success, but in many cases removal of the embryo has been found to prevent or retard mobilization and adversely to affect development of the requisite enzymes.

Let us look first at those seeds in which the cotyledons are the storage organs. A number of researchers have used pea seeds (a hypogeal germinator) for their experiments. Unfortunately, variable, sometimes contradictory results have been found, possibly because different varieties have been employed or, more likely, because the experimental conditions were not standardized. Respiration, differentiation of mitochondria, and other sub-cellular changes are all impaired in detached pea cotyledons [10, 112]. The cotyledons therefore seem to require a factor from the axis for the development and maintenance of respiratory metabolism — this can be supplied even by a small piece of attached axial tissue [112, 119]. As far as protein and starch mobilization in peas are concerned it is unclear if the presence of the axis is necessary. In some cases protein and starch breakdown proceed equally well in detached and attached cotyledons [10] but in others both the development of proteinase activity and

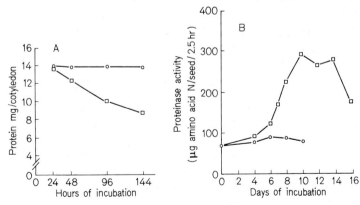

Fig. 7.11A and B. The influence of the attached axis on protein breakdown and extractable proteinase activity in *Pisum sativum*. (A) Protein breakdown in the cv. Burpeeana. After Chin et al., 1972[20]. (B) Proteinase activity in the cv. Alaska. After Yomo and Varner, 1973 [118]. o——o: in detached cotyledons; □——□: in cotyledons with attached axis

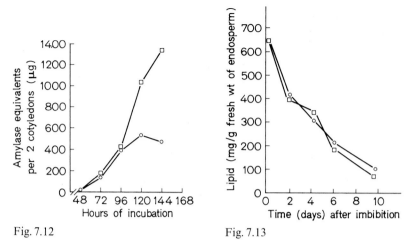

Fig. 7.12   Fig. 7.13

Fig. 7.12. The development of α-amylase activity in cotyledons of *Phaseolus vulgaris*. o——o: in detached cotyledons; □——□: in cotyledons with attached axis. From Dale, 1969[29]

Fig. 7.13. Fat breakdown in the endosperm of castor bean. o——o: endosperms without embryo; □——□: endosperms of intact seed. After Marriott and Northcote, 1975[81]

protein breakdown are prevented by isolating the cotyledons [20, 45, 118] (Fig. 7.11). The consequences of cotyledon excision for α-amylase activity include inhibition [79, 112], promotion [118], and no appreciable effect whatsoever [10]. Excised cotyledons of *Phaseolus vulgaris* also develop less α-amylase than do the attached organs but this effect is manifest only 96 h after excision (Fig. 7.12). Thus, the initiation of the enzyme can occur without the axis but continued production after four days requires its presence. Unfortunately, these particular studies on amylases were not accompanied by investigations on starch breakdown in the seeds so it is difficult to relate the enzymology to other biochemical changes. We might note again though, that starch breakdown in pea cotyledons continues unchanged even when they are detached [10].

Turning to other epigeal germinators we find similar evidence for the role of the axis. In the course of normal growth of *C. maxima* the activities of extractable proteinase and dipeptidase in the cotyledons increase and, of course, the protein is hydrolysed. But when the axis is removed these three processes all slow down [91, 105, 116]. *Cucurbita* stores a large proportion of its reserves as triglycerides, and enzymes involved in fat metabolism, especially isocitrate lyase, also show dependence upon the axis [92]. Stimulatory effects of the axis on mobilization in the cotyledons have been described for two other epigeally germinating species, *Gossypium* and *Helianthus annuus*. In *Gossypium*, distal halves of the cotyledons severed from the axis show little lipase activity [14]. Reducing sugar, presumably derived from stored fat, reaches higher concentrations in the attached cotyledons of *Helianthus* than in detached ones [42].

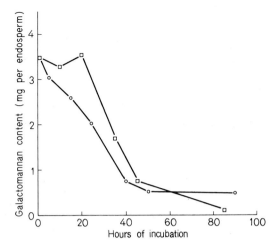

Fig. 7.14. Galactomannan mobilization in *Trigonella foenum-graecum*. ○——○ : endosperms without embryos; □——□ : endosperms of intact seeds. Based on Reid and Meier, 1972[99]

In addition to the above findings there is nevertheless a body of evidence which raises doubts concerning the relationship between axis and cotyledons in food mobilization. We should recall, in this context, that α- and β-amylase have been reported to reach higher levels in detached pea cotyledons [118], that starch and protein breakdown occur in detached pea cotyledons [10] and that α-amylase development in *P. vulgaris* cotyledons is unaffected by excision, at least over the first four days. Similarly, removing the axis has no deleterious effect on isocitrate lyase activity in cotyledons of *Arachis hypogaea* (peanut) and the enzyme reaches a level identical to that in intact seeds [80].

Studies on seeds which store reserves in non-embryonic tissues also yield evidence for and against the concept that the embryo initiates and controls mobilization. Two species of *Pinus, Pinus ponderosa* and *Pinus sylvestris,* exhibit a positive role of the embryo on enzymes of the storage tissue – in these cases haploid megagametophytic tissue which stores fat, protein and a little starch. In previously chilled (i.e. stratified) *P. ponderosa* isocitrate lyase activity in the storage tissue rises during the first four days after commencement of imbibition, visible germination occurring on the second day. When the embryo is removed at the start of imbibition the level eventually reached by the enzyme is decreased by about 60%; removal of the embryo after two days has no such effect [13]. In *P. sylvestris* the level of α-amylase in the gametophytic tissue is greater when the embryo is present than when it is not and there is, incidentally, a further beneficial effect of removing the seed coat [86].

The castor bean has a relatively bulky, oily endosperm surrounding the fragile embryo. A high rate of fat (oil) breakdown is measurable in this storage tissue whether or not the embryo is present (Fig. 7.13). Similarly, a number of enzymes including fatty acyl co-enzyme A dehydrogenase, isocitrate lyase, and fructose-1,6-diphosphatase, all of which feature in fat mobilization, appear in the endosperm irrespective of the absence of the embryo, though somewhat higher activity is attained when the embryo remains inside the endosperm [54].

The endosperm of the legume *Trigonella foenum-graecum* (fenugreek) stores galactomannan as a prominent part of its reserves and this is hydrolysed by enzymes secreted from the peripheral layer of cells, the aleurone layer. Unlike the aleurone layer of many cereals there is no evidence that the one in *Trigonella* is under the control of the embryo; enzyme secretion and galactomannan breakdown occur rapidly in endosperm from which the embryo has been removed (Fig. 7.14). It should be noted, however, that the important storage reserves of fenugreek are fat and protein deposited within the cotyledons; the role of galactomannans might be to conserve water within the seed rather than to provide carbohydrate for the embryo (see Chap. 6).

7.2.2. The Mechanism of Axial and Embryonic Control

Returning now to those seeds which are reputed to exhibit embryonic or axial control of enzyme activity and food mobilization we can examine the possible mechanisms of these effects. They fall essentially into two groups: (1) "Hormonal" control; (2) Sink effects.

(1) "Hormonal" Control

Stimulated, no doubt, by the elucidation of the mechanism in cereals, several attempts have been made to demonstrate that a factor from the embryo or axis is also implicated in other seeds. This has been approached in two ways. Firstly, the promotive capacity of extracts or diffusates of embryos and axes has been investigated; secondly, it has been argued that if a known hormone such as gibberellin or cytokinin can replace the beneficial effect of the embryo or axis this raises the possibility that an endogenous hormone might naturally be involved.

In only a few instances has the possibility of a diffusible or extractable factor been demonstrated. Protein hydrolysis in detached cotyledons of *C. maxima* is somewhat enhanced when these organs are incubated in a liquid medium together with excised embryonic axes [116]. The inference from this result is that the axis produces a "hormonal" factor which promotes metabolism in the cotyledons. A "hormonal" factor also seems to diffuse into the incubation medium bathing *P. ponderosa* embryos since this medium, when applied to isolated megagametophytic tissue, stimulates 44% more isocitrate lyase than in the control tissue [13]. Similarly, a promotive extract obtained from *Ricinus* (castor bean) embryos increases the activity of extractable fructose-1,6-diphosphatase in isolated endosperm of this seed [100]. It should be recognized, however, that these effects are nowhere near as dramatic as the action of cereal embryonic diffusates and extracts upon their endosperm. Moreover, some of the experiments whose results have just been described were not rigorously carried out and are open to serious criticisms (see below, Sect. 7.2.3). It can be stated, in summary, that no embryonic or axial "hormonal" influence comparable to that exhibited by certain cereal grains has yet been found in other seeds.

Accompanying these studies on diffusible or extractable factors have been those on the actions of applied plant growth regulators. Not surprisingly, one which has been used extensively is gibberellin. This, of course, is because it is so impressively involved in cereal grains. In one seed — castor bean — applied GA apparently induces greater fat breakdown in the endosperm when the amount of fat is expressed on a fresh weight basis [81]. But GA also causes an increase in fresh weight [82] and a recalculation to take this into account actually reveals no GA stimulation of fat disappearance from the endosperm. Gibberellin, further, has no effect on the activity of acid and alkaline lipases in castor bean endosperm though it does increase the activity of isocitrate lyase and hydroxyacyl co-enzyme A dehydrogenase [82]. These results highlight a point which is worth emphasizing; that is, there is often no coincidence between enzyme levels and reserve breakdown. Higher or lower activities of enzyme do not necessarily accompany higher or lower rates of food mobilization.

Lipase of *Gossypium* cotyledons [14] and acid lipase in *Pyrus malus* (apple) [101] show small increases in response to applied GA. The rising activity in acid lipase of *Pyrus* seems simply to coincide with visible germination and growth of the embryos, thus casting doubt on any direct action of GA upon enzyme level. Promotive effects of gibberellins, relatively small in magnitude, have been reported for α-amylase in isolated pea cotyledons [79, 112] but there are also reports which discount any such action of GA in this species and in *P. vulgaris* [39, 103, 118].

Evidence for a diffusible or extractable "hormonal" factor in embryos and axes has been mentioned above with respect to three seeds — *C. maxima, Ricinus communis,* and *P. ponderosa.* If these promotive factors were gibberellins one would expect applied GA to substitute for the embryo or axis. In *Cucurbita* and *Pinus* the axis and embryo cannot be replaced by GA and thus the stimulatory factor in the embryo is not likely to be GA alone. Isolated endosperm of *Ricinus,* however, does respond to GA by showing enhanced activity of fructose-1,6-diphosphatase, as it does to embryonic extracts.

The effects of exogenous gibberellin thus leave us with a rather confused picture as some storage tissues react to it while others do not. But certainly, even when there is a positive response, it does not measure up to that displayed by wheat or barley aleurone cells which can show several hundred-fold stimulations of enzyme levels. Probably the "best" promotion induced by applied GA, measuring goodness in quantitative terms, is seen in *Corylus* cotyledons where there is about a five-fold stimulation of isocitrate lyase activity [95]. Endogenous gibberellin probably features in the normal germination and growth processes of this seed so that such stimulatory effects on isocitrate lyase may be important.

Other hormones which have been tested extensively for their axis or embryo "replacement value" are the cytokinins. Some conflicting results emerge from experiments on pea cotyledons where effects upon α-amylase range from inhibition [102], through a small stimulation [112], to complete replacement of the axis by a combination of zeatin riboside and GA [79]. The cytokinin, kinetin, enhances extractable α-amylase activity in *P. vulgaris* cotyledons and promotes reducing sugar formation in *H. annuus* cotyledons [42]. Again, the effects are relatively small and even in the controls (i.e. not treated with kinetin) appreciable amounts of α-amylase and reducing sugar both appear.

Three enzymes in isolated, cytokinin-treated cotyledons of *C. maxima* — isocitrate lyase, proteinase and dipeptidase — all reach levels comparable to those found in cotyledons of intact embryos [91, 92, 105]. Since cytokinin (benzyladenine) thus apparently replaces the lost axis the conclusion has been made, and quoted in the research literature, that the axis in *Cucurbita* embryos produces cytokinin which regulates enzyme activity or formation. No direct confirmation of this conclusion has yet been made; there is no experimental evidence that the axis does indeed export cytokinins to the cotyledons let alone at levels similar to those which have been applied from an exogenous source. Moreover, the experiments using external cytokinin are themselves open to criticism on grounds that are discussed later (Sect. 7.2.3).

Our conclusions regarding the role of plant hormones in enzyme activity and food mobilization unfortunately again acquire a rather negative flavour since the part played by cytokinins is somewhat unclear.

(2) Sink Effects

Another way in which the embryo or axis might regulate food mobilization is by acting mainly as a sink. This implies that the storage tissues may be quite autonomous, that they may possess or be able to synthesize all the requisite enzymes for conversion of the food reserves into transportable chemical compounds, without needing some factor from the embryo or axis. Alternatively, the embryo or axis could provide an initial trigger before autonomy is achieved. The embryo and axis could, in either case, exert their major promotive effect simply by drawing off the products of food breakdown from the storage tissue, thus maintaining the concentrations below certain critical levels. If the products of reserve food mobilization begin to accumulate, feedback phenomena arrest further food transformation perhaps through effects on reaction equilibria or on enzyme action and synthesis themselves.

It seems very likely that this must be an important action of the embryo or axis; let us now see the experimental evidence directly bearing upon it. Synthesis of fatty acyl co-enzyme A dehydrogenase, fructose-1,6-diphosphatase and isocitrate lyase in castor bean endosperm is initiated in the absence of the embryo but the final levels attained by these enzymes is lower than in the endosperm of the intact seed. The isolated endosperm, unlike that in the seed, accumulates a relatively high concentration of glucose and, to a lesser extent, sucrose [54] and it may be that this accumulation prevents the continued increase in enzyme levels that occurs in the intact seed. Interestingly, of the three endosperm enzymes, isocitrate lyase is least affected by removing the embryo and it is also the enzyme which responds least to added glucose. Isocitrate lyase in the cotyledons of castor bean, and also of *Cucurbita*, is, however, partially suppressed by glucose, though surprisingly high concentrations (0.1 M) must be applied [77]. The development of this enzyme in *Pinus pinea* seeds is lowered by the fatty acids, octanoic and oleic acids [115]. One can envisage that if these acids accumulate in the seed the glyoxylate cycle enzyme would then be "switched off."

Inhibition by accumulation of food reserve breakdown products is also suggested by work on detached pea cotyledons in which the build-up of amino acids is associated with the repression of proteinase activity. In fact, feeding intact germinated seeds with a mixture of amino acids (1% casein hydrolysate) causes low levels of proteinase in the cotyledons [118]. Nevertheless, the relationship between endogenous amino acid levels and proteinase activity in peas is somewhat unclear. This is because attached pea cotyledons have been found to contain high concentrations of soluble nitrogenous compounds (i.e. including amino acids) and yet they continue to mobilize stored protein while detached cotyledons, with lower soluble nitrogen, do not [45].

7.2.3. Conclusions and Appraisals

The reader may be disappointed at not having received a clear answer to the question of how food mobilization in non-cereal seeds is regulated. Part of the cause of our uncertainty is indicated at the beginning of this account; that is, cereal grains are an "easier" experimental system! When we try to manipulate other seeds in the same manner as cereals we cannot avoid producing damaged cotyledons and endosperms from which the testa often has had to be removed. The result is storage tissue in a physiological condition very different from that in the intact seed. Whether meaningful comparisons can be made between such tissues and those in the seed is a debatable point.

But these inherent difficulties are only a partial cause of the uncertain picture that has emerged, for in many cases the experimenters have introduced factors which exacerbate the situation. For example, detached cotyledons and endosperms are often fully or partly immersed in a liquid medium and then compared,

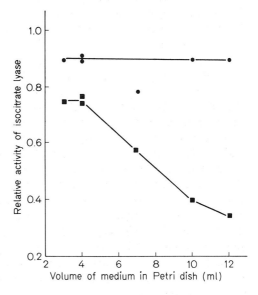

Fig. 7.15. The effect of incubation conditions on the apparent axis effect in *Cucumis sativus* cotyledons. Detached cotyledons and intact seeds were incubated for three days after which time their levels of isocitrate lyase were measured. ●——●: cotyledons with axis attached; ■——■: detached cotyledons. From P. Slack, 1976—personal communication

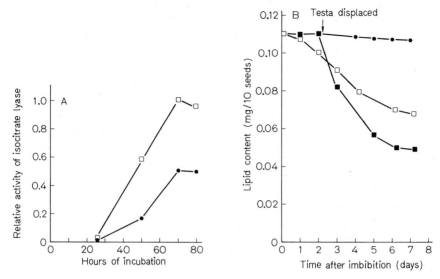

Fig. 7.16A and B. The effect of testa removal on enzyme activity and fat mobilization in *Cucumis sativus* cotyledons. (A) Isocitrate lyase activity in isolated cotyledons. □——□: without testa; ●——●: with testa. (B) Fat mobilization. □——□: isolated cotyledons without testa; ●——●: isolated cotyledons with testa; ■——■: intact seed (the time of natural displacement of the testa is indicated). From P. Slack, 1976 — personal communication

for food mobilization activity, with cotyledons or endosperm in the intact seed or seedling; the latter, are, of necessity, held under favourable growing conditions which, in most cases, do not include total immersion in liquid! Submergence of cotyledons certainly alters the patterns of enzyme activity and food mobilization, most probably because of the resultant paucity of oxygen (Fig. 7.15). When steps are taken to secure comparable aeration the reputed differences between attached and detached cotyledons almost disappear. This sensitivity of cotyledons to incubation conditions might well account for many of the reports in the literature on inhibition of enzymes in axis-less cotyledons. It may also be responsible for the disparity between experimental results of different researchers who often have not been cautious enough about their experimental conditions. Differences in oxygen availability due to varying degrees of immersion might even develop during a single experiment. For example, although detached cotyledons and intact seeds might both start an experiment in the same volume of liquid, this soon drops as the seed germinates and grows. The detached cotyledons meanwhile remain immersed while those on the young seedling are now above liquid level. Placing embryonic axes in the same vessel as detached cotyledons (to test for diffusing substances) can have the same effect; the liquid level drops as the axes consume water for their growth and the cotyledons thus become better aerated.

Availability of oxygen is indeed a factor which plays an important role in food mobilization in some seeds. In *Cucumis* for example, shortly after germination a peg of tissue develops on the lower surface of the hypocotyl which serves to lever off the testa (Fig. 4.13A). When this occurs rapid hydrolysis of the fat reserves begins. This is probably a causal relationship since artificial removal of the testa prior to axis growth, even from detached cotyledons, promotes enzyme production and fat breakdown (Fig. 7.16A and B).

Many of the experiments with applied hormones fail to take account of the effects of these substances on growth. The cytokinins are especially well known to promote cotyledon expansion which, of course, requires the synthesis of new cell wall material. Fat breakdown occurs rapidly in cotyledons treated with cytokinin but there is virtually no decrease in dry weight. What we are witnessing is probably a transfer of carbon from triglyceride to cellulose, i.e. from source to sink. It is not surprising that cytokinin accelerates fat hydrolysis and isocitrate lyase activity but it is questionable that it is acting directly on those processes. It may be acting simply to create an internal sink through its effect on cell expansion.

We are thus left with indefinite conclusions concerning the control mechanisms in dicot seeds. It seems likely that for some years hence the cereal grain will continue as the best-understood regulatory system in the mobilization of food reserves.

Some Articles of General Interest

1. Briggs, D.E.: Hormones and carbohydrate metabolism in germinating cereal grains. In: Biosynthesis and its Control in Plants. Milborrow, B.V. (ed.). London: Academic Press, 1973, pp. 219–277
2. Jones, R.L.: Gibberellins: their physiological role. Ann. Rev. Pl. Physiol. **24**, 571–598 (1973)
3. Mann, J.D.: Mechanism of action of gibberellins. In: Gibberellins and Plant Growth. Krishnamoorthy, H.N. (ed.). New Delhi: Wiley Eastern Ltd., 1975, pp. 239–287
4. Palmer, G.H.: The industrial use of gibberellic acid and its scientific basis—a review. J. Inst. Brew. **80**, 13–30 (1974)
4a. Palmer, G.H., Bathgate, G.N.: Malting and brewing. In: Advances in Cereal Science and Technology. Pomeranz, Y. (ed.). St. Paul: Am. Assoc. Cereal Chem., 1976, pp. 237–324
5. Varner, J.E.: Gibberellin control of a secretory tissue. In: The Chemistry and Biochemistry of Plant Hormones. Recent Advances in Phytochemistry. Runeckles, V.C., Sondheimer, E., Walton, D.C. (eds.). New York: Academic Press, 1974, Vol. VII, pp. 123–130
6. Yomo, H., Varner, J.E.: Hormonal control of a secretory tissue. In: Current Topics in Developmental Biology. Moscona, A.A., Monroy, A. (eds.). New York: Academic Press, 1971, Vol. VI, pp. 111–144

References

7. Amrhein, N.: Planta (Berl.) **118**, 241–258 (1974)
8. Armstrong, J.E., Jones, R.L.: J. Cell Biol. **59**, 444–455 (1973)
9. Ashford, A.E., Jacobsen, J.V.: Planta (Berl.) **120**, 81–105 (1974)
10. Bain, J., Mercer, F.V.: Australian J. Biol. Sci. **19**, 85–96 (1966)
11. Bennett, P.A., Chrispeels, M.J.: Pl. Physiol. **49**, 445–447 (1972)
12. Ben-Tal, Y., Varner, J.E.: Pl. Physiol. **54**, 813–816 (1974)
13. Bilderback, D.E.: Physiol. Plantarum **31**, 200–203 (1974)
14. Black, H.S., Altschul, A.M.: Biochem. Biophys. Res. Commun. **19**, 661–664 (1965)
15. Brown, H.T., Morris, G.H.: J. Chem. Soc. **57**, 458–528 (1890)
16. Carlson, P.S.: Nature (London) **237**, 39–41 (1972)
17. Chandra, G.R.: Nature (London) **248**, 161–162 (1974)
18. Chandra, G.R., Duynstee, E.E.: Biochim. Biophys. Acta **232**, 514–523 (1971)
19. Chen, R.-F., Jones, R.L.: Planta (Berl.) **119**, 207–220 (1974)
20. Chin, T., Poulson, R., Beevers, L.: Pl. Physiol. **49**, 482–489 (1972)
21. Chrispeels, M.J.: Biochem. Biophys. Res. Commun. **53**, 99–104 (1973)
21a. Chrispeels, M.J., Baumgartner, B., Harris, N.: Proc. Natl. Acad. Sci. **73**, 3168–3172 (1976)

22. Chrispeels, M.J., Varner, J.E.: Pl. Physiol. **42**, 398–406 (1967)
23. Chrispeels, M.J., Varner, J.E.: Pl. Physiol. **42**, 1008–1016 (1967)
24. Chrispeels, M.J., Tenner, A.J., Johnson, K.D.: Planta (Berl.) **113**, 35–46 (1973)
25. Clutterbuck, V.J., Briggs, D.E.: Phytochemistry **12**, 537–546 (1973)
26. Colborne, A.J., Morris, G., Laidman, D.L.: J. Exp. Botany **27**, 759–767 (1976)
27. Collins, G.G., Jenner, C.F., Paleg, L.G.: Pl. Physiol. **49**, 404–410 (1972)
28. Crozier, A., Kuo, C.C., Durley, R.C., Pharis, R.P.: Can. J. Botany **48**, 867–877 (1970)
29. Dale, J.E.: Planta (Berl.) **89**, 155–164 (1969)
30. Doig, R.I., Colborne, A.J., Morris, G., Laidman, D.L.: J. Exp. Botany **26**, 387–398 (1975)
31. Doig, R.I., Colborne, A.J., Morris, G., Laidman, D.L.: J. Exp. Botany **26**, 399–410 (1975)
32. Eastwood, D., Laidman, D.L.: Phytochemistry **10**, 1275–1284 (1971)
33. Eastwood, D., Laidman, D.L.: Phytochemistry **10**, 1459–1467 (1971)
34. Eastwood, D., Tavener, R.J.A., Laidman, D.L.: Nature (London) **221**, 1267 (1969)
35. Evins, W.H.: Biochem. J. **10**, 4295–4303 (1971)
36. Filner, P., Varner, J.E.: Proc. Natl. Acad. Sci. **58**, 1520–1526 (1967)
37. Firn, R.D.: Planta (Berl.) **125**, 227–233 (1975)
38. Firn, R.D., Kende, H.: Pl. Physiol. **54**, 911–915 (1974)
38a. Gale, M.D., Marshall, G.A.: J. Exp. Botany **37**, 729–735 (1973)
39. Gepstain, S., Ilan, I.: Pl. Cell Physiol. (Tokyo) **11**, 819–822 (1970)
40. Gibson, R.A., Paleg, L.G.: Biochem. J. **128**, 367–375 (1972)
41. Gibson, R.A., Paleg, L.G.: Australian J. Pl. Physiol. **2**, 41–49 (1975)
42. Gilad, T., Ilan, I., Reinhold, L.: Israel J. Botany **19**, 447–450 (1970)
43. Goldstein, L.D., Jennings, P.H.: Pl. Physiol. **55**, 893–898 (1975)
44. Groat, J.I., Briggs, D.E.: Phytochemistry **8**, 1615–1627 (1969)
45. Guardiola, J.L., Sutcliffe, J.: Ann. Botany (London) **35**, 791–807 (1971)
46. Haberlandt, G.: Ber. Deut. Botan. Ges. **8**, 40–48 (1890)
47. Hardie, D.G.: Phytochemistry **14**, 1719–1722 (1975)
48. Harvey, B.M.R., Oaks, A.: Planta (Berl.) **121**, 67–74 (1974)
49. Hayashi, T.: J. Agr. Chem. Soc. Japan **16**, 531–538 (1940)
50. Higgins, T.J.V., Zwar, J.A., Jacobsen, J.V.: Nature (London) **260**, 166–169 (1976)
51. Ho, D.T.-H., Varner, J.E.: Proc. Natl. Acad. Sci. **71**, 4783–4786 (1974)
52. Hock, B.: Planta (Berl.) **93**, 26–38 (1970)
53. Honigman, W.A., Taiz, L.: Am. J. Botany **62**, suppl., 9 (1975)
54. Huang, A.H., Beevers, H.: Pl. Physiol. **54**, 277–279 (1974)
55. Ingle, J., Hageman, R.H.: Pl. Physiol. **40**, 672–675 (1965)
56. Jacobsen, J.V., Knox, R.B.: Planta (Berl.) **112**, 213–224 (1973)
57. Jacobsen, J.V., Knox, R.B.: Planta (Berl.) **115**, 193–206 (1974)
58. Jacobsen, J.V., Varner, J.E.: Pl. Physiol. **42**, 1596–1600 (1967)
59. Jacobsen, J.V., Zwar, J.A.: Australian J. Pl. Physiol. **1**, 343–356 (1974)
60. Jacobsen, J.V., Zwar, J.A.: Proc. Natl. Acad. Sci. **71**, 3290–3293 (1974)
61. Jacobsen, J.V., Scandalios, J.G., Varner, J.E.: Pl. Physiol. **45**, 367–371 (1970)
62. Johnson, K.D., Chrispeels, M.J.: Planta (Berl.) **111**, 353–364 (1973)
63. Johnson, K.D., Kende, H.: Proc. Natl. Acad. Sci. **68**, 2674–2677 (1971)
64. Jones, R.L.: Planta (Berl.) **85**, 359–375 (1969)
65. Jones, R.L.: Planta (Berl.) **87**, 119–133 (1969)
66. Jones, R.L.: Planta (Berl.) **88**, 73–86 (1969)
67. Jones, R.L.: Pl. Physiol. **47**, 412–416 (1971)
68. Jones, R.L.: Planta (Berl.) **103**, 75–109 (1972)
69. Jones, R.L.: Pl. Physiol. **52**, 303–304 (1973)
70. Jones, R.L., Armstrong, J.E.: Pl. Physiol. **48**, 137–142 (1971)
71. Jones, R.L., Chen, R.-F.: J. Cell Sci. **20**, 183–198 (1976)
72. Jones, R.L., Price, J.M.: Planta (Berl.) **94**, 191–202 (1970)
73. Keates, R.A.B.: Nature (London) **244**, 355–356 (1973)
74. Khan, A.A., Verbeek, R., Waters, E.C., Jr., Van Onckelen, H.A.: Pl. Physiol. **51**, 641–645 (1973)

75. Kirsop, B.H., Pollock, J.R.A.: J. Inst. Brew. **64**, 227–233 (1958)

76. Koehler, D.E., Varner, J.E.: Pl. Physiol. **52**, 208–214 (1973)

77. Lado, P., Schwendiman, M., Marrè, E.: Biochim. Biophys. Acta **157**, 140–148 (1968)

78. Laidman, D.L., Colborne, A.J., Doig, R.I., Varty, K.: In: Mechanisms of Regulation of Plant Growth. Bieleski, R.L., Ferguson, A.R., Cresswell, M.M. (eds.). Wellington: Royal Soc. of N. Z., 1974, Bulletin **12**, 581–590

79. Locker, A., Ilan, I.: Pl. Cell Physiol. (Tokyo) **16**, 449–454 (1975)

80. Marcus, A., Feeley, J.: Biochim. Biophys. Acta **89**, 170–171 (1964)

81. Marriott, K.M., Northcote, D.H.: Biochem. J. **148**, 139–144 (1975)

82. Marriott, K.M., Northcote, D.H.: Biochem. J. **152**, 65–70 (1975)

83. McNeil, M., Albersheim, P., Taiz, L., Jones, R.L.: Pl. Physiol. **55**, 64–68 (1975)

84. Muto, S., Beevers, H.: Pl. Physiol. **54**, 23–28 (1974)

85. Naylor, J.M.: Can. J. Botany **44**, 19–32 (1966)

86. Nyman, B.: Physiol. Plantarum **25**, 112–117 (1971)

87. Ogawa, Y.: Pl. Cell Physiol. (Tokyo) **7**, 509–517 (1966)

88. Paleg, L.G.: Pl. Physiol. **35**, 293–299 (1960)

89. Paleg, L.G., Coombe, B.G., Buttrose, M.S.: Pl. Physiol. **37**, 798–803 (1962)

90. Palmiano, E.P., Juliano, B.O.: Pl. Physiol. **49**, 751–756 (1972)

91. Penner, D., Ashton, F.M.: Pl. Physiol. **42**, 791–796 (1967)

92. Penner, D., Ashton, F.M.: Biochim. Biophys. Acta **148**, 481–485 (1967)

93. Phillips, M.L., Paleg, L.G.: In: Proc. 7th Intern. Conf. on Pl. Growth Sub., Carr, D.J. (ed.). Canberra, Australia 1972, pp. 396–406

94. Phinney, B.O.: In: Plant Growth Regulation. Klein, R.M. (ed.). Ames, Iowa: Iowa State Univ. Press, 1959, pp. 489–501

95. Pinfield, N.J.: Planta (Berl.) **82**, 337–341 (1968)

96. Pollard, C.J.: Pl. Physiol. **44**, 1227–1232 (1969)

97. Pollard, C.J., Nelson, D.C.: Biochim. Biophys. Acta **244**, 372–376 (1971)

98. Presley, H.J., Fowden, L.: Phytochemistry **4**, 169–175 (1965)

99. Reid, J.S.G., Meier, H.: Planta (Berl.) **106**, 44–60 (1972)

100. Scala, J., Patrick, C., Macbeth, G.: Phytochemistry **8**, 37–44 (1969)

101. Smolenska, G., Lewak, S.: Planta (Berl.) **116**, 361–370 (1974)

102. Sprent, J.: Planta (Berl.) **81**, 80–87 (1968)

103. Sprent, J.: Planta (Berl.) **82**, 299–301 (1968)

104. Sundblom, N.-O., Mikola, J.: Physiol. Plantarum **27**, 281–284 (1972)

105. Sze, H., Ashton, F.M.: Phytochemistry **10**, 2935–2942 (1971)

105a. Taiz, L., Honigman, W.A.: Pl. Physiol. **58**, 380–386 (1976)

106. Taiz, L., Jones, R.L.: Planta (Berl.) **92**, 73–84 (1970)

107. Taiz, L., Jones, R.L.: Am. J. Botany **60**, 67–75 (1973)

108. Tavener, R.J.A., Laidman, D.L.: Phytochemistry **11**, 989–997 (1972)

109. Van Onckelen, H.A., Verbeek, R., Khan, A.A.: Pl. Physiol. **53**, 562–568 (1974)

110. Varner, J.E.: Advan. Exp. Med. Biol. **62**, 65–78 (1975)

111. Varner, J.E., Mense, R.M.: Pl. Physiol. **49**, 187–189 (1972)

112. Varner, J.E., Balce, L.V., Huang, R.C.: Pl. Physiol. **38**, 89–92 (1963)

113. Varty, K., Laidman, D.L.: J. Exp. Botany **27**, 748–758 (1976)

114. Vigil, E.L., Ruddat, M.: Pl. Physiol. **51**, 549–558 (1973)

115. Vincenzini, M.T., Vincieri, F., Vanni, P.: Pl. Physiol. **52**, 549–553 (1973)

116. Wiley, L., Ashton, F.M.: Physiol. Plantarum **20**, 688–696 (1967)

117. Yomo, H.: Hakko Kyokaishi **18**, 600–602 (1960) [cited in Chem. Abstracts **55**, 26145 (1961)]

118. Yomo, H., Varner, J.E.: Pl. Physiol. **51**, 708–713 (1973)

119. Young, J.L., Huang, R.C., Vanecko, S., Marks, J.D., Varner, J.E.: Pl. Physiol. **35**, 288–292 (1960)

120. Yung, K.H., Mann, J.D.: Pl. Physiol. **42**, 195–200 (1967)

Glossary and Index of English and Botanical Names

Author Index

Numbers in bold type indicate the pages of the reference citations

Subject Index

Physiology and Biochemistry of Seeds in Relation to Germination

In Two Volumes

Volume 2
J.D. Bewley, M. Black

Viability, Dormancy, and Environmental Control

1982. 153 figures. XII, 375 pages. ISBN 3-540-11656-7

Contents: Viability and Longevity. – Dormancy. – The Release from Dormancy. – The Control of Dormancy. – Perspective on Dormancy. – Environmental Control of Germination. – Glossary and Index of English and Botanical Names. – Author Index. – Subject Index.

The book provides a comprehensive, up-to-date treatment of selected aspects of seed physiology and biochemistry, including viability, longevity, dormancy, and the environmental control of germination. Discussion is at a level which will be useful to advanced students as well as to research workers and lectures. The subject is treated in a critical manner, ensuring that areas of ignorance, over-assumption or incomplete knowledge are fully discussed, and modern developments, discoveries and concepts receive thorough consideration. Together with Volume 1, this book represents the most wide-ranging advanced text available in this subject area.

L. van der Pijl

Principles of Dispersal in Higher Plants

3rd revised and expanded edition. 1982. 30 figures. X, 215 pages. ISBN 3-540-11280-4

Contents: Introduction. – General Terminology. – The Units of Dispersal. – The Relation Between Flowers, Seeds and Fruits. – Ecological Dispersal Classes, Established on the Basis of the Dispersing Agents. – Dispersal Strategy and the Biocoenosis. – Establishment. – The Evolution of Dispersal Organs in General. – Ecological Developments in Leguminous Fruits. – Dispersal and the Evolution of Grasses. – Man and His Plants in Relation to Dispersal. – References. – Subject Index. – Index of Scientific Plant Names. – Index of Scientific Animal Names.

From the reviews:
"The third edition of this classic work has been completely revised and expanded to include research results obtained over the last 15 years. As in previous editions, the emphasis remains on principles and ecology (especially synecology), evolution and establishment after transport. ... the present work is an invaluable addition to the literature on reproductive biology of plants... Few botanists today are better qualified than van der Pijl to write on dispersal (and pollution) biology... an excellent up-to-date treatment of a long neglected subject... this splendid volume is unlikely to be surpassed for quite some time..." *Science*

Springer-Verlag
Berlin
Heidelberg
New York
Tokyo

Encyclopedia of Plant Physiology

New Series

Editors: A. Pirson, M. H. Zimmermann

Springer-Verlag
Berlin
Heidelberg
New York
Tokyo